T0248621

Proceedings of the 30th International Geological Congress

Volume 19

Geochemistry

## Proceedings of the 30th International Geological Congress

Volume 1 : Origin and History of the Earth
Volume 2 : Geosciences and Human Survival, Environment, and Natural Hazards
Volume 3 : Global Changes and Future Environment
Volume 4 : Structure of the Lithosphere and Deep Processes
Volume 5 : Contemporary Lithospheric Motion / Seismic Geology
Volume 6 : Global Tectonic Zones / Supercontinent Formation and Disposal
Volume 7 : Orogenic Belts / Geological Mapping
Volume 8 : Basin Analysis / Global Sedimentary Geology / Sedimentology
Volume 9 : Energy and Mineral Resources for the 21st Century / Geology of Mineral Deposits / Mineral Economics
Volume 10 : New Technology for Geosciences
Volume 11 : Stratigraphy
Volume 12 : Paleontology and Historical Geology
Volume 13 : Marine Geology and Palaeoceanography
Volume 14 : Structural Geology and Geomechanics
Volume 15 : Igneous Petrology
Volume 16 : Mineralogy
Volume 17 : Precambrian Geology and Metamorphic Petrology
Volume 18.A : Geology of Fossil Fuels - Oil and Gas
Volume 18.B : Geology of Fossil Fuels - Coal
Volume 19 : Geochemistry
Volume 20 : Geophysics
Volume 21 : Quaternary Geology
Volume 22 : Hydrogeology
Volume 23 : Engineering Geology
Volume 24 : Environmental Geology
Volume 25 : Mathematical Geology and Geoinformatics
Volume 26 : Comparative Planetology / Geological Education / History of Geosciences

PROCEEDINGS OF THE

# 30TH INTERNATIONAL GEOLOGICAL CONGRESS

BEIJING, CHINA, 4 - 14 AUGUST 1996

## VOLUME 19

# GEOCHEMISTRY

EDITORS:
XIE XUEJIN
*INSTITUTE OF GEOPHYSICAL AND GEOCHEMICAL EXPLORATION, BEIJING, CHINA*

CRC Press
Taylor & Francis Group
Boca Raton London New York

CRC Press is an imprint of the
Taylor & Francis Group, an **informa** business

First published 1997 by VSP

Published 2019 by CRC Press
Taylor & Francis Group
6000 Broken Sound Parkway NW, Suite 300
Boca Raton, FL 33487-2742



First issued in paperback 2019



ISBN 13: 978-0-367-44796-0 (pbk)
ISBN 13: 978-90-6764-267-5 (hbk)



**Visit the Taylor & Francis Web site at**
**http://www.taylorandfrancis.com**

**and the CRC Press Web site at**
**http://www.crcpress.com**

# CONTENTS

# FLUID ROCK INTERACTION

*Proc 30th Int'l. Geol. Congr.*, Vol. 19, pp. 3-25
Xie Xuejin (Ed)
VSP 1997

# Hydrothermal Stability of Berthierine: Experimental and Modeling Studies

Zhihong (John) Zhou, William D. Gunter
Alberta Research Council, P.O. Box 8330, Edmonton, Alberta, Canada T6H 5X2
Donald R. Peacor
Department of Geological Sciences, University of Michigan, Ann Arbor, MI 48109

## Abstract

Autoclave experimental results show that berthierine is not stable in hydrothermal solutions at 250°C. In neutral to alkaline solution, the principal decomposition product of the berthierine is a swelling clay. The swelling clay produced in this reaction is Fe-rich and trioctahedral and is classified as Fe-saponite. In acidic solution, Fe-chlorite is the major reaction product. A geochemical computer program, PATHARC, was used to simulate the hydrothermal reactions of berthierine. The model predicted all major reactions observed in the experiments, including the destruction of berthierine, the formation of Fe-saponite in neutral to alkaline solutions, and the formation of Fe-chlorite in acidic solutions.

There is a significant volume increase for the berthierine to Fe-saponite reaction. This volume increase will decrease the porosity and permeability of the hydrocarbon reservoirs. The degree of the damage will depend on the mineralogy, clay concentration, and initial porosity of the hydrocarbon reservoirs. The more berthierine in the reservoir and the less the initial porosity, the greater the damage. There should be no volume increase from the berthierine to Fe-chlorite reaction. The potential of fines migration will likely increase as the surface coating berthierine disintegrates or is replaced by loosely adhering Fe-saponite and analcime. However, the Fe-saponite may have a limited potential for osmotic swelling.

## INTRODUCTION

Berthierine often occurs as a diagenetic mineral which is stable in low temperature reducing environments [1]. Curtis et al. [2] noted that Fe-chlorite replaces berthierine as temperatures rise. Velde [1] indicated that this transformation occurs at about 100°C. However, Jahren and Aagaard [3] described pore-filing berthierine in Jurasic clastic reservoirs from offshore Norway that was undergoing transformation to chamosite (an Fe-chlorite) at about 70°C.

Hydrothermal stability of berthierine has implications not only in sedimentary diagenesis but also in hydrocarbon recovery processes. Thermal recovery processes are currently used to produce hydrocarbons from some of the world's largest heavy oil and bitumen reservoirs (e.g. Athabasca and Cold Lake, Alberta, San Joaquin Valley, California, and Duri, Indonesia). The Cold Lake reservoir, Alberta contains $3.5 \times 10^{10}$ $m^3$ of bitumen in place [4] and is the most active area for *in situ* thermal recovery in Canada. Recent studies indicate that berthierine is among the most abundant clay minerals in the Clearwater Formation at Cold [5]. It is commonly recognized that hydrothermal reactions or "artificial diagenesis" occurring during thermal recovery can cause formation (permeability) damage [6-10]. Therefore, it is of practical significance to determine the stability of the berthierine, to characterise its reaction products, and to determine their effect on the recovery process.

Perhaps due to its variable chemical composition and small grain size, the thermodynamic properties of berthierine available in the literature are poorly known; kinetic data does not exist. Thus, prediction into its fate during thermal recovery would bear a large uncertainty to say the least. For this reason, the Cold Lake berthierine was purified and subjected to a series of autoclave experiments [11]. Thermodynamic properties for a chemically simple berthierine were calculated. In this study, berthierine-rich bulk sands are used in autoclave experiments in order to determine the effect of matrix sands on the hydrothermal stability of berthierine. The reaction paths during thermal recovery are modeled using the PATHARC geochemical program. The potential of formation damage caused by berthierine reactions are discussed.

## MATERIAL AND METHODS

### *Materials*

The geological occurrence and origin of the Cold Lake berthierine have been discussed in detail by Hornbrook and Longstaffe [5]. The berthierine at Cold Lake occurs in the Lower Cretaceous Clearwater Formation which are deltaic siliciclastic sands. The berthierine formed during early diagenesis. Petrographically, the Cold Lake berthierine is present as coatings on the sand grains. The samples used in this study were from Cold Lake 6-21-64 4W4 at 439.7m. The unconsolidated sands were bitumen saturated, which was removed by a cold extraction technique using methylene chloride. The bitumen-extracted bulk sands were used for autoclave tests. Four solutions were used in the autoclave tests:

Solution 1: 0.1 M $NaHCO_3/CO_2$ solution;
Solution 2: 0.1 M $NaHCO_3$/0.08 M NaOH solution;
Solution 3: 0.03 M $Na_2SiO_3$/0.1 M NaCl solution;
Solution 4: 0.02 M HCl solution.

### *Experimental Methods*

The hydrothermal stability of the berthierine was studied through autoclave experiments. Sixteen autoclave experiments were conducted but only four of them will be presented here (Table 1). For each autoclave run, 2.0 to 2.5 grams of bulk sands and 100 mL of solution were mixed by sonification. The pH of the suspension was measured before it was placed in a 300 mL stainless steel autoclaves. After the suspension was added, the autoclave were purged and pressure tested with nitrogen gas before being heated. All 4 tests were conducted at 250°C for a period of 6 weeks.

When an elevated $CO_2$ pressure was required, the autoclave was connected to a delivery system to allow $CO_2$ pressurization at temperature. The solution was heated first to reach temperature equilibrium (250°C). Then, a $CO_2$ pressure was applied on top of the steam pressure at the temperature. When the pressure equilibrated (approximately 15 minutes), the valves were closed and the autoclaves weighed. Total weights for each autoclave were recorded before and after heating to ensure no leaks developed throughout the experiment.

After the specified run time, the autoclaves were removed from the oven, air cooled for 10 minutes and quenched to room temperature by submerging in cold water. This quench procedure was used to minimize the precipitation of quartz and other minerals during cooling. Sub-samples for water analysis were removed from the aqueous layer, filtered through 0.45μm filters and appropriately preserved for cation and anion analysis. Measurements for pH were made on separate aliquots. The solids and remaining water were transferred into dialysis tubing, washed with de-ionized water and freeze dried for X-ray diffraction (XRD), scanning electron microscopy (SEM) with energy-dispersive X-ray analyzer (EDX). The post-run bulk-sands were then further separated into clay and non-clay fractions. The clay fraction was subjected to detailed XRD and analytical electron microscopy (AEM) analyses.

The pre- and post-run solutions were analyzed for major elements (Na, K, Ca, Mg, Al, Fe, and Si) by inductively coupled plasma (ICP). Chloride was determined by a potentiometric titration with $AgNO_3$. Total inorganic carbon (TIC) was determined using a Dohrman total carbon analyzer. Total dissolved solids, ion balance and saturation index of quartz were calculated using a geochemical model, SOLMINEQ.88 [12].

The transmission electron microscopy (TEM) analysis was done in the Electron Microscopy Laboratory, the University of Michigan at Ann Arbor. The TEM is equipped with a quantitative energy dispersive X-ray analyzer (EDX) and therefore can be used as an analytical electron microscope (AEM). The clays were first buried in an epoxy and an micro thin section was cut after the epoxy was hardened. Then the thin section was ion-milled to produce ultra thin edges on which the TEM and AEM analyses were performed In a given analysis, the TEM module was usually used first in which the mineral structure was determined by lattice image and electron diffraction patterns; then AEM analysis was done on the same thin edge. Thus, we have coupled structural and chemical information on the mineral being analyzed

## EXPERIMENTAL RESULTS

### *The Pre-run Materials*

The sand grains are ubiquitously coated by diagenetic clay minerals (Fig. 1). The coating material has a corn-flake morphology. The coating is usually 3-5 μm in thickness. XRD analysis shows that the sands are mainly quartz and albite, but a berthierine peak is visible at 7 12Å (Fig. 2). Detailed XRD analysis of the clay fraction shows that it is mainly composed of berthierine but also contains a significant amount of a mixed smectite/chlorite clay (S/C) and a small amount of quartz and albite (Fig. 3). The presence of the S/C in the sample is best shown by the EG-treatment of the Ca-saturated clay fraction which shifts the 14 Å peak to 16 Å. Upon heat treatment, the 14 Å peak gradually decreased to 12.65 Å at 550°C. The chemical composition of berthierine were determined using an analytical electron microscope (AEM) and is shown in Table 2. The chemical composition was calculated on the basis of 28 octahedral and tetrahedral charges.

Because the EDX cannot differentiate ferrous Fe and ferric Fe, the $Fe^{2+}/Fe^{3+}$ ratio in the berthierine is not known. However, because berthierine is a trioctahedral mineral, all the Fe was considered as ferrous.

***The Post-run Materials***

The compositions of the post-run solutions are shown in Table 3; and XRD results are summarized in Table 4. Highlights of the major changes in each group are discussed here.

*$NaHCO_3/CO_2(aq)$ Solution (Near Neutral).* $NaHCO_3$ solution was equilibrated with a $CO_2$ pressure equivalent of 2.5 bar at 25°C. This was used to buffer the solution pH (25°C) at 6.30. There were little change in solution pH during the hydrothermal reaction (Table 3). Dissolved silica was close to quartz solubility. Berthierine was largely consumed and the albite to quartz peak ratio increased during the reaction (Fig. 4). The C/S phase cannot be detected in the post-run samples because it was either consumed in the reaction or transformed to other clay minerals. Large amounts of analcime are present and were probably produced at the expense of albite. Most analcime crystals are larger than 2μm in diameter and therefore they can only seen in the bulk XRD traces. A new swelling clay has been formed and predominates the clay fractions of the both samples. The swelling clay has well developed (*00l*) diffraction peaks and shows a clear shift upon ethylene glycol treatment (Fig. 5).

*$NaHCO_3/NaOH$ Solution (Alkaline).* The solution had a measured pH (25°C) of 10.49. The solution pH decreased significantly after the hydrothermal reactions. Berthierine and C/S were reacted out completely. The reaction products are mainly 17 Å swelling clay, with a very small amount of illite. The hydrothermal swelling clay produced in the reaction has a distinct morphology (Fig. 6). The albite/quartz peak ratio has increased compared with the pre-run sample. Euhedral albite crystals can be seen in the post-run samples (Fig. 6).

*$Na_2SiO_3/NaCl$ Solution (Highly Alkaline).* These tests were conducted in 0.30 M $Na_2SiO_3$/0.1 M NaCl solution with a buffered pH (25°C) of 11.46. However, both the pH and dissolved silica in the pre-run solution were greater than the nominal value. The dissolved Si was measured at 2030 mg/l or 0.034 M and the pH was measured at 12.20. After the reaction, the solution pH decreased significantly; the silica concentration increased. Berthierine and C/S were consumed and the Fe-rich swelling clay and analcime were formed. The albite/quartz peak area ratio did not change significantly. In addition, a small amount of stilpnomelane was detected. This hydrated layer silicate mineral is not common but has been found in alkaline hydrothermal solutions.

*HCl Solution (Acidic).* The solution had a measured pH of 2.20. Solution pH increased by 3.5 units after the hydrothermal reaction. Berthierine content decreased significantly but some survived the reaction. Large amounts of swelling clay and chlorite were produced. The co-existence of berthierine, the swelling clay, and chlorite is successfully predicted by the PATH modeling (see following). The albite/quartz peak area ratio may have decreased.

***Nature of the clay minerals formed during hydrothermal reactions***

In addition to analcime, the reaction products contain two major clay minerals: the 17 Å swelling clay and a chlorite. The swelling clay has variable morphology and is Fe-rich in composition and trioctahedral in crystal structure. In order to model the hydrothermal reactions of berthierine-rich sands, efforts were made to document the chemical composition of the swelling clay and the chlorite. Table 5 shows the chemical composition of the swelling clay as determined by AEM. It can be noted that 1) total exchangeable cations are high (M=1.6 based on 20 oxygens and 4 hydroxides). From a mineralogical point of view, the swelling clay should be classified as a vermiculite. The operation definition of smectite is its layer charge, lying between 0.20 to 1.20 (for every 22 oxygens), and that of vermiculite between 1.2 to 1.8 [13]. However, it is possible that the apparent high layer charge is due to contamination by salts deposited on the clay surface. In addition, upon ethylene glycol treatment, the (001) peak of the swelling clay was clearly shifted to 17 Å, which is characteristic of smectite. It should be noted that none of the known vermiculites contains significant ferrous Fe in their structure [13]. Although trioctahedral Fe-smectite is rare, both natural and synthetic trioctahedral Fe-smectite have been reported in literature and they are tentatively classified as Fe-saponite [14]. Based on this, the swelling clay formed in our hydrothermal tests will still be called Fe-saponite.

The composition of the chlorite is shown in Table 6. The analysis indicates the consistent presence of K. This particular sample was K-saturated. Thus it is possible that the K is adsorbed onto the chlorite surface. Based on their chemical composition (Tables 2, 5 and 6), the structural formula for berthierine, Fe-saponite, and Fe-chlorite are assigned as:

Berthierine - $(Fe_{1.5}Al_{0.7}Mg_{0.65})(Si_{1.6}Al_{0.4})O_5(OH)_4$

Fe-saponite - $Na_{0.8}(Fe_{1.9}Al_{0.2}Mg_{0.9})(Si_3Al)O_{10}(OH)_2$

Fe-chlorite - $(Fe_{2.6}Al_{2.0}Mg_{0.80})(Si_{3.2}Al_{0.8})O_{10}(OH)_8$

***Summary of experimental results***

The major observations from the autoclave experiments are: 1) berthierine is consumed in solutions with a wide range of composition (pH). In most cases, it is consumed completely in six weeks of experiment; 2) Fe-saponite is the predominant reaction product in neutral to basic solutions; 3) Fe-chlorite is the major reaction product in acidic solution; 4) analcime is an important reaction product in some runs; and 5) quartz to albite ratio in the bulk-sands samples may be decreased in all but acidic solutions.

## GEOCHEMICAL MODELLING OF THE HYDROTHERMAL REACTIONS

***Thermodynamic Properties of Berthierine, Fe-saponite, and Fe-chlorite***

The hydrothermal stability of the Cold Lake berthierine and its decomposition reactions can be estimated through geochemical modeling. The prerequisite condition for such a modeling is that the thermodynamic properties of the reactants and the products are known. The major reactant minerals are quartz, albite, and berthierine, and the major product minerals are Fe-saponite, analcime, and chlorite. Among which, the thermodynamic properties of quartz, albite, analcime are well known, but those of berthierine, Fe-smectite, and Fe-chlorite are not known. We adapted the method of Tardy and Garrels [15] to estimate the free energy of formation of berthierine, Fe-smectite, and Fe-chlorite. The third law entropies and heat capacity constants of these minerals were estimated using the method of the correpondance of structures [16]. Table 7 summarizes the estimated thermodynamic properties of the three minerals.

*Reaction Simulation with PATHARC*

With the estimated thermodynamic properties of berthierine, Fe-saponite and Fe-chlorite, a geochemical computer program, PATHARC [17], was used to simulate the hydrothermal reactions of berthierine-rich sands. Based on thermodynamic equilibrium and mass balance, PATHARC calculates the solution composition and mineral assemblages as reactions progress towards equilibrium. The solid/solution ratio and the solution compositions are identical to those used in the experiments. The solid composition is estimated based on the results of XRD analysis. The starting material is assumed to contain 60 wt% of quartz, 30 wt% of albite and 10 wt% of berthierine. The C/S phase was not included because its chemical composition and thermodynamic properties are not known.

*Run 1: The Reaction of Bulk Sands in $NaHCO_3/CO_2$ Solution at 250°C.* The modeling results are shown in Fig. 7. The scale on the X-axis, Log (Reaction Progress), is a complex function of reaction time; the scale on Y-axis represents the amount of minerals per 1 kg of water. It should be noted that the quartz and albite have a different scale from the rest of the minerals. All of the berthierine and a small portion of quartz were dissolved during the reaction. The albite content fluctuated slightly but its net change was minimal. A large amount of Fe-saponite and a lesser amount of Fe-chlorite are predicted to form as the berthierine is dissolved. The relative abundance of the Fe-saponite and Fe-chlorite fluctuated with reaction progress. This is because of a number of minerals (e.g. antigorite, talc, and siderite) whose contents were very small (therefore they are not plotted on Fig. 7). However, the appearance/disappearance of these minerals affected the overall equilibrium in the system and caused the fluctuation in Fe-saponite and Fe-chlorite contents. The modeling results are consistent with experimental observation in two major aspects. First, both the experiment and prediction shows that the berthierine was consumed and replaced by Fe-saponite. Second, the albite/quartz ratio was predicted to decrease, as seen in the experiment. The formation of Fe-chlorite, as predicted by modeling, was not observed in the experiment. The PATHARC model does not predict the formation of analcime, which was observed in experiments.

*Run 2: The Reaction of Bulk Sands in $NaHCO_3/NaOH$ Solution at 250°C.* The PATHARC simulation predicts that all of the berthierine in the sands will be destroyed during the reaction (Fig. 8). A small fraction of quartz was also dissolved. Albite contents fluctuated but the net change is minimal. The fluctuation in albite content explains the euhedral albite crystals observed in the SEM (Fig. 6). A large amount of Fe-saponite and a lesser amount of Fe-chlorite were formed. A small amount of paragonite was formed during the reaction but was consumed at the end of the reaction. A very small amount of ferrous oxide, serpentine and talc appeared and then disappeared during the reaction (modelled). They are not plotted in Fig. 8 but were responsible for the fluctuation of Fe-saponite and Fe-chlorite contents between -4.5 and -4.0 of the Log (Reaction Progress). The modeling results are comparable to the experimental results except that the formation of chlorite is not observed in the experiment.

*Run 3: The Reaction of Bulk Sands in $Na_2SiO_3/NaCl$ Solution at 250°C.* The PATH.ARC simulation predicts that the berthierine will be consumed completely in the reaction at 250°C (Fig. 9). The quartz content showed a small decrease. The albite content showed a slight increase. The major reaction product was again Fe-saponite, which is consistent with the experimental result.

*Run 4: The Reaction of Bulk Sands in HCl Solution at 250°C.* The PATHARC simulation predicts very complex hydrothermal reactions in this system (Fig. 10). The reaction was allowed to proceed until the Log(Reaction Progress) equalled -3.6 so that the terminal pH value would correspond to the observed pH value. In this interval, 2.5 grams of albite and 1.25 grams of berthierine would be consumed and about one gram of quartz would be formed in the reaction. A large amount (1.4 grams) of Fe-chlorite was predicted to form. Kaolinite would be formed initially with Fe-chlorite but was consumed later on in the reaction when Mg-smectite was saturated and started to precipitate. The solution pH increased slowly from 2 to 5.5 in this period. The experimental observation fits the modeling results. First, about half of the berthierine remained after the reaction. Secondly, a significant amount of smectite and chlorite were produced. Thirdly, the quartz/albite ratio increased after the reaction. It is important to note that the solution pH at the end of the first interval was about 5.5, similar to the measured post-run value. The results suggest that of berthierine-rich oil sands/HCl solution reaction did not reach equilibrium during the six weeks of reaction time. Should the reaction last longer, the terminal solution pH would rise to 6.7 and the stable mineral assemblage would be quartz, albite, Fe-chlorite and paragonite.

***Summary of Modeling Results***

All the major experimental observations were simulated by the PATHARC modeling, which are: 1) berthierine is not stable under the experimental conditions and will be consumed by hydrothermal reactions; 2) Fe-saponite is the predominant reaction product in neutral to alkaline solutions; and 3) Fe-chlorite is the predominant reaction product in acidic solution. However, the formation of analcime may be slightly underestimated by PATHARC modeling, whereas the formation of Fe-chlorite is overestimated by PATHARC modeling. These inconsistencies may be minimized when the thermodynamic and kinetic properties of the concerned minerals are improved. It is interest to note that mica, kaolinite, and other minerals can be formed and then dissapear again in the reactions.

## DISCUSSION

***Hydrothermal Stability of Berthierine***

The experimental results shown here confirm that berthierine can be transformed into chlorite in hydrothermal solution as suggested in the literature (1-3). However, this reaction is only significant in acidic solution and will be overtaken by the berthierine to Fe-saponite reaction in neutral to alkaline solutions. The berthierine-to-chlorite transformation may not significantly affect reservoir quality because berthierine and chlorite have similar chemical composition and physical properties.

To our best knowledge, the berthierine to Fe-saponite reaction has not been reported in literature. The experimental results shown here strongly indicate that berthierine would be decomposed to form a Fe-rich swelling clay in neutral to alkaline solution

The experimental data is strongly supported by PATHARC geochemical modeling. Based on the estimated thermodynamic data, the PATHARC program successfully predicts that berthierine is an unstable phase relative to Fe-saponite and Fe-chlorite. The PATHARC program also predicts that the Fe-saponite is the predominant reaction product in most hydrothermal solutions tested in this study.

***The effect of berthierine decomposition on reservoir quality***

A direct result of mineral reaction in a hydrocarbon reservoir is the change in its porosity and permeability. Usually, a coreflood experiment has to be conducted to evaluate this effect. At present, some insight on this effect may be obtained by considering the volume changes caused by berthierine decomposition. Based on reaction stoichiometry and specific gravity of reactant and product minerals, the berthierine to Fe-saponite reaction is accompanied by a volume change of 50% [11]. It should be noted that this volume increase does not include the osmotic component of clay swelling, which would take place in fresh water and cause further formation damage. This indicates that a significant increase in solid volume can be expected for the berthierine to Fe-saponite reaction. The reaction will undoubtedly reduce the porosity and permeability of the hydrocarbon reservoirs. However, the magnitude of this effect on reservoir porosity and permeability would depend on the clay concentration and initial porosity of the reservoirs. However, it should be emphasized again that this conclusion is based on the experiments and PATHARC modeling, both of which were conducted at 250°C. If the temperature is lower, the reaction may not take place or only proceed very slowly.

As described above, the Cold Lake berthierine occurs mainly as surface coatings of the sand matrix in the reservoirs. As such, the colloidal and hydrodynamic forces required to dislodge the berthierine particles would be large. However, after hydrothermal reaction, the berthierine will be destroyed and the produced analcime and Fe-saponite will loosely sit on sand grains or have accumulated in the pores. Fe-saponite occurs as elongated fibres and laths. Thus, the potential for fines transport may be increased. Clay swelling is a common problem in hydrocarbon reservoirs [18]. The swelling behavior of Fe-saponite is unknown. Most of the saponites in nature are Mg-rich end-members. Assuming that the swelling sensitivities of the Fe-saponite is comparable to Mg-saponite, the hydrothermal Fe-saponite would have a relative small swelling potential when compared with other smectites. In addition, the layer charge of the Fe-saponite found in the autoclave tests are very high. This would again limit its osmotic swelling. We conclude that the Fe-saponite only has a limited potential for osmotic swelling.

## CONCLUSIONS

Based on the experimental results and PATHARC modeling, we conclude:

1. The Cold Lake berthierine is not stable in hydrothermal solutions over a wide range in pH at 250°C. Both the experimental and modeling results show it will decompose in hydrothermal reactions. In neutral to alkaline solution, the principal decomposition product of the berthierine is a swelling clay, with accessory analcime. The swelling clay produced in this reaction is Fe-rich and trioctahedral and is classified as Fe-saponite. However, the layer charge of Fe-saponite is very high. In acidic solution, Fe-chlorite is the major reaction product.

2. There is more than a 50% volume increase for the berthierine-to-saponite reaction. This volume increase will decrease the porosity and permeability of the hydrocarbon reservoirs. The degree of the damage will depend on the reservoir mineralogy, clay concentration, and initial porosity of the hydrocarbon reservoirs. The more berthierine in the reservoir and the less the initial porosity, the greater the damage. There should be no volume increase from the berthierine to Fe-chlorite reaction.
3. The potential of fines migration will likely increase as the surface coating berthierine is broken down or replaced by loosely distributed Fe-saponite and analcime.
4. The Fe-saponite may have a limited potential for osmotic swelling. The impairment of permeability by smectite swelling will be limited.

## ACKNOWLEDGEMENTS

We thank Dr. F.J. Longstaffe of the University of Western Ontario for his comprehensive work on the occurrence and origin of the Cold Lake berthierine, which led us to develop the current study. We thank Dr. J. Dudley of Imperial Oil and Mr. Andrew Fox of Amoco for their enthusiasm and support for this project. Bernice Kadatz and Shauna Cameron of the Alberta Research Council (ARC) performed autoclave experiments, and Dr. E.H. Perkins provided assistance in PATH modeling. Funding by the AOSTRA/ARC/CANMET/Industry Research Program is gratefully acknowledged.

## REFERENCES

1. B. Velde. *Clay Minerals: A Physico-Chemical Explanation of Their Occurrence*. Developments in Sedimentology, Elsevier, Amsterdam & New York, pp. 40, 1985.
2. C.D. Curtis. Clay mineral precipitation and transformation during burial diagenesis, *Philosophical Transactions of the Royal Society of London*, **A315**, 91-105, 1985.
3. J.S. Jahren and P. Aagaard. Compositional variations in diagenetic chlorites and illites, and relationships with formation-water chemistry, *Clay Minerals*, **24**, 159-170, 1989.
4. D.M. Wightman, B. Rottenfusser, J. Kramers, R. Harrison. Geology of Alberta oil sands deposits In *AOSTRA Technical Handbook on Oil Sands, Bitumens and Heavy Oils*, L.G. Hepler and C. Hsi (eds.), AOSTRA, 203-258, 1989.
5. E.R.C Hornibrook and F.J. Longstaffe. Berthierine from the Lower Cretaceous Clearwater Formation, Alberta, Canada. *Clays and Clay Minerals*, **44**, 1-21, 1996.
6. J.A. Boon, T. Hamilton, L. Holloway and B. Wiwchar. Reaction between rock matrix and injected fluids in Cold Lake oil sand - potential for formation damage, *Journal of Canadian Petroleum Technology*, **22**, 55-66, 1983.
7. W-L. Huang and J.M. Lango. Experimental studies of silicate-carbonate reactions II. Applications to steam flooding of oil sands, *Applied Geochemistry*, **9**, 523-532, 1994.
8. W.D. Gunter and G.W. Bird. $CO_2$ production in tar sand reservoirs under *in situ* steam temperature: reactive calcite dissolution, *Chemical Geology*, **70**, 301-311, 1988.
9. Sedimentology Research Group. The effects of *in situ* steam injection on Cold Lake oil sands, *Bulletin of Canadian Petroleum Geology*, 29, 447-478, 1981.
10. I. Hutcheon. A review of artificial diagenesis during thermal enhanced recovery, *Mem. Am. Ass. Petrol. Geol*, 37, 413-429, 1984.

11. Z. Zhou, W.D. Gunter, B. Kadatz, and S. Cameron. Hydrothermal stability of the clay minerals from the clearwater reservoirs at Cold Lake, Alberta, *Heavy Crude and Tar Sands - Fueling for A Clean and Safe Environment, the 6th Unitar International Conference on Heavy Crude and Tar Sands*, UD DOE, 27-35, 1995.
12. Y.K. Kharaka, W.D. Gunter, P.K. Aggarwal, E.H. Perkins, and J.D. DeBraal. *SOLMINEQ.88: A Computer Program for Geochemical Modelling of Water-Rock Interactions*, U.S. Geological Survey, Water-Resources Investigations Report 88-4227, 1988.
13. De La Calle and H. Suquet. Vermiculite, in *Hydrous Phyllosilicates*, S.W. Bailey (ed), MSA Reviews in Mineralogy, **19**, 455-496, 1988.
14. N. Guven. Smectites in *Hydrous Phyllosilicates*, S.W. Bailey (ed), MSA Reviews in Mineralogy, **19**, 497-560, 1988.
15. Y. Tardy and R.M. Garrels. A method of estimating the Gibbs energies of formation of layers silicates, *Geochimica et Cosmochemica Acta*, **38**, 1101-1116, 1974.
16. H.W. Nesbitt. Estimation of the thermodynamic properties of Na-Ca- and Mg-beidellites, *Canadian Mineralogist*, **15**, pp. 22-29, 1977.
17. E.H. Perkins and W.D. Gunter. *A Users Manual for β PATHARC.94*, Alberta Research Council, ENVTR 95-11, 1995.
18. Z. Zhou, W.D. Gunter, B. Kadatz, and S. Cameron. Effect of clay swelling on reservoir quality, *The Journal of Canadian Petroleum Technology*, 35, 18-23, 1996.

**Table 1: Materials and Conditions for Autoclave Tests**

| Run ID | Solution Composition | Temp °C | Duration (weeks) | Prerun pH 25°C | Postrun pH 25°C | Postrun pH 250°C |
|---|---|---|---|---|---|---|
| B1 | 0.1M $NaHCO_3$ 2.5 Bar $pCO_2$ | 250 | 6 | 8.41 | 6.38 | 6.38 |
| B2 | 0.1M $NaHCO_3$ 0.08M NaOH | 250 | 6 | 10.5 | 9.96 | 7.76 |
| B3 | 0.03M $Na_2SiO_3$ 0.1M NaCl | 250 | 6 | 12.27 | 11.59 | 9.49 |
| B4 | 0.01M HCl | 250 | 6 | 2.21 | 5.52 | 5.52 |

**Table 2: Analytical Electron Microscope Analysis of the Cold Lake Berthierine**

| | Si | Al | Fe | Mg | Cr | Mn | Ti | Ca | K | Na |
|---|---|---|---|---|---|---|---|---|---|---|
| 1 | 6.254 | 4.414 | 5.901 | 2.490 | .000 | .000 | .000 | .000 | .174 | .789 |
| 2 | 6.476 | 4.119 | 6.119 | 2.141 | .195 | .317 | .000 | .000 | .000 | .000 |
| 3 | 6.446 | 4.539 | 5.803 | 1.883 | .166 | .223 | .000 | .000 | .285 | .000 |
| 4 | 6.634 | 3.816 | 5.968 | 2.083 | .000 | .124 | .417 | .000 | .000 | .000 |
| 5 | 6.892 | 3.871 | 5.350 | 2.469 | .000 | .114 | .238 | .000 | 000 | 000 |
| 6 | 6.763 | 4.062 | 5.889 | 2.093 | .052 | .108 | .000 | .000 | .000 | .426 |
| 7 | 6.190 | 4.433 | 6.472 | 2.310 | .000 | .082 | .000 | .066 | .079 | .000 |
| 8 | 6.271 | 4.434 | 6.139 | 2.556 | .000 | .054 | .000 | .000 | .095 | .000 |
| 9 | 6.040 | 4.556 | 5.992 | 3.094 | .000 | .000 | .000 | .000 | .000 | .000 |
| 10 | 6.460 | 4.259 | 5.974 | 2.596 | .000 | .000 | .000 | .000 | .244 | .000 |
| Av. | 6.443 | 4.250 | 5.962 | 2.372 | .041 | .102 | .066 | 007 | .088 | .122 |
| Std | 0.252 | 0.256 | 0.270 | 0.330 | .072 | .098 | .137 | 020 | .105 | .256 |

**Table 3: Composition of Post-run Solutions**

| Sample ID | B1 | B2 | B3 | B4 |
|---|---|---|---|---|
| Temperature C | 250 | 250 | 250 | 250 |
| Time (weeks) | 6 | 6 | 6 | 6 |
| pH | 6.38 | 9.98 | 11.70 | 5.52 |
| pH (TIC) | 6.77 | 9.96 | 11.59 | ---- |
| Na | 1910 | 2970 | 3300 | 239 |
| Na×10[#] | 2200 | 3800 | 3490 | 230 |
| K | 48 | 122 | 120 | 64.5 |
| Ca | 6.8 | 0.2 | -0.01 | 24.1 |
| Mg | 0.3 | <0.1 | -0.1 | 0.4 |
| Al | 0.4 | 0.2 | 0.3 | 0.1 |
| Fe | 1.7 | 0.2 | 0.3 | 0.1 |
| Mn | 0.3 | <0.01 | <0.01 | <0.01 |
| Cl** | | | 3600 | 431 |
| TIC (measured) | 1369 | 1233 | 30.3 | 3.15 |
| TIC (corrected)* | 2046 | | | |
| Si | 204 | 1050 | 1122 | 356 |
| SiO2 | 436 | 2246 | 2400 | 762 |
| TDS* | 7443 | 12403 | 9762 | 1526 |
| Ion Balance (%)* | -4.59 | -16.76 | -12.82 | 7.66 |
| S.I. Quartz | 0.264 | -0.68 | -131 | 0.247 |

*values from SOLMINEQ.88 calculation; **from titration, [#]calculated from the samples diluted 10 times at quenching time. Solute concentrations are all in mg/L.

**Table 4: Normalized Peak Intensities of Minerals in Post-run Clay Fractions (%)***

| | Smec. | C/S | Chl. | Illite | Berth. | Anal | Quartz | Albite |
|---|---|---|---|---|---|---|---|---|
| Prerun | | 4.42 | | | 94.26 | | 0.79 | 0.54 |
| Post-run B1 | 61.60 | | | 6.39 | 8.29 | @ | 14.51 | 9.21 |
| Post-run B2 | 63.95 | | | 15.07 | 2.44 | | 10.55 | 8.00 |
| Post-run B3 | 74.28 | | | 6.95 | | @ | 12.40 | 5.62 |
| Post-run B4 | 47.51 | | 2.49 | | 30.66 | | 18.29 | 1.05 |

*The pre and post runs of the "bulk sands" were separated into the <2 μm and >2 μm fractions, only the <2 μm fractions are shown here. Post-run B3 contains 0.74 % of stilpnomelane. @Found in bulk sample, but not in the clay fractions.

**Table 5: Analytical Electron Microscope Analysis of Fe-Saponite in Post-run Sands**

| ID | Si | Al | Fe | Mg | Cr | Mn | Ti | Ca | K | Na |
|---|---|---|---|---|---|---|---|---|---|---|
| 1 | 5.875 | 2.575 | 4.007 | 1.451 | .055 | .054 | .000 | .149 | .087 | 1.270 |
| 2 | 5.757 | 2.638 | 4.100 | 1.404 | .041 | .051 | .048 | .182 | 063 | 1.206 |
| 3 | 6.460 | 2.405 | 3.405 | 1.391 | .000 | .072 | .000 | .150 | .053 | 0.855 |
| 4 | 6.248 | 2.293 | 3.877 | 1.191 | .000 | .000 | .000 | .186 | 094 | 1.525 |
| Av. | 6.085 | 2.478 | 3.847 | 1.359 | .024 | .044 | .012 | .167 | .074 | 1.214 |
| Std | 0.282 | 0.137 | 0.267 | 0.100 | .025 | .027 | .021 | .017 | .017 | 0.239 |

**Table 6: Analytical Electron Microscope Analysis of Fe-Chlorite in Post-run Sands**

| ID | Si | Al | Fe | Mg | Cr | Mn | Ti | Ca | K | Na |
|---|---|---|---|---|---|---|---|---|---|---|
| 1 | 6.595 | 5.650 | 4.962 | 1.281 | .000 | .000 | .000 | .000 | .182 | .000 |
| 2 | 6.162 | 5.758 | 4.938 | 1.644 | .000 | .126 | .000 | .000 | .153 | .511 |
| 3 | 6.065 | 4.820 | 5.968 | 2.483 | .000 | .000 | .000 | .000 | .379 | .000 |
| 4 | 6.048 | 5.879 | 5.559 | 1.241 | .000 | .000 | .000 | .152 | .267 | .000 |
| 5 | 6.789 | 6.234 | 3.769 | 1.042 | .000 | .124 | .000 | .000 | .268 | .000 |
| 6 | 7.060 | 5.711 | 3.845 | 0.917 | .000 | .000 | .186 | .000 | .357 | .000 |
| 7 | 6.477 | 5.598 | 4.783 | 1.431 | .000 | .000 | .107 | .000 | .441 | .000 |
| 8 | 6.753 | 5.526 | 4.150 | 1.841 | .000 | .000 | .000 | .000 | .427 | .000 |
| Av. | 6.275 | 5.381 | 4.517 | 1.341 | .000 | .036 | .040 | .015 | 328 | .048 |
| Std | 0.543 | 0.632 | 0.825 | 0.511 | .000 | .055 | .064 | .046 | .122 | .153 |

**Table 7: Thermodynamic Properties of Berthierine, Fe-Saponite, and Fe-Chlorite**

| | Berthierine | Fe-Saponite | Fe-Chlorite |
|---|---|---|---|
| $\Delta G_f°$ (J/mol) | -7,106,189 | -5,140,755 | -7,166,774 |
| $\Delta H_f°$(J/mol) | -7,723,808 | -5,501,572 | -7,783,055 |
| S° (J/k-mol) | 544.45 | 364.13 | 538.70 |
| a | 675.39 | 431.20 | 705.01 |
| b x $10^3$ | 241.75 | 112.87 | 179.7 |
| c x $10^{-5}$ | -157.10 | -89.54 | -155.14 |

Berthierine: $(Fe_3Mg_{1.3}Al_{1.4})(Si_3Al)O_{10}(OH)_8$
Fe-Saponite: $Na_{.8}(Fe_{1.9}Al_{.2}Mg_{.9})(Si_3Al)O_{10}(OH)_2$
Fe-Chlorite: $(Fe_{2.6}Al_2Mg_{.8})(Si_{3.2}Al_{.8})O_{10}(OH)_2$

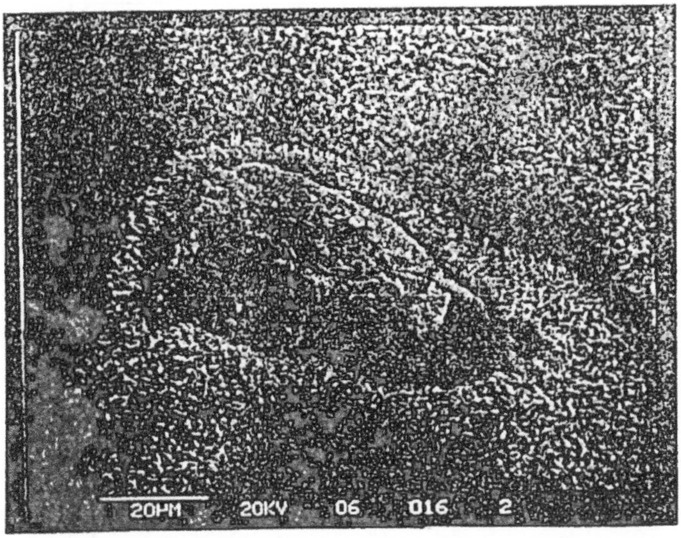


**Fig. 1** Berthierine coatings on sand grains at the Clearwater Formation, Cold Lake, Alberta.

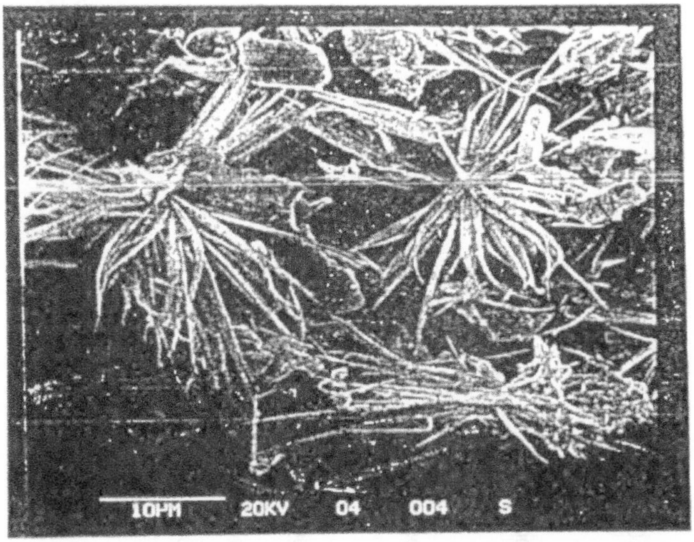


**Fig. 6** Scanning electron micro-photo of swelling clays formed during hydrothermal reaction of oil sands in $Na_2HCO_3$/NaOH Solution. The swelling clays have distinct morphology of radiant fibers. The rectangular grains are albite.

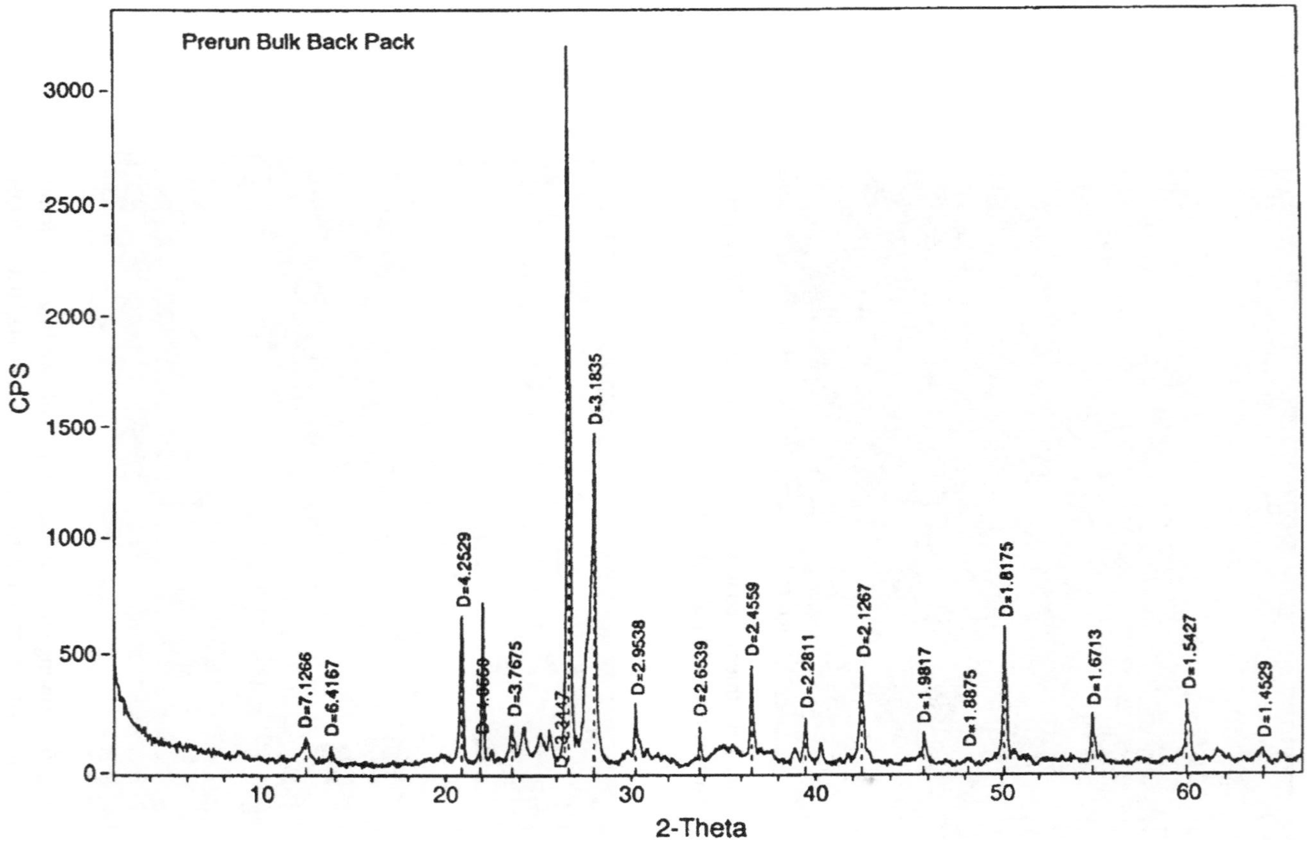


**Fig. 2** X-ray Diffraction (XRD) trace of the oil sands from Cold Lake, Alberta. All the XRD analysis were done with Cu K-alpha radiation.

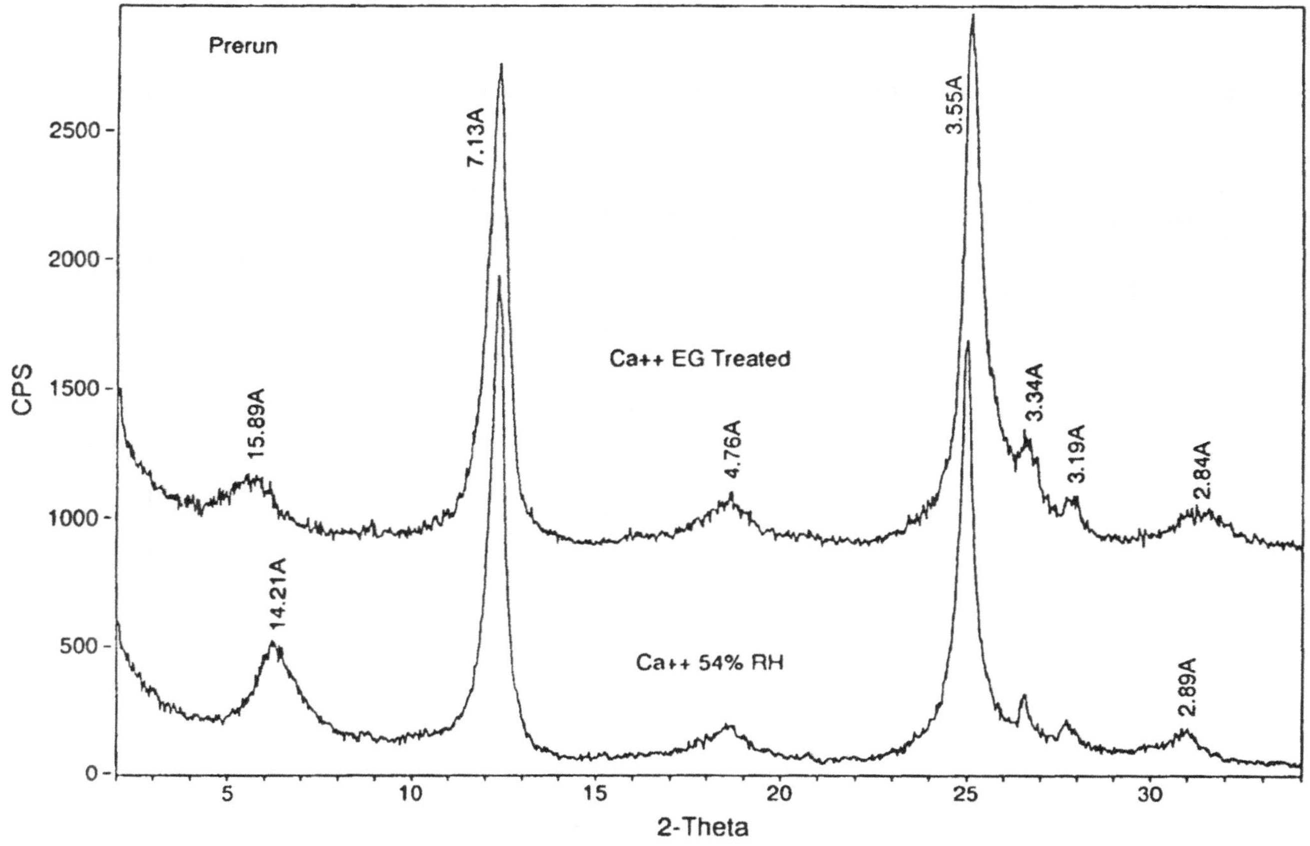


**Fig. 3** X-ray Diffraction (XRD) traces of clay fraction from Cold Lake oil sands. Ca-54%RH: Ca-saturated and equilibrated with 54% RH; Ca-EG: Ca-saturated and ethylene glycol treated. Berthierine (7.13 and 3.55Å) is predominant.

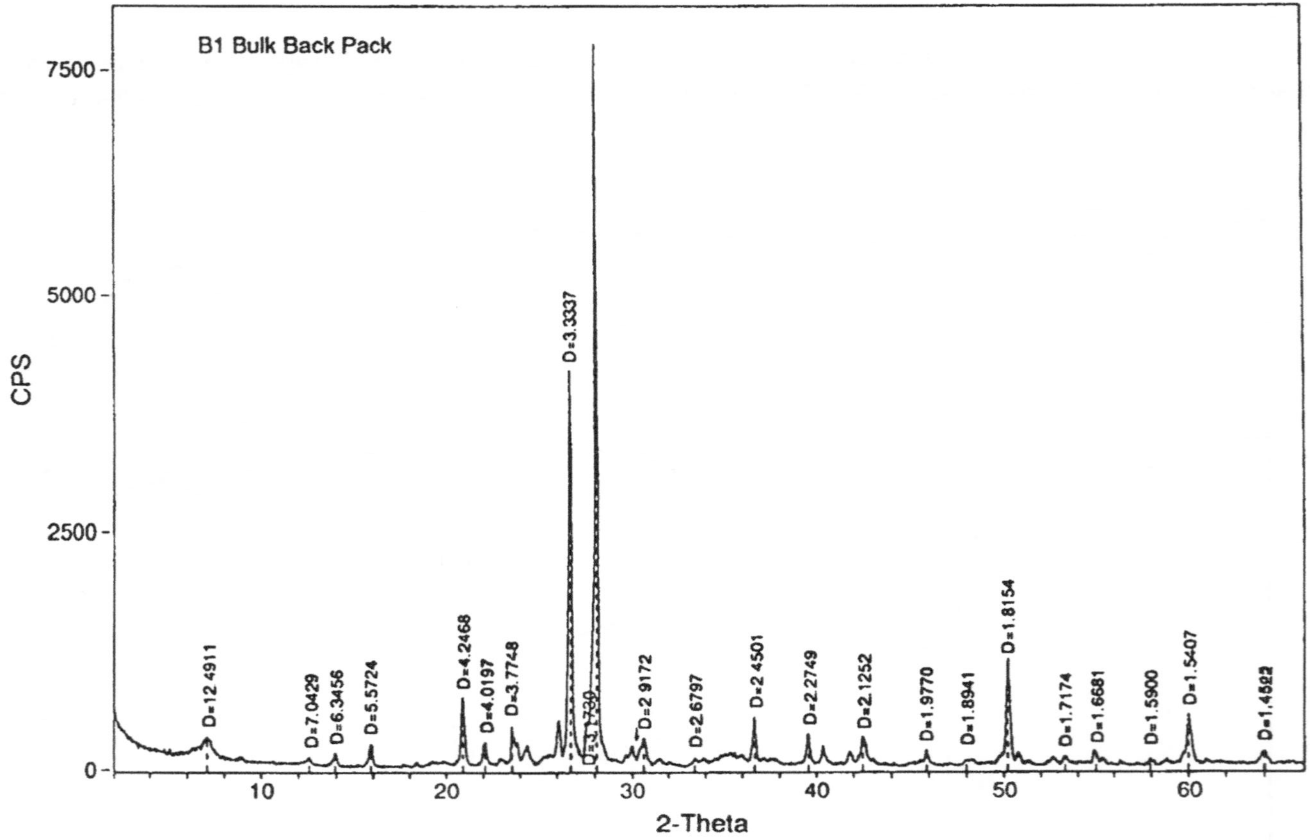


**Fig. 4 X-ray Diffraction (XRD) trace of post-run oil sands. The albite/quartz peak intensity ratio is higher than that of the pre-run sample.**

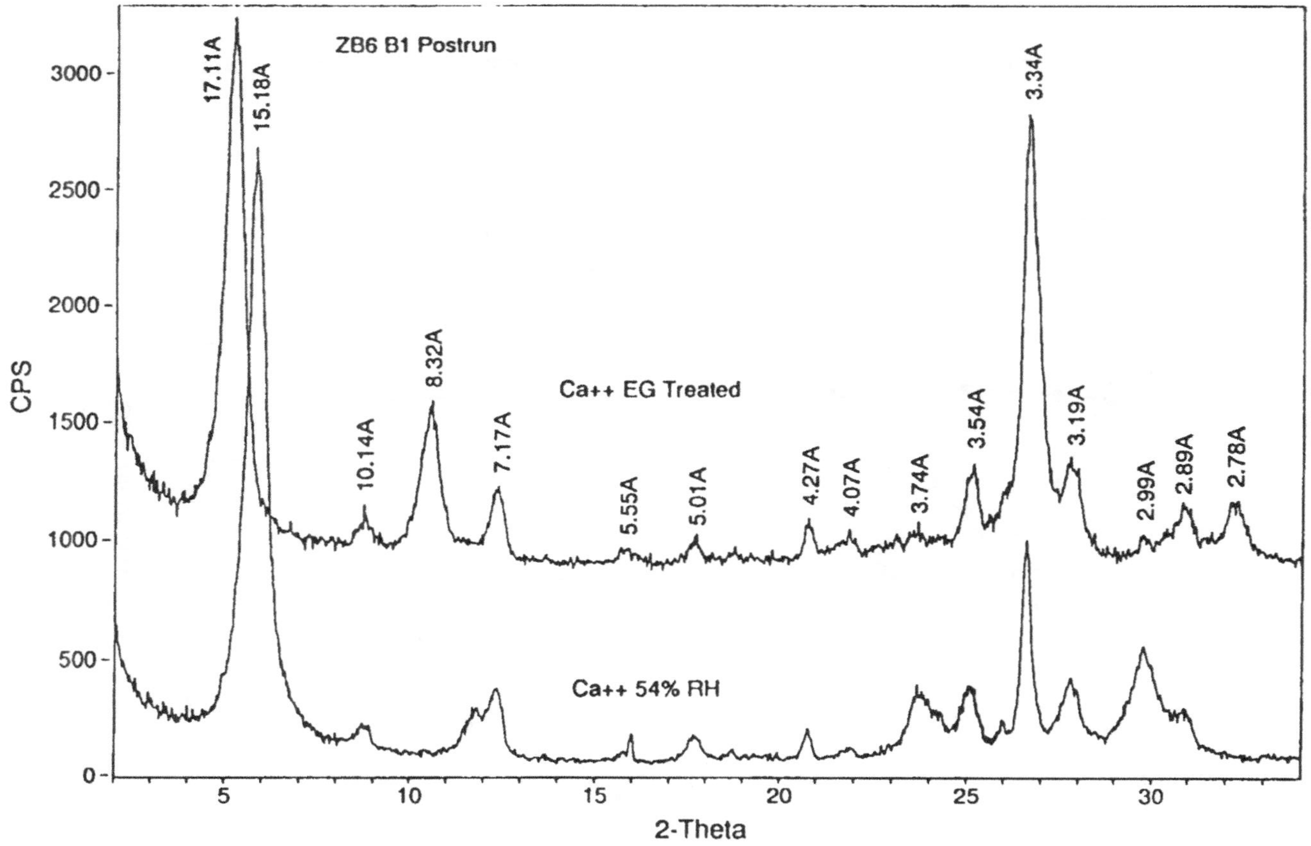


**Fig. 5** X-ray Diffraction (XRD) traces of the clay fraction in the post-run oil sands XRD Traces of ZB6 Clay Fraction. The 17 Å swelling clay is predominant. The swelling clay has well developed *(00l)* diffraction bands. Less amounts of berthierine (7Å & 3.5Å), quartz, albite, and illite (10Å & 5Å) are present.

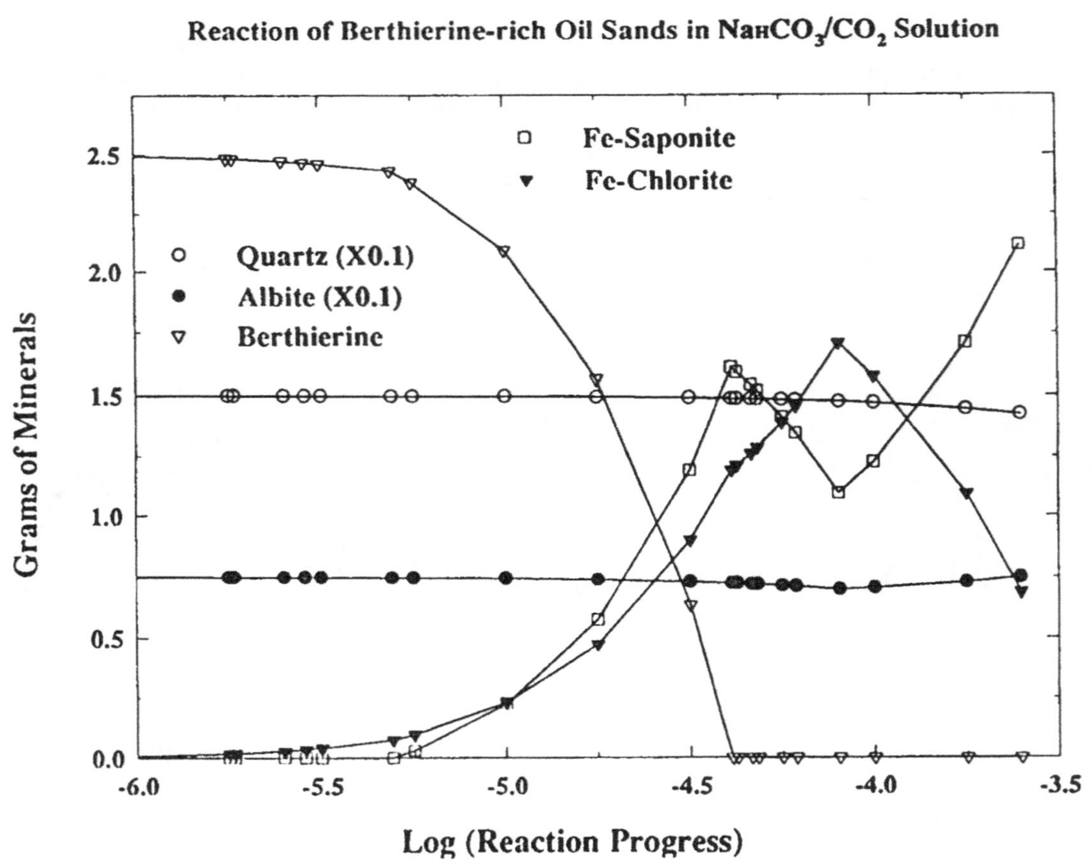


**Fig. 7** PATHARC Run 1: Reaction of bulk sands in $NaHCO_3/CO_2$ solution at 250°C

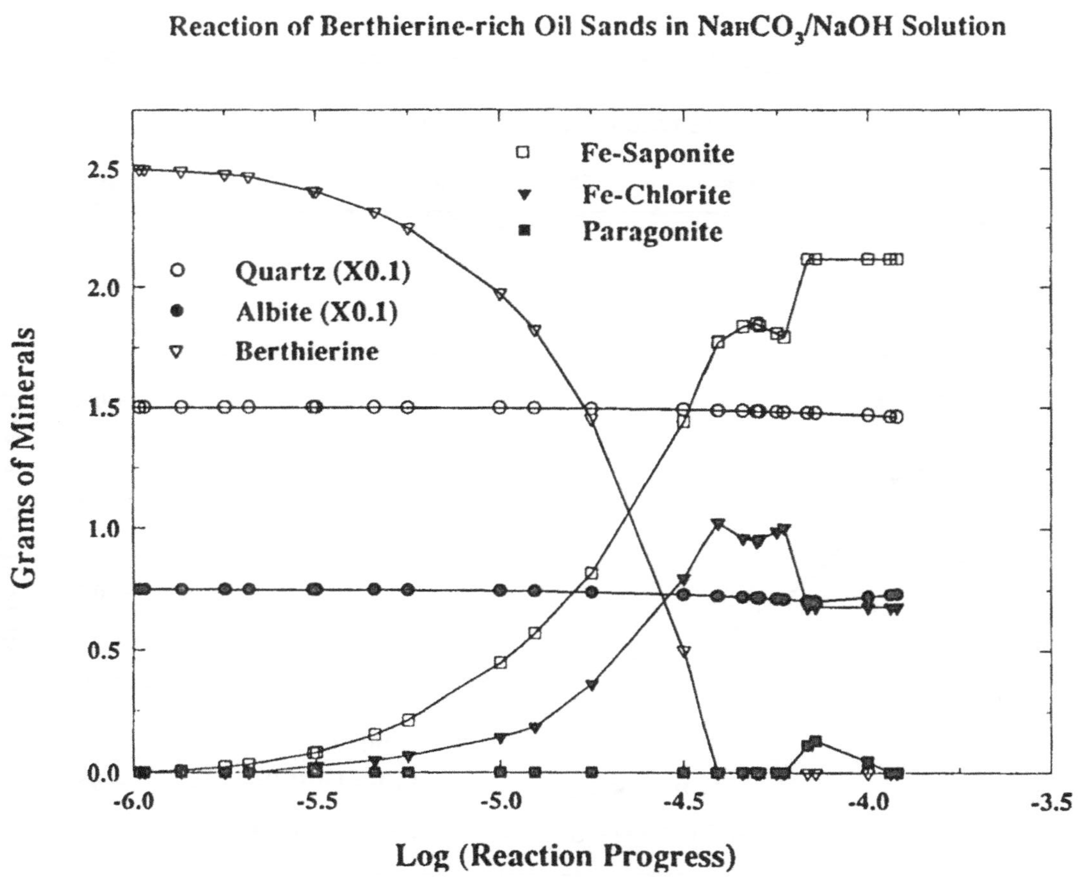


**Fig. 8** PATHARC Run 2: Reaction of bulk sands in $NaHCO_3$/NaOH solution at 250°C

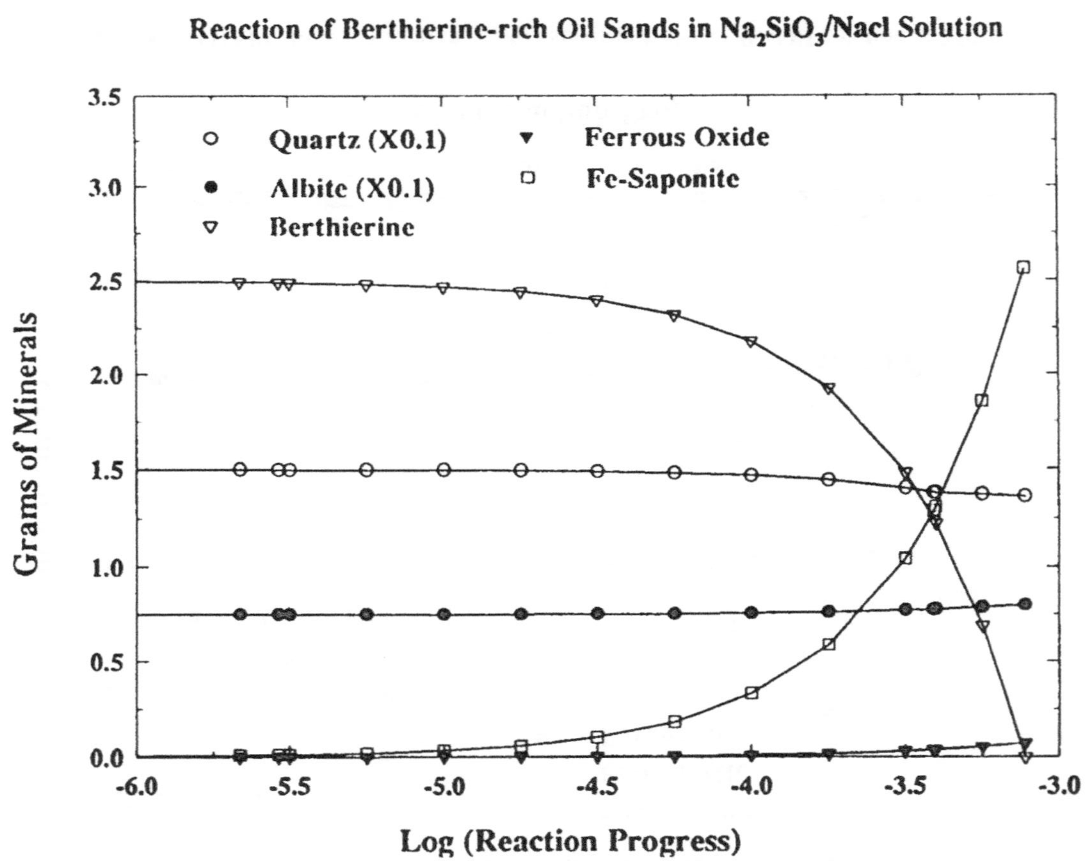


**Fig. 9** PATHARC Run 3: Reaction of bulk sands in $Na_2SiO_3$/NaCl solution at 250°C

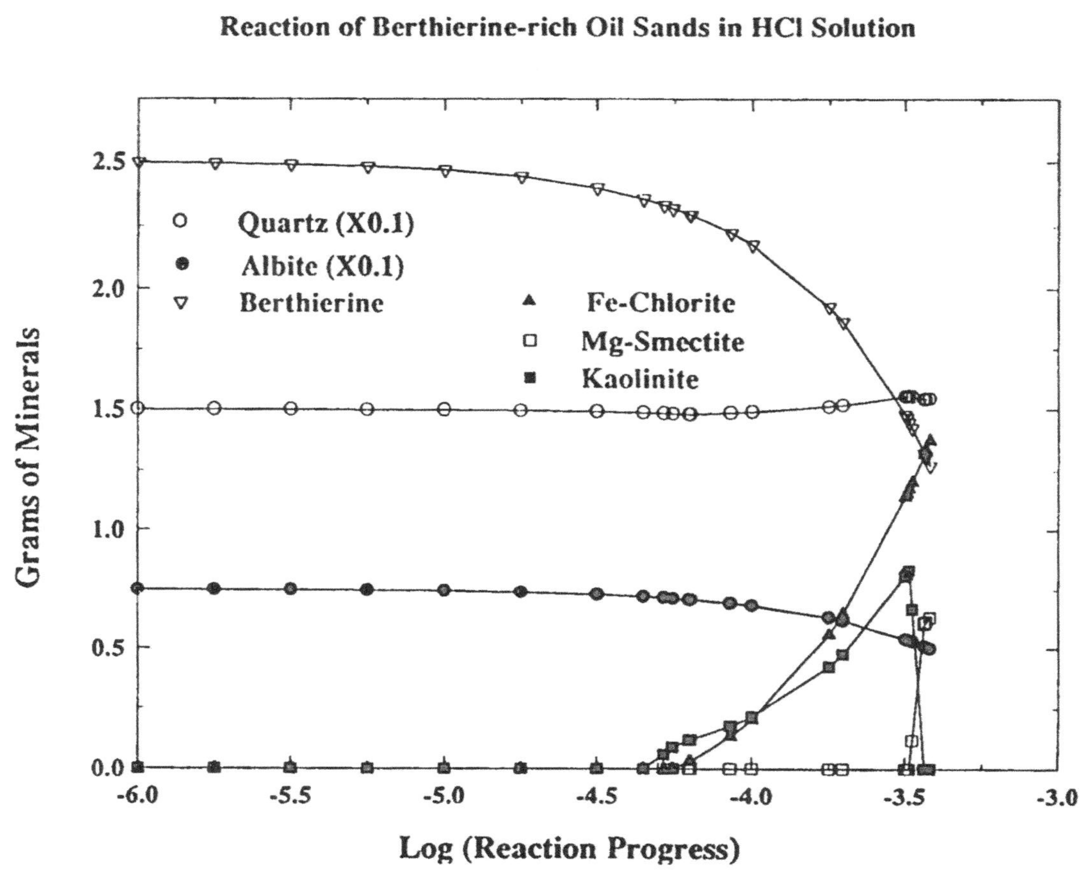


**Fig. 10** PATHARC Run 4: Reaction of bulk sands in HCl solution at 250°C until pH = 5.5.

# GEOCHEMICAL KINETICS

*Proc. 30th Int'l. Geol. Congr.*, Vol. 19, pp. 29-47
Xie Xuejin (Ed)


# KINETICS OF MINERAL DISSOLUTIONS IN OPEN-FLOW SYSTEMS AND NON-LINEAR DYNAMIC BEHAVIORS IN THE FLUID/SOLID INTERFACE

RONGHUA ZHANG and SHUMIN HU
*Open Research Lab. of Geochemical Kinetics, Ministry of Geology and Mineral Resources. Chinese Academy of Geol. Sci., Institute of Mineral Deposits, Baiwanzhuang Road 26, Beijing 100037, China.*
TONG JIANCHANG
*Computer center, Peking Univ. 100875, Beijing, China*

**Abstract**

In the recent years, we have obtained plenty of data of the mineral dissolution rates in aqueous solutions based on the kinetic experiments of the open-flow system. The experiments suggest that the kinetic processes of the mineral dissolutions have non-linear behavior, particularly in the reaction system far removed from equilibrium state. It has been found that the reaction rate is only stable in a short period of time, when the experimental conditions are fixed. The time for keeping the steady state of the kinetic process depends on the different minerals, the different solutions, and the physical conditions. The rates of the mineral dissolutions were measured by using a packed bed reactor and in an open-flow system. The experiments in the mineral - aqueous solution system indicate that the output concentrations of the dissolving species of the reactant products from the open flow system are not stable in some cases, in a very long term experiment, while maintaining the constant chemical and hydrodynamic conditions. The observation of the output products is carried by using an electronic conductivity detector, and recording the electronic conductivity of the output solutions continuously. And also we take the liquid example for chemical analysis. The experimental results also indicate that the reaction rates were weakly waving within a long time period. The phenomena of the non-linear behavior have been found in the fluorite-HCl-H2O system, such as the multi - steady states, self - transient of the state, and complex chaotic oscillation of the concentrations of the dissolving species in the output solutions. When change the input from water to HCl - aqueous solution in a continuous flow system, the reaction system will have a response in the output solutions for this change. This is a step change of an input solution, which can drive a non-linear response of the output solution. For instance, a temporary complex oscillation of the output solution appears in that case. This study suggests a solid-liquid interface pattern to illustrate this non-linear behavior. This model describes the coupling processes of adsorption , surface reaction, surface ion exchange and diffusion or transportation processes. And also, the multi-steps of the reaction and intermediate products on surface involve in the kinetic process, which can derive a non-lincar kinetic processes.

*Keywords: Mineral dissolution, Chemical kinetics, Non-linear behavior*

## INTRODUCTION

In the recent more than ten years, the dissolution rates of a lot of minerals in hydrothermal solutions have been tested [23,13,14,11,12,6]. We have measured the dissolution rates of many minerals: fluorite, albite, calcite, dolomite, zeolite, etc. in aqueous solutions and in the open-flow systems at elevated temperatures, T (up to 300°C) and pressures, P (up to 13.8 MPa). The experiments provided a lots of new data of mineral dissolution rates, and also found that the mineral dissolution has non-linear behavior in an open system as operating the experiments at far from equilibrium state [26, 27, 28, 29, 30, 31]. Some scientists reported that both dissolution and precipitation show strong nonlinear dependence on the $\Delta G$ of the reactions between the mineral and aqueous solutions [16,2,31]. Authors found that the reaction system of fluorite in the HCl - H2O solution has the nonlinear behavior when the reaction system removes from the close equilibrium state to a far equilibrium state. And also, authors found that this system has very complex nonlinear dynamic processes, such as multi-steady states, self-transition and more complex oscillation etc., while in a far equilibrium state of an open system.

## EXPERIMENTAL APPROACH

### *Non-Ideal Flow System*

Open flow-through reaction systems differ fundamentally from the static autoclave systems in that, matter and energies are continuously put into and removed from the systems. Open flow through reactors are ideally suited to the description of steady state and non-steady state kinetic processes and the determination of the mechanism of controlling reaction rates. Since the conditions are uniform over the reactor volume, the expression for a steady state of the reaction flow system can be expressed as follows:

$$F_{A,inp} = F_{A,out} + (-r_{AF})V_R \qquad 1.$$

where, F refers to the mole flow rate, [A, inp] means the input of A, [A, out] means the output of A. -r stands for dissolution rate of the mineral AF. This equation shows that: one can define a practical steady state in a flow system as that portion of rate vs time is linear ( i.e. output concentrations of dissolving species are time - invariant ).

$$F_{A,inp} = F_{A,out} + (-r_{AF})V_R + \frac{\partial N_A}{\partial t} \qquad 2.$$

where $\frac{\partial N_A}{\partial t}$ means the rate of accumulation of A within reactor. If $\frac{\partial N_A}{\partial t} = 0$, it is a steady state of the system. Oppositely, it is non-steady state system [7]. The continuous flow reactors can be divided into ideal and non-ideal flow systems according to fluid flow patterns. The ideal flow reactors consist of: plug flow reactor (PFR) and continuous stirred tank reactor, (or called continuous flow and well mixed reactor, CSTR ). The design of the experiment for dissolution study can be carried out by using the above two sorts of reactors with the different purposes. We use non-ideal flow reactor for fluorite dissolution study ( packed bed reactor ), which offers plug flow system superimposed by an axial dispersion. This non - ideal flow reactor can be used to simulate natural flow system and hydrothermal systems.

### *Method*

*Experimental Equipments*

The experimental equipments were constructed of titanium vessel, liquid pump, back pressure regular etc. (See Fig.1. In ref. [26,27]). And also, we improve the previous research approach, 1) we use a pump with high flow rate that the velocity ranges from 0.5 to 8 ml/min ; 2) before take solution sample, we drive the aqueous solution to pass through an electronic conductivity detector ( Wascan tech. Inc. Model 213a ) and watch the changes of the conductivity of the output solution, at any time. As measuring the reaction rates at a steady state of the reaction system for an interest, we need to know when the steady state was achieved. In any one run, the flow rate and the input solutions (e.g. pH) were held constant for a long enough time that a steady state conditions were closely achieved. So we identify the steady state of the reaction system by recording the electronic conductivity of the output solution at a variety of time. As the system achieve a steady state, the electronic conductivity of the output solutions will maintain a constant value in a long term ( for instance, in several days for fluorite - aqueous solution system). How long the time for the system to remain in a steady state depends on the hydrodynamic and chemical conditions. The reaction - flow system is also connected with computer system. The temperature and the electronic conductivity of output solutions of the system at a constant pressure are all recorded by the computer system continuously. The changes of the experimental system can be also found through a monitor.

The mineral fractions were cleaned ultrasonically in acetone for 1 hours to remove surface fines. The homogeneity of the fluorite sample was checked by X - ray diffraction (XRD) of a powder as well as by scanning electron microscope. Finally, the fluorite samples for test of the dissolution rates were analyzed chemically.

Surface areas of the natural mineral powder of the three kind of particle size were measured using the BET method, which is the continuous flow chromatography method using nitrogen and helium gas. Input solution to the experimental systems were made from deionized and degassed water. The mineral particles were put inside the tubular vessel, that is called a packed bed reactor ( PBR ). This tubular packed bed reactor with non-ideal flow was used to carry out our mineral dissolution study [27].

The most experiments on mineral dissolution included fluorite, albite, dolomite, etc., are operated by using packed bed reactor with requisite feed system, T, P and flow rate (U) control. We vary the experimental parameters, T, pH, flow rate during the course of each run, in fixed surface area of mineral sample and at constant pressure. We change the different surface area for different runs. After sampling the output solution, we analyze the composition of output solutions, by using inductively coupled plasma spectrophotometer (ICP-ES) and chromatograhpy (DIONEX 2020), and also measure the pH values of the aqueous samples. Normally, the mineral dissolution rate is expressed as -r . In our experimental flow system (PBR), the reaction rate can be written as:

$$-r = C/\bar{t} \qquad 3.$$

where define $C = C_{i,out} - C_{i,inp}$ , $C_{i,out}$ = concentration of I at outlet. and $C_{i,inp}$ = concentration of component i at inlet, $\bar{t}$ is the average residence time.

## EXPERIMENTAL RESULTS

Many scientists reported, the initial dissolution rate of a mineral is rapid, and decreases over time in the kinetic experiments, until the reaction rate becomes stable. After the initial non-steady state dissolution, the steady-state kinetic process is proceeding [12,8,30,28,31].

Usually, in the initial state of mineral dissolution the reaction rate is fast that over the first few days the dissolution rates fall progressively, presumably due to the removal of disturbed materials or fine particles and adjustment of mineral surface to the aqueous solution [18,5]. How long the initial state takes that depends on physico-chemical conditions, mineral nature and the chemistry of the input solutions. After this initial state, we can begin to measure the rate of mineral dissolution. As mentioned above that the measured data of the dissolution rates were stable in a short period of time.

One can define a practical steady state in such reactor as that portion of r and t plots that is linear. Therefore, the data of the mineral dissolution rate in a flow system has been obtained. Also the nonlinear kinetic phenomena have been found while long time operating the experiments. As fixed the physico-chemical condition for the reaction system of fluorite-aqueous solutions, the dissolution rates (r) were not stable and changed with total accumulation of the reaction time,($\Sigma$ t ) within a long time term of dissolution. This is found by continuous recording the electronic conductivity, $\lambda$, of the output solutions through a electronic conductivity detector. And also, the chemical instability of the reaction products has also been found by applying the time resolved spectra of FT-IR to test output solutions [10].

The reaction rate is not stable in a long time term of the experiment that the reaction rates wave; self-transition of the state happens, etc., in some cases. When change the input solution to the flow system, the nonlinear response of the output solution occurs in the flow system, such as very complex oscillation, multi-steady state, a temporary chaotic oscillation of the output solutions ( shown as the changes of the concentrations of dissolving species of i, Ci, or its the electronic conductivities $\lambda$ ).

### *Nonlinear Kinetics in Dissolution*

The measurement of the dissolution rates of minerals shows that the reaction rates have a nonlinear behavior with how far the distance removed form the equilibrium state of the reaction system. This is the reaction rates are strongly depending on the (Cs - C) or (Ks-Q), or chemical affinity of the reaction of $\Delta$G (Fig. 1).

The Ks - Q value shows the reaction system how far from the equilibrium state. The slope of the curves plotting between log r and log (Ks-Q) expresses that the variation of rate order is nonlinear, during the system removes from a close equilibrium state to a far from equilibrium state.

The general rate expression is described as follows:

$$-r = k\ Sa\ (Ks - Q)^{\alpha} \qquad 4.$$

where Ks is the solubility product, Q refers to the activity product, α is the reaction order, k stands for the rate constant, Sa is the surface area of mineral particles in a unit volume of aqueous solution. The Fig. 1 expresses the log r varies with log (Ks-Q), in which the slope of the curves is changing from high value to zero when the system is going to a far removed equilibrium state, i.e. the log(Ks-Q) decreases.

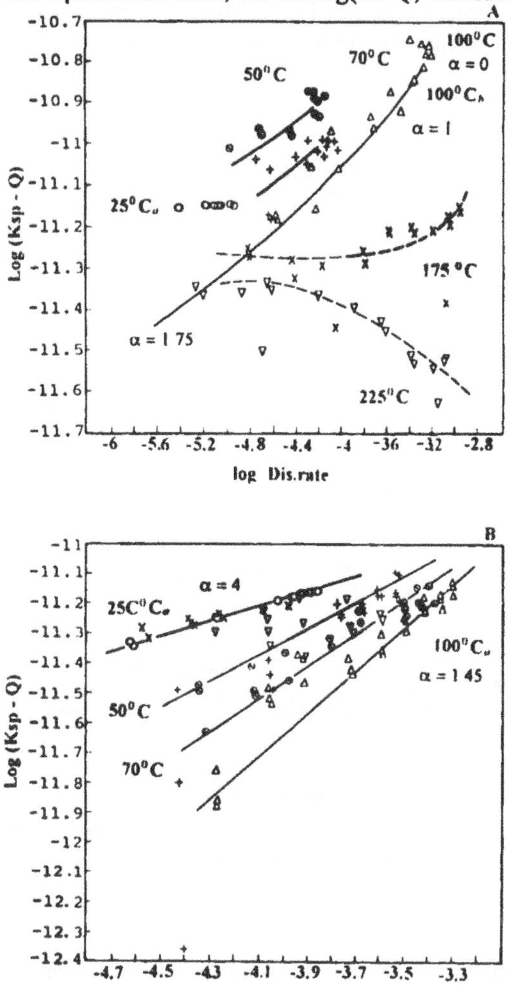


Fig.1. The logarithm of the dissolution rate of fluorite in aqueous solution vs log (Ks-Q). A, particles in size fraction of 18-35 mesh; B, in 60-80 mesh.

*Reaction Rate Wave*

It is often seen that the many studies for the rates of mineral dissolution are performed by using the mathematic average rates of the mineral dissolutions. Author had found the reaction rate variation in a long-term dissolution as fixed physico-chemical conditions, in 1986 ( Pr. Zhang visiting Princeton University), see Fig.2. Fig.2 expresses that the reaction rate of fluorite in water varies with the accumulation time, in the case of 25 and 100°C, in the flow system.

As keep operating the reaction system in long-term dissolution, 200 - 400 day, we found that the reaction rate also waved automatically. See Fig.3. It is well known that surface motion and deformation may be induced by chemical and physical constraints, e.g. etch pits occur caused by chemical reaction [31]. These instabilities may lead to ordered behavoir with spacial and temporal patterns, or more complex dynamics [24].

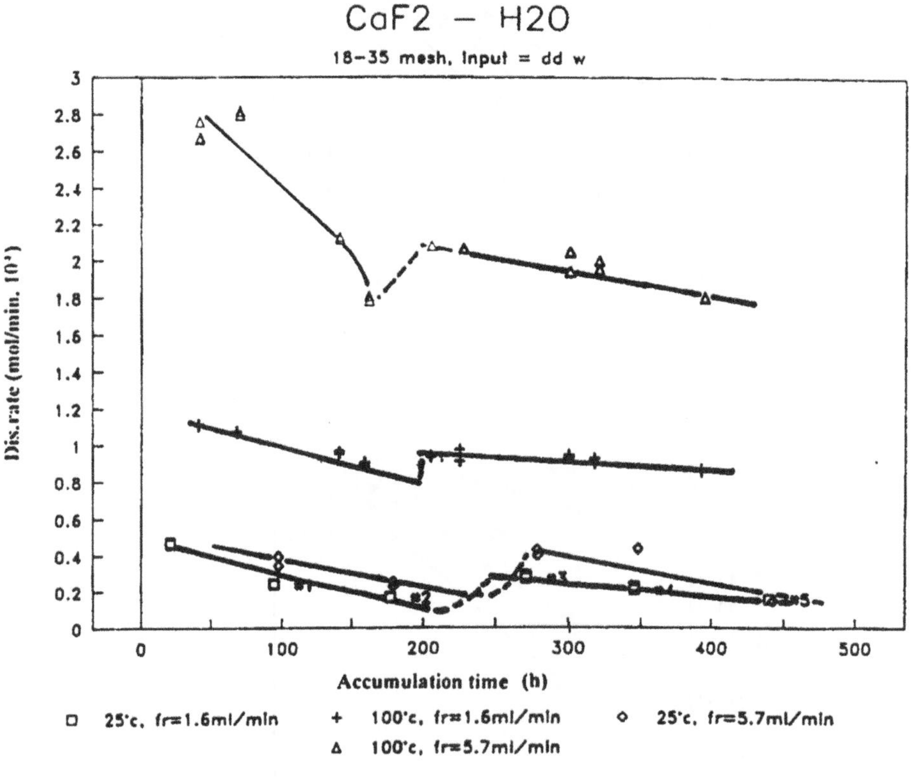


Fig.2 The variation of the dissolution rates in fluorite - water system as a function of the accumulation time.

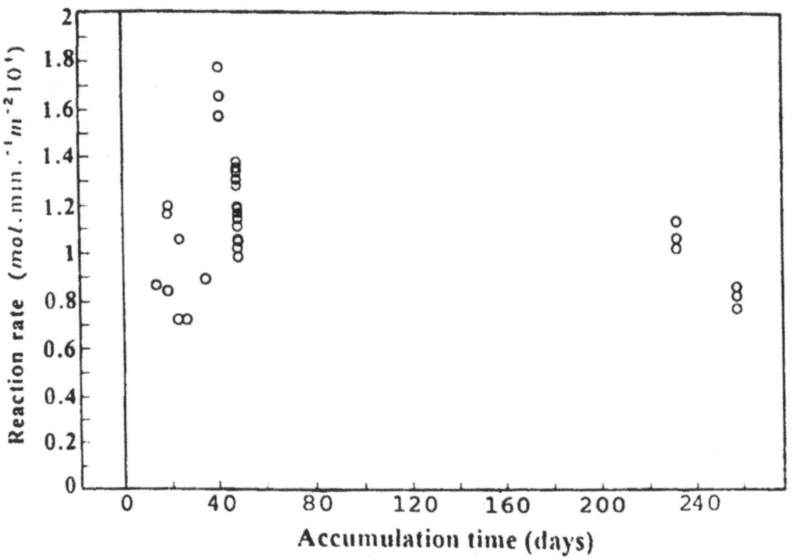


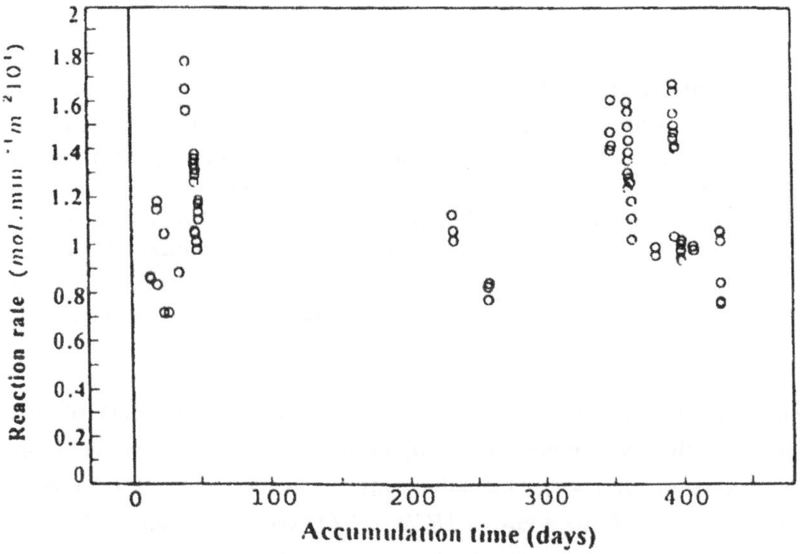


Fig.3 The reaction rate of fluorite in HCl-H2O solution in long-term period

The flow system of the packed bed reactor has a residence time distribution function dF(t), shown in Fig. 4, which expresses a nonideal flow system. For a continuous flow system F(t) the volume fraction of the fluid at the outlet that has remained in the system for a time less than t. For a constant density flow system, F(t) is also the weight fraction of the effluent with an age less than t. Fig. 4 shows that a step change of input tracer.

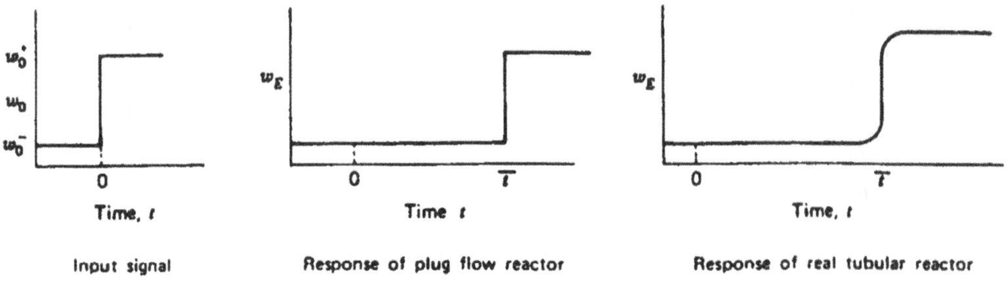


Fig.4 Generalized response of an arbitrary reactor to a step change in input tracer concentration.

When change input water to input $HCl - H_2O$ solution to fluorite - water system, we find a response curve for this a step change of the input solution. Usually, the response curve is similar to that shown in Fig.4. But, in some cases of the long-term dissolution, it has been found that the response curves are nonlinear, complex oscillation, shown in Fig.5. Fig.5 and Fig.6 are all expressed by recording of the electronic conductivity of the output solution from the reactor. Fig.5 shows probably the multi-steady states. Fig.6 shows the response curves of the conductivity of output solutions, there is a complex oscillation of the output solutions, while a step change of the input from water to HCl solution.

In the Fig.6 the response curve shows a temporary chaotic oscillation of the output solution in conductivity and in the concentrations of $Ca^{2+}$, $F^-$, $Cl^-$, which occurs when change input water to $HCl - H_2O$ solution ( pH 3.26. flow rate 2.25 ml/min.). Note that the time evolution of the $CaF_2 - HCl - H_2O$ system, as running the experiment in long term dissolution (2 months to 1 year), provides an opportunity to find the nonlinear response curves.

The time evolution reveals the changes of spectral compositions of the reaction products by using the time resolution of the FT-IR . The chemical compositions of the output solutions were also analyzed continuously. We found that the multi-quasi steady state, spontaneous transitions with abrupt jump and the excitation in output concentrations, chemical hysteresis, in $CaF_2 - HCl - H_2O$ system, as maintained constant pH of input HCl solution, and constant T and P.

## MINERAL -AQUEOUS SOLUTION INTERFACE

The Cl, F, Ca concentrations and pH of the output solutions vary by flow rates. In some cases, the output Cl concentrations are lower than input Cl, which is an evidence that $Cl^-$ ions could accumulate on the mineral surface, as well as $H^+$ ions. These facts prove that the proton $H^+$ and $Cl^-$ ions can be adsorbed on mineral surface. And also, a surface ion exchange occurs on mineral surface: $2H^+ = Ca^{2+}$ .

New methology develops for study on the solid - liquid interface in the past ten years [9,4]. Based on X - ray photoelectron spectroscopy study and FT-IR spectrum study of the surface of the reacted fluorite and output solutions, it has been found that the surface compositions are changeable, and new product $Ca(F,Cl)_2$ formed on mineral surface.

We comprise XPS spectra between the fresh surface and the reacted surface of fluorite, and found that the Cl peak appears in the XPS spectra of the reacted fluorite. And the kinetic energy Ek is different with that of pure component $CaCl_2$.

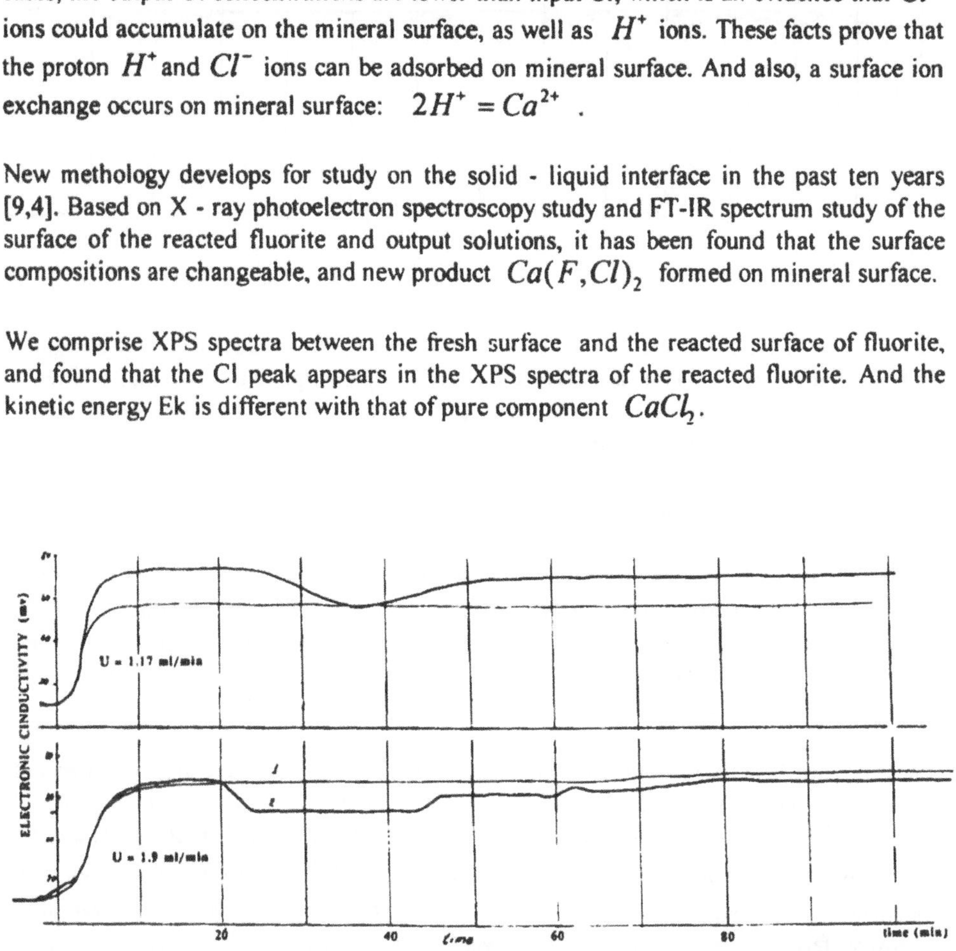


Fig.5 the response curves of the conductivity of the CaF2-HCl-H2O system.
The response curves show one step change to the system from input water to input HCl solution, pH=3.26, in the case: 25°C, pressure is 1.38 Mpa, flow rate range at 1.17 - 1.9 ml/min. And the accumulation time is about 2 weeks.

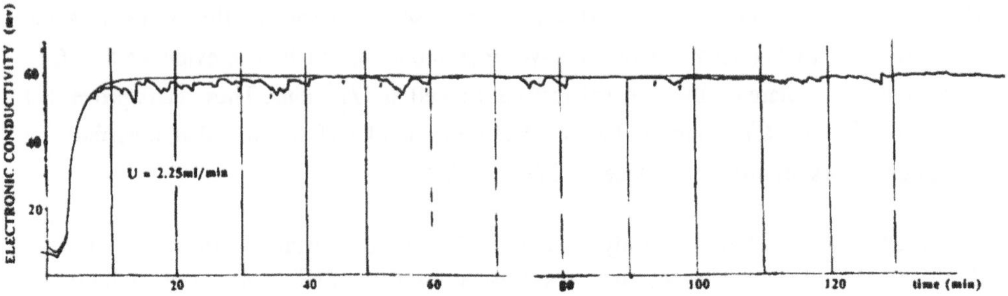


Fig.6 One response curve shows the chaotic oscillation in the conductivity of the output solution. The response curves are derived from one step change to the system, i.e. from input water to input HCl solution. In the case: pH = 3.26, flow rate is 2.25 ml/min, the accumulation time is about 40 days.

When a lot of fluorite crystals react with HCl solution in a flow system, we take a piece of crystal per day, then we analyze the surface by using the XPS spectra. The Cl peak in XPS spectra and its intensity increases with the accumulation time. We can calculated the percentage of the ion covered on surface and the ratio $F^- / Cl^-$ on the mineral surface, which change with time evolution. Our combined XPS spectra and dissolution rate measurements of the fluorite - HCl-H2O system show that the surface concentrations of Cl and F vary with time, $\Sigma$ t, for detail, see reference [30].

Concerning the temporary chaotic oscillation as shown in Fig.6, the chemical analyses tell us that Ca, F, and Cl vary with time in different frequency, that is due to the existence of time variables in the heterogeneous system and the surface chemical reactions: H and Cl adsorption on fluorite surface, $Ca^{2+} / 2H^+$ and $F^- / Cl^-$ ion exchanges on surface. The surface ion exchanges cause that the proton $H^+$ and $Cl^-$ ions can penetrate into mineral surface, and beneath the surface layer, while the reaction time accumulates in a long term period. These facts indicate that there are many variable intermediate products in the aqueous solution-solid interface of the fluorite - HCl solution system: $Ca^{2+}, F^-, CaF^+, Ca(F,Cl)_2, CaCl^+, Cl^-$ that provide the data and facts for modeling the nonlinear kinetics of mineral dissolution in a flow system.

**DISCUSSION**

Concerned above complex liquid-solid reaction process with non-linear behavior, it could compose the adsorption rates of several aqueous species, their desorption rates and the release rate of Calcium ion into the solution. Based on the surface reactions and surface ion exchanges, we can describe the kinetic process by a group reaction equations as the equation group [ A ].

On the other hand, the chaotic dynamics theory considers that there will be a chaos occurring in a reaction process, if it consists of more than 2 variable intermediate products in the reaction [25,15]. The following reactions could occur in the interface:

$$CaF_2 + H^+ + 2* \xrightarrow{k_1} CaF^+_{ad} + HF^0 + *$$

$$H^+ + * \underset{k_-}{\overset{k_2}{=}} H^+_{ad}$$

$$CaF^+_{ad} + H^+_{ad} \xrightarrow{k_3} Ca^{2+} + HF^0 + 2*$$

$$2* + H_2O \underset{k_{-4}}{\overset{k_4}{=}} H_2O_{ad}$$

$$CaF^+_{ad} + Cl^- + * \underset{k_{-5}}{\overset{k_5}{=}} Ca(F,Cl)_{2,ad}$$

$$Ca(F,Cl)_{2,ad} + H^+_{ad} \xrightarrow{k_6} CaF^+_{ad} + H^+ + Cl^- + * \qquad \text{A.}$$

All experimental data that includes the spectra of FT-IR for the mineral-liquid interface, the output concentrations of an aqueous species i, Ci, and the electronic conductivity of the output solution, λ were recorded by means of direct hookup to a microcomputer continuously. Assume that the concentrations of the aqueous species $Ca^{2+}, F^-, CaF^+, Ca(F,Cl)_2, CaCl^+, Cl^-$ are the variables, and remain the constant input HCl solution in the open flow system as a dynamic system. And then, send for a fractal, mathematic analysis.

Now, define $\Theta_1, \Theta_2, \Theta_3, \Theta_4$ to express the % of the surface covered by the surface adsorption state of $CaF^+, H^+, H_2O$ and the intermediate product $Ca(F,Cl)_2$ on mineral surface separately. Than we can set up a group of rate equations as follows:

$$\frac{d\Theta_1}{dt} = \overline{k_1}(1-\Theta_1-\Theta_2-\Theta_3-\Theta_4)^2(H^+) - \overline{k_3}\Theta_1\Theta_2 + \overline{k_{-5}}\Theta_4 + \overline{k_6}\Theta_2\Theta_4$$
$$-\overline{k_5}(1-\Theta_1-\Theta_2-\Theta_3-\Theta_4)\Theta_1(Cl^-)$$

$$\frac{d\Theta_2}{dt} = \overline{k_2}(1-\Theta_1-\Theta_2-\Theta_3-\Theta_4)(H^+) - \overline{k_3}\Theta_1\Theta_2 - \overline{k_{-2}}\Theta_2 - \overline{k_6}\Theta_2\Theta_4$$

$$\frac{d\Theta_3}{dt} = \overline{k_4}(1-\Theta_1-\Theta_2-\Theta_3-\Theta_4)^2 - \overline{k_{-4}}\Theta_3$$

$$\frac{d\Theta_4}{dt} = \overline{k_5}(1-\Theta_1-\Theta_2-\Theta_3-\Theta_4)(Cl^-)\Theta_1 - \overline{k_6}\Theta_2\Theta_4 - \overline{k_{-5}}\Theta_4$$

$$\frac{d(Ca^{2+})}{dt} = \overline{k_3}\Theta_2\Theta_2 - f(Ca^{2+}) \qquad \text{B}$$

Where, f refers to flow rate, $(Ca^{2+}),(H^+),(Cl^-)$ stands for the concentrations of the aqueous species separetely. This group of rate equations can be simplified as define : $\overline{k_1}(H^+)=k_1$ ,$\overline{k_2}(H^+)=k_2$, $\overline{k_3}=k_3$, $\overline{k_4}=k_4$, $X=(Ca^{2+})$, $\overline{k_5}(Cl^-)=k_5$, $\overline{k_6}=k_6$, $\overline{k_{-2}}=k_{-2}$, $\overline{k_{-4}}=k_{-4}$, $\overline{k_{-5}}=k_{-5}$. Therfore, this group of rate equations includes the variables: $\Theta_1,\Theta_2,\Theta_3,\Theta_4,X$. When estimate its non-linear characteristics , we need to indentify the coefficients of $k_{1,}k_2,k_{-2},k_3,k_4,k_{-4},k_5,k_{-5},k_6,f$. And also we need to determine the initial values of the variables such as: $\Theta_1^0,\Theta_2^0\ \Theta_3^0,\Theta_4^0,X^0$. Concerning the group A of rate equations, if $\frac{dx}{dt}\neq 0$, based on the experimental data, we can estimate the value range for coefficient k. And then, we find the approval values for the coefficient of $k_{1,}k_2,k_{-2},k_3,k_4,k_{-4},k_5,k_{-5},k_6,f$ , and simulate the conditions for the oscillation of X (Here, X refers to the concentration of $Ca^{2+}$).

If $\frac{dX}{dt}=0$, we choose the approval values of the coefficients of $k_{1,}k_2,k_{-2},k_3,k_4,k_{-4},k_5,k_{-5},k_6,f$. and the boundary condition for this group of rate equations, then simulate the nonlinear behavior of theses variables of $\Theta_1,\Theta_2,\Theta_3,\Theta_4,X$ in the time evolution.

We resolve this group differential equations and do numerical simulation, when estimate the coefficient k i.e. the rate constant for each reaction in the group reactions by using the experimental data. The numerical simulations have obtained a lots of interesting results.

One can find the variables are changing with time, showing complex oscillation in the dynamic process. In the other cases, it has found that the multi-steady states occur in the dynamic process. See Fig.7 & 8.

The group of reactions (A) can be simplified in order to illustrate the nonlinear kinetic phenomena in principle, as follows:

$$CaF_2 + H^+ + 2* \xrightarrow{k_1} CaF_{ad}^+ + HF^0 + *$$

$$H^+ + * \underset{k_{-2}}{\overset{k_2}{=}} H_{ad}^+$$

$$CaF_{ad}^+ + H_{ad}^+ \xrightarrow{k_3} Ca^{2+} + HF^0 + 2*$$

$$2* + H_2O \underset{k_{-4}}{\overset{k_4}{=}} H_2O_{ad} \qquad \text{C.}$$

Based on this group of reactions, we set up the another group rate equations as follows:

$$\frac{d\Theta_1}{dt} = \overline{k_1}(1-\Theta_1-\Theta_2-\Theta_3-\Theta_4)^2(H^+) - \overline{k_3}\Theta_1\Theta_2$$

$$\frac{d\Theta_2}{dt} = \overline{k_2}(1-\Theta_1-\Theta_2-\Theta_3-\Theta_4)(H^+) - \overline{k_3}\Theta_1\Theta_2 - \overline{k_{-2}}\Theta_2$$

$$\frac{d\Theta_3}{dt} = \overline{k_4}(1-\Theta_1-\Theta_2-\Theta_3-\Theta_4)^2 - \overline{k_{-4}}\Theta_3$$

$$\frac{d(Ca^{2+})}{dt} = \overline{k_3}\Theta_1\Theta_2 - f(Ca^{2+}) \qquad \text{D.}$$

In the same way, we simulate the dynamic process in order to interpret a lots of experimental results. Fig.7 and 8 are the phase space diagram in which one dimension is specified by its concentrations of the variable (Y axis) and the time (X axis).

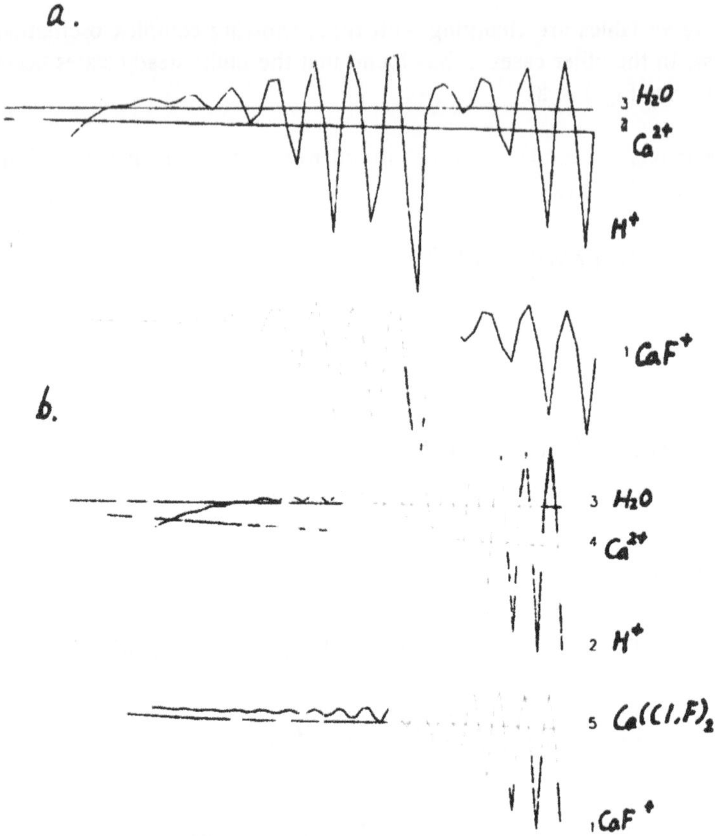


Fig.7a. The variable of CaF+, H+, H2O and Ca in the system varies with time. The mathematical simulation is based on the simplified group of rate equations.
Fig.7b The dynamic process of the variable of $\Theta_1,\Theta_2,\Theta_3,\Theta_4,X$

Fig.7a shows the variables of CaF+, H+, H2O and Ca in the fluorite - HCl -H2O system which vary with time. The mathematical simulation is based on the simplified group of rate equations (D). Fig.7b expresses the dynamic process of the variable of. $\Theta_1,\Theta_2,\Theta_3,\Theta_4,X$. From the diagram Fig.7a and 7b, you may find that the variables move in the phase diagram which is nonlinear dynamic process as sensitivity to initial conditions. When we change the initial values, we can get the quite different moving path of the variables. Fig.7b shows that the different initial values for equation group B lead to have the different dynamic behaviors. Thus facts show that these variables inhibit the characteristics of the chaotic dynamics [1].

The mathematical analysis for the solutions of the group of the rate equations has found that the group of equations (D) has the different solutions. At first, we simplify the group equation (D) , then get the following form:

$$\frac{dx}{dt} = k_1(1-x-y-z)^2 - k_3xy$$

$$\frac{dy}{dt} = k_2(1-x-y-z) - k_3xy - k_{-2}y$$

$$\frac{dz}{dt} = k_4(1-x-y-z)^2 - k_{-4}z \qquad \text{E.}$$

As resolving the steady state solutions of the equation group (E), see reference [30,31], for detail. Then, this equation obtained as follows:

$$x^3 + Ax^2 + Bx + C = 0$$

while,

$$A = -1 + \frac{k_2}{k_1} + \frac{2k_{-2}}{k_3} + (\frac{k_2}{k_1})^2(\frac{k_4}{k_{-4}})$$

$$B = (\frac{k_{-2}}{k_3})^2 + \frac{k_2^2}{k_3k_1} + \frac{k_2k_{-2}}{k_1k_3} - \frac{2k_{-2}}{k_3}$$

$$C = -(\frac{k_{-2}}{k_3})^2 \qquad \text{F.}$$

As numerical analysis of this group equation (F), reform the group equation (5), then it has obtained as follows:

$$s^3 + Ps + Q = 0 \qquad 6$$

while,

$$\Delta = (\frac{Q}{2})^2 + (\frac{P}{3})^3 \qquad 7$$

Now, we find that the equation ( 5 ) has three steady state resolutions $x_1^*, x_2^*, x_3^*$, if $\Delta <$ 0, and the equation ( 5 ) has one steady state solution $x_1^*$ , if $\Delta > 0$ .

After analyzing of the stability of the steady state, find: both kinds of steady-state solutions are unstable. In the phase space, a surface of a ball evolves from the outside to the inside in the system, only one steady state is inside of the ball. So a periodic solution may exist in the system. In the case of three steady state solutions, the three steady-state solutions are unstable, and the surface of the ball does not reach infinite. So it has a chaotic solution in the system.

Our experiments demonstrate the system may exhibit the multi-steady state and bifurcation. As plotting concentration time historical diagram, we found that the intermediate products have the different characteristics of their oscillations, as shown in Fig.7a & 7b. In the stereographic plot for $H^+, CaF^+, CaCl^+$ variables, shown in Fig.8.

The Fig.8 is a phase space diagram. The phase space of a dynamic system is a mathematical space with orthogonal coordinate directions representing each of the variables need to specify the instantaneous state of the system. These variables consist of $CaF^+_{ad}, H^+_{ad}, Ca^{2+}, Ca(F,Cl)_{2,ad}, CaCl^+, etc.$

The interfacial kinetics between carbonate and aqueous solutions have been also found to have non-linear behavior: chemical wave ( reaction rate r, or concentration of dissolving species Ci) due to the coupling processes of the temperature effect on r and mass transfer (diffusion and flowing), e.g. the exothermal reaction of calcite dissolution at T > 200°C, in a far from equilibrium system. Authors will have another paper to discuss the calcite dissolution process.

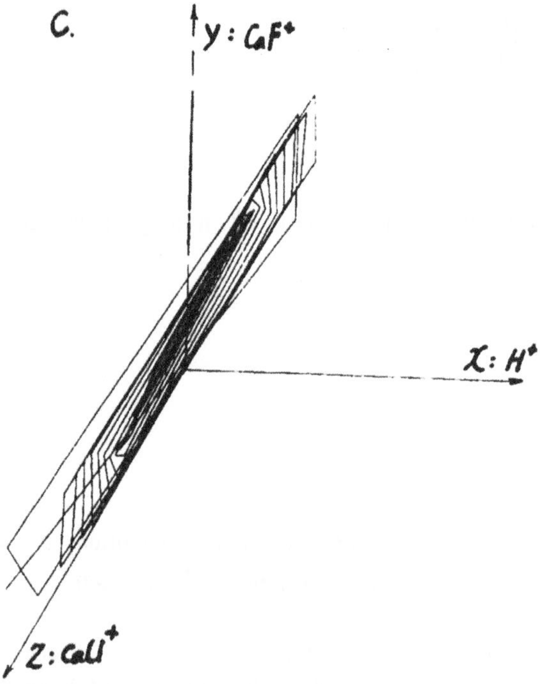


Fig.8 The phase space diagram: the X, Y, and Z axis express the variables of $H^+, CaF^+, CaCl^+$, based on the equation group B.

## CONCLUSION

Our kinetic study reveals the non-linear dynamic behavior in fluorite-HCl-H2O system. This non-linear dynamic process appears in the different scale, in mineral-solution interface, in the reaction and flow system. This study suggests a solid-liquid interface pattern to illustrate this non-linear dynamic process. This model describes the coupling processes of adsorption , surface reaction, surface ion exchange and diffusion or transportation processes. And also, the multi-steps of the reaction and the intermediate products on surface involve in the kinetic process, which can provide the information for modeling non-linear kinetics on interface. The solid-liquid interface pattern model could only explain the part of the non-linear dynamic processes in the flow system. Furthermore, we consider that the complex non-linear dynamic process occurring in the heterogeneous, pororious, flow system accompanying reaction, need a mathematic model of the chaotic and non-linear dynamics in the furthermore study.

## *ACKNOWLEDGMENT*

This research was supported by the basic research grant (8502215) of the Ministry of Geology and Mineral Resources. We are very grateful to National Science and Technology Committee for giving support. This work was also supported by Chinese Natural Science Foundation, 49070179, 49473196. We thank the friends in the Open Research Laboratory of Geochemical Kinetics to support this research work. We acknowledge Pr. Jiang for helping abvice and discussion.

## REFERENCES

1. Baker G.L., and Gollub J.P., 1996, Chaotic Dynamics: an introduction, Cambridge University press.
2. Brantley S.L., 1992, Kinetics of dissolution and precipitation- experimental and field results, Water-rock interaction, Kharaka, & Maest (eds) Balkema, Rotterdam.
3. Casey H.W., 1987, Heterogeneous kinetics and diffusive boundary layers: the example of reaction in a fracture, J. Geophysic R., 92, 8007 - 8014.
4. Casey W.H, Hochella JKr.M.F., And Westrich H.R., 1993, The surface chemistry of manganiferous silicate minerals as inferred from experiments on tephroite (Mn2SiO4), Geochim. Et Cosmichim. Acta Vo.57, p 785-793.
5. Chou L. And Wollast R., 1984, Study of the weathering of albite at room temperature and pressure with a fluidized bed reactor. Geochim. Cosmochim. Acta 48, 2205-2218.
6. Dove P.M., And Crerar D.A., 1990, Kinetics of quartz dissolution in electrolyte solutions using a hydrothermal mixed flow reactor, Geochim. Cosmochim Acta, v.54, p955-969.
7. Fogler H.S.,1986, Elements of chemical engineering, Princeton Hall, Englewood Cliffs.
8. Ganor J., Mogollon J.L., and Lasaga A.C., 1995, The effect of pH on kaolinite dissolution rates and on activation energy, Geochim Cosmochim. Acta, Vol.59, no. 6, p 037-1052.

9. Hochella, Jr. M. F., And White A. F., 1990, ed: Mineral - water interface geochemistry. Min. Rev. Vol. 23, Min. Soc. Am.
10. Hu S., and Zhang R., 1995, Ft-IR and UV spectrum of interface in the mineral/water interaction, Water-rock interaction, Kharaka & chudaev (eds) Balkema, Rotterdam.
11. Knauss K.G., And Wolery T.J., 1988, The dissolution kinetics of quartz as a function of pH and time at 70°C, Geochim. Cosmochim. Acta 52, p43-55 , 1983.
12. Knauss K.G., Nguyen, and Weed H.C., 1993, Diopside dissolution kinetics as a function of pH, CO2, temperature, and time, Geochim. et Cosmochim. vol. 57, p 285 - 294.
13. Lasaga A.C., 1981, Rate laws of chemical reactions, and Transition state theory. In Kinetics of Geochemical Processes (ed: Lasaga A.C. And Kirkpatrick), 1-67 and 135-169,Rev.Mineral.8.
14. ------,1984,Chemical kinetics if water-rock interactions, J. Geophys. Res. 89, 4009-4025.
15. Levine H., 1990, Interfacial patterns: A unified viewpoint, Chaos (eds) Campbell D.K. AIP New York.
16. Nagy K.L. and Lasaga A.C., 1992, Dissolution and precipitation kinetics of gibbsite at 80°C and pH 3: The dependence on solution saturation state. Geochim. Cosmochim. Acta 56, 3093-3111.
17. Paces T., 1973, Steady-state kinetics and equilibrium between ground water and granitic rock. Geochim. Cosmochim. Acta 37, 2641-2663.
18. Petrovich J..D. , 1981, Kinetics of dissolution of mechanically comminuted rock - forming oxides and silicates: I. Deformation and dissolution of quartz under laboratory conditions, Geochim. Et Cosmochim., vol. 55, p3273 - 3286.
19. Plummer R.L., Parkhurst D.L. And Wigley T.M.L., 1979, Critical review of the kinetics of calcite dissolution and precipitation, Am. Ch. Soc: Washington, D.C., 537-573.
20. Posey-Dowty J., Crerar D.A., Hellmann R., And Chang C., 1986, Kinetics of mineral - water reactions : theory, design and application of circulating hydrothermal equipment, Amer. Mineral., vol.71, 85-94.
21. Richardson C. K., and Holland H. D., 1979, The solubility of fluorite in hydrothermal solutions, an experimental study, Geoch. Et Cosmochom., v 43, p 1313 - 1325.
22. Rickard T.D. And Sjoberg S.E., 1983, Mixed kinetic control of calcite dissolution rate, Amer. J. S., 283, 815-830.
23. Rimstidt J.D. And Barnes H.L., 1980, The kinetics of silica -water reactions, Geochim. Cosmochim. Acta, 44, 1683-1699.
24. Sanfeld A., and Steinchen A., 1984, Mechanical instability and dissipative structures at liquid interfaces, Chemical instabilities (eds) Nicolis G., and Baras F., Reidel Publishing company.
25. Schmitz R.A., D'Netto G.A., Razon J.R., and Brown J.R., 1984, Theoretical and experimental studies of catalytic reactions, Chemical instabilities (eds) Nicolis G., and Baras F., Reidel Publishing company.
26. Zhang R., Posey-Dowty J., Hellmann R., Borcsik M., Crerar D., And Hu S., 1989,Kinetic study of mineral - water reactions in hydrothermal Flow systems at elevated temperatures and pressures, Science in China B, 11. 1212-1222.(English abstract)

27. Zhang R., Posey-Dowty J., Hellmann R., Borcsik M. Crerar D., And Hu S., 1990. Kinetics of mineral - water reactions in hydrothermal flow systems at elevated temperatures and pressures, Science in China (series B) Vol.33, No.9, 1136-1152.( English edition )
28. Zhang R., Hu S., Hellmann R., And Crerar D., 1992, book: Chemical kinetics of minerals in hydrothermal systems and mass transfer, Science published House, Beijing ( Chinese with a detail abstract in English ).
29. Zhang R., And Hu S., 1992, Chemical kinetic control on mineral dissolution in hydrothermal flow system, Abstract, The second International thermodyna ics of natural processes, Novosibirck, Russian.
30. Zhang R., and Hu S., 1996, Kinetics of fluorite dissolution in open-flow system and surface chemistry, Science in China (D), Vol.26, no.1. p 41-51.
31. Zhang Ronghua, Hu Shumin, Jian Lu, Tong Jianchang et al, 1997, Mineral-fluid reaction in open system and non-linear dynamic process, Sience publising house, Beijing.

*Proc 30th Int'l. Geol. Congr.*, Vol. 19, pp. 49-61
Xie Xuejin (Ed)
VSP 1997

# Kinetic Measurements of Mineral Dissolution By Using FT-IR and UV Spectrum, and XPS

SHUMIN HU and RONGHUA ZHANG
*Open research laboratory of Geochemical Kinetics, Ministry of Geology and Mineral Resources. Chinese Academy of Geological Sciences, Baiwanzhuang Road #26, 100037, Beijing, China.*

**Abstract**

The kinetics of the interaction between fluorite and aqueous solution could be investigated by applying of FT - IR spectroscopy. UV spectroscopy and X-ray photoelectron spectroscopy(XPS). Authors observe the variation in FT - IR spectra and the XPS spectra of the reaction products with time evolution and investigate the modifications of the surface structure and composition of the mineral after reacted with a liquid phase. It has been found that the interface kinetics conducted the spatial and temporal non-linear characters that reveal the chemical instability of the reaction between fluorite and aqueous solutions.

Key words: FT-IR spectrum. XPS. Chemical kinetics. Solid/liquid interface. Real time measurement

## INTRODUCTION

Greatly improved characterizations of chemical speciation in aqueous fluids and in the solid-liquid interfaces are required to further understand geochemical processes involving the earth fluids and kinetics of water-rock interactions. We propose direct spectroscopic investigations of the fluorite -HCl -H2O system. The experimental study of the reaction kinetics, especially the study of heterogeneous reaction and mineral/aqueous solution interface, is carried out by using new technique of a microscopic methods instead of the macroscopic approach. That is a new development of applying the FT-IR spectrum, UV spectrum and XPS spectrum methods and dynamic methods to observe the solid surface, the liquid phase in reaction system and the solid-liquid interface. These new techniques and methods have advantage in discovering the new kinetic mechanism of the heterogeneous reactions and in understanding the surface chemistry in microscopic scale [3].

Interface of mineral/aqueous phase has been observed by using new developing technique. Many scientists report about these new developments (Hochella et al, 1990). With this purpose, authors operate the kinetic experiments of mineral - water interaction in a flow system by using a packed bed reactor. The flow-reactor system contains a tubular pressure vessel, in which mineral particles were put, that authors described in their papers, please see ref. in detail [8,9,11].

The aqueous phase products of the output solutions from a flow - reactor can be observed and measured by using the FT - IR spectroscopy continuously. The reacted mineral surface could be also observed by using FT - IR spectroscopy. Authors also utilize UV and Raman spectroscopy as well as XPS, to analyze the surface after testing dissolution rate. Authors found that the composition and speciation of the output solution from fluorite - HCl - H2O systems, and of the mineral surface change with reaction time, i.e. found that chemical instability in the interfacial processes of the reaction systems. The real time measurement and dynamic technique of FT - IR spectroscopy help us to discovery the temporary and transition state compounds in the interface of fluorite and HCl solutions. which might be a factor caused the chemical instability of the fluorite and HCl - H2O solution interaction.

## IR- SPECTRA OF THE AQUEOUS PHASE

FT -IR spectrum ranges from 400 to 4000 wavenumber (cm -1). Water has very strong absorbance in the spectrum range.

### *IR spectra of water*

IR spectroscopy is not well suited for direct determination of salute structure in aqueous solutions because of the IR spectrum of the water itself. The fundamental vibrations, as well as overtones and combination bands of water, extend from roughly 1600 cm-1 in IR to 7400 cm-1 in NIR [2]. The measurement by using IR spectrum is difficult for aqueous solution, when an aqueous species has absorbance in the same spectrum range as water, with a purpose in distinguishing the difference between water and aqueous species of low concentration solution [1,2].

At first, we need to prepare the different kinds of the aqueous solutions as standards, e.g., the different concentrations and pH. Observe their FT-IR spectra, and distinguish the different characters of their spectra. Our combined the IR spectra of the different solutions and their chemical compositions suggest that the distributions of the different aqueous species and the complex in aqueous phase.

As put a small film of the output solution from reaction system on the surface of optical glass ( optical material is TII): the infrared spectra, the absorbance of water, CaF2 and CaCl2 aqueous solutions, can be identified, see Fig.1.

However, the study of the infrared absorption of aqueous species, metal-ligand vibration in aqueous phase is greatly precluded because of strong water absorption in the infrared region. Some spectra of the aqueous species were measured using D2O solution for reduce interference caused by vibration overtone of H2O in NIR.

### *IR spectra of the fluorite-HCl-H2O system*

In order to know what kinetic process happen in the system of fluorite - HCl - H2O, we can analyses the FT-IR spectra from the reaction system. For example, within a concentration range from 5 to 30 ppm of Calcium, an aqueous phase could be distinguished by using a high resolution FT - IR: one of the peak position of absorbance characteristic bands of water centered at 702.3 cm-1. See Fig.1  In Fig.1, FT - IR spectra of aqueous solutions are drown as 5 curves: only one is water spectrum. The others are CaF2 saturation solution (SA) , 1 m CaCl2 solutions and aqueous product from reactor.

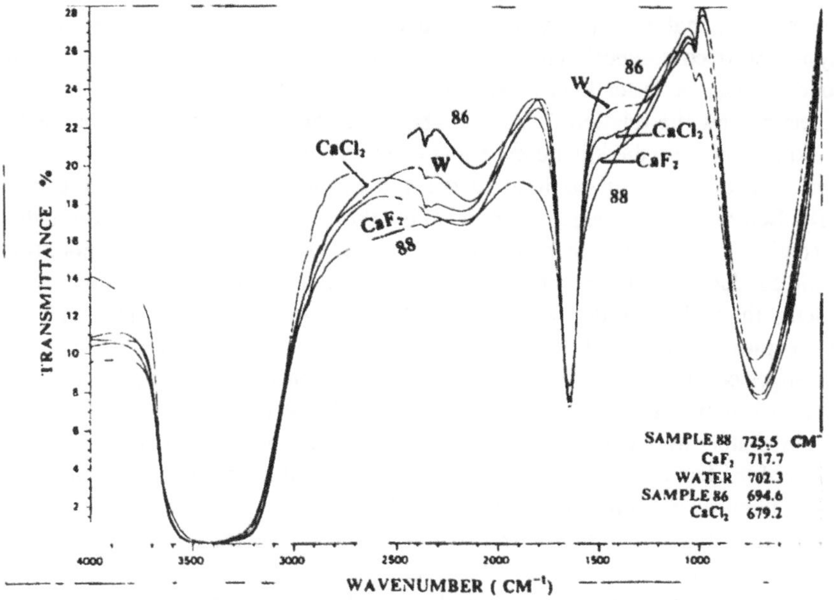


Fig.1 FT - IR spectra of aqueous solution: there are 5 curves, only one is water spectrum. The other are CaF2 saturation solution (SA), 1 m CaCl2 solutions and aqueous product from reactor.

The peak of the band of 1m $CaCl_2$ solution is at 679.2 cm-1. The peak of the band of $CaF_2$ saturation solution is at 717.7 cm-1. These absorbance bands for $CaF_2$ and $CaCl_2$ solutions do not show the positions of the true vibrational spectra of the aqueous species. The harmonious vibration frequency $\omega_e$ of CaF and CaCl are 587.1 and 369.8 cm -1 separately. The vibration frequencies $\nu_3$ of metal compounds $CaF_2$ and $CaCl_2$ are 553.7 and 402.3 cm -1. These peak positions of the bands, shown in Fig.1, are the inclined vibration of the water affected by aqueous species or complex: $CaF_2, CaF^+, Ca(F,Cl)_2, CaCl_2, CaCl^+$ etc. Probably, there is a complex of $Ca(F,Cl)_2$ in water. The reaction products of the output aqueous solution flowing from the fluorite - HCl - water system have special IR absorbance bands, which are different with them of CaF2 and CaCl2 solution. Comparison with spectrum of the CaF2 saturation solution, their peak positions of spectra of the reaction products from fluorite /

HCl solution interaction, turn to at the lower wavenumber. Their wavenumbers decrease by increasing of the Cl concentration in the aqueous phase. The peak positions and the bands of reaction products vary in various reaction conditions: pH, temperature, flow rates, etc. And also, they were found to be changed with the changeable dissolution rates.

*Dynamic observation*

Dynamic method is that the spectra of the aqueous solutions of the output from the reaction system were continuously observed and recorded in every hour, every minute and every second ( it can be faster than a second by using rapid scan technique ). And the dynamic spectra show the relation of wavenumbers and the reaction time of the reaction system. Authors operate the kinetic experiments in a continuous flow system by using a Packed Bed Reactor. Pump in solution to the system, and collect the output solution continuously. Before take the aqueous sample, allow the output solution passes through an electronic conductivity detector, then record the conductivity in real time. By monitoring the electronic conductivity of the output solutions from the fluorite - HCl solution system during a long time observation, we could find a variation of dissolution rates and temporary complex oscillation of the output solution concentration. While the temporary chemical changes occurring in the mineral / water interaction inside of the pressure vessel in the reaction - flow system, the output solution could be observed continuously and analyzed immediately by aid fo FT - IR. The behavior of spectrum intensities and the wavenumbers of the bands of the output solutions were found to change with time, which indicates the chemical instability in the fluorite - HCl - H2O system in some cases. For instance, see Fig.2. Fig.2 represents the real time FT - IR spectra of aqueous solution, collected output solutions from CaF2 - HCl - H2O flow systems, in 3 hours and in the case: input pH = 3.26, pressure inside of vessel = 13.8 MPa, flow rate = 1.97 ml/min. The analytical method is to put a drop of aqueous product taken from reaction system on an optical glass, made a thin liquid film, then observe. It has been found that one peak of the bands changes from 660 cm-1 change to 770 cm-1 of the output solutions in 3 hours, as we continue to observe the output solutions, which IR spectra represent the complex dynamic process in the interfacial kinetics of $CaF_2 - HCl - H_2O$ system.

Fig.2 also shows the instability of the system that the mixture compounds of the solute and the reaction products in the aqueous phase are characterized of the variation of the wavenumbers with time. The inclinations of water vibration spectrum due to aqueous species, as mentioned before, vary with respect to time, which indicates that the interaction among the mixed aqueous species is still processing.

It is well known that the differences of the IR spectra between the compounds of calcium with F, Cl, Br, and I , and the compounds hydrogen with F, Cl, Br and I are all expressed in that their wavenumbers. And also, the vibration frequencies vary with the atomic number. Therefore, for the halides of hydrogen and halides of calcium, their harmonic vibration frequency is progressively changing, and decreases in the sequence of F->Cl->Br->I. Fig.2 shows that an important bond frequency of the mixed speciation in the aqueous phase decreases, or its wavenumber increases. This fact expresses that probably $CaF^+$ is getting to dominate the bond peak in the wavenumber range between 600 - 800 cm-1. In some cases, the two bond peaks in the spectrum range were found of the reaction products [10,12,13,14,15]. The wavenumber changes would be derived from the following reactions:

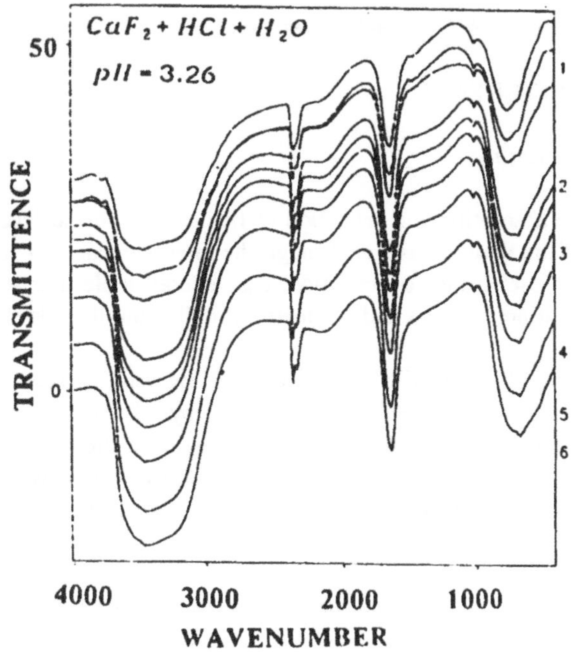


Fig.2 The real time FT - IR spectra of aqueous solution, collected output solutions from CaF2 - HCl - H2O flow systems, in 3 hours and in the case: input pH = 3.26, pressure inside of vessel = 13.8 MPa, flow rate = 1.97 ml/min.

$$CaF^+_{ad} + CaCl^+_{ad} = Ca(F,Cl)^0_{2,ad} + Ca^{2+}$$

$$CaF^+_{ad} = CaF^+_{aq} + *$$

$$CaCl^{+}_{ad} = CaCl^+_{aq} + *$$

$CaF^+_{ad}$ and $CaCl^+_{ad}$ are the adsorption species on fluorite surface. $CaF^+_{aq}$ and $CaCl^+_{aq}$ stand for species in the aqueous solution, * refers to the active center on mineral surface. Assume that these reactions occur in the interface between mineral and liquid phase, then aqueous phase will have the changeable speciation with time. And also this reaction occurs in the aqueous phase:

$$Ca(F,Cl)^0_{2,aq} + Ca^{2+} = CaF^+_{aq} + CaCl^+_{aq}$$

and $Ca(F,Cl)^0_{2,aq}$ may not be stable, therefore, more $CaF^+$ species will remain in the aqueous phase due to $CaCl^+ \rightarrow Ca^{2+} + Cl^-$ reaction.

## MINERAL - LIQUID INTERFACE

### *IR spectra of the mineral surface*

It is well known that the absorption and reflection of IR beam on the surface of the solid phase and the surface of the liquid phase, or through the interface, can be used to identify the surface structure and the surface composition, e.g., overall molecular symmetry and coordination number; ligation numbers and symmetry of individual ligands etc.

Usually, the method of in situ DRFTIRS is used to treat the solid surface. The fluorite surfaces after testing the dissolution rates were analyzed by FT - IR in situ DRFTIRS. Then it was found that the surfaces of the reacted fluorites have the different spectra as they were taken from the different positions inside of pressure vessel, or were collected at the different accumulation time of the reaction. FT - IR spectra demonstrate the surface speciation changes with the position inside of vessel and with reaction time.

The modification of the liquid-solid interfaceis has been proved by the following experiment. The fluorite reacted with HCl solution (pH = 0.5) in 9 days, in a continuous stirred flow reactor, at 25°C and 1 atm pressure. As we collected the crystal sample every day and then analyzed the surface of the reacted fluorite by IR spectroscopy. We use fresh fluorite as standard sample. Using spectral subtraction of the two sorts of spectra, before and after reaction of fluorite, obtain the informations about the modifications of the interface of fluorite and aqueous phase. The chemical modification and new surface complex on fluorite will result the frequency shift and band broadening, and lead the distorting of band shape of the original spectra. The spectral subtraction between fresh fluorite and reacted fluorite indicates that there are some new bands and peaks appear in the spectra, shown in Fig. 3. We can indentify what new complex and compounds occur on surface, and what a stretching and binding vibrations could be observed to conduct the absorbance bands in the spectral subtraction. And also the 8 different spectra of reacted surface express in the time evolution sequence from the second day to the ninth day. There are three different peak positions (band) in the wavenumber from 1200 to 1500 cm -1, in which one could be the surface complex $Ca(F,Cl)_2$. Within 9 days, the two peaks of the bands within the range from near 1360 to 1460 cm -1, gradually changed to one peak of the bands at 1415 to 1420 cm -1 after the third day, which prove that the surface complex or compounds varied with time.

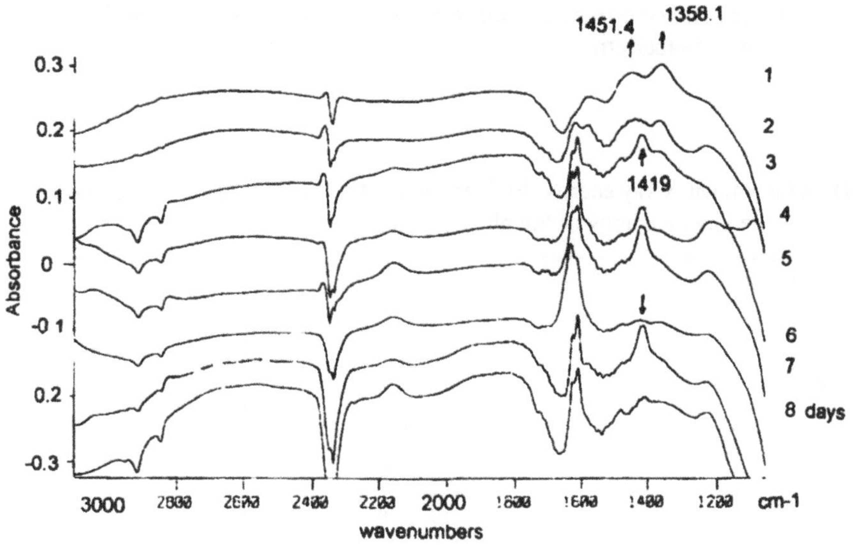


Fig.3 Spectral subtraction of the fluorite surface before and after reaction within 9 days. The top spectrum is of the second day. The bottom spectrum is of the ninth day

Authors have already found that the bulk concentrations, i.e. reacted aqueous products in the vertical vessel change with position from bottom to the top of the vessel. Note that we pump in solution from bottom up to the top in the reaction flow system. Now, it has been found that their surface compositions of the mineral particles vary in vertical distance of the vessel, in which, their spectra change from bottom to the top.

The Fig.4 expresses the spectra of the surface of reacted fluorite in the different position inside of the vessel, after fluorite reacted with HCl solution ( pH = 3.26) in the condition of flow reactor at temperature from 25 - 100°C and at 13.6 MPa.

*XPS study on the interface of fluorite-aqueous phase*

XPS is the most useful surface technique for delineating the chemistry of the surface. A source of X-ray is directly on mineral sample surface: it ionizes all ions on the mineral surface [7]:

$$A_{surface} + h\nu \rightarrow A^{+}_{surface} + e \qquad 1.$$

where e is the ejected photoelectron. The kinetic energy of the ejected in equation (1) is given by the Einstein photoelectric equation:

$$h\nu = E_b + E_k \qquad 2.$$

where, $h\nu$ is the known X-ray energy, Eb is the ionization or binding energy of the electrons, Ek is the kinetic energy of the photoelectrons.

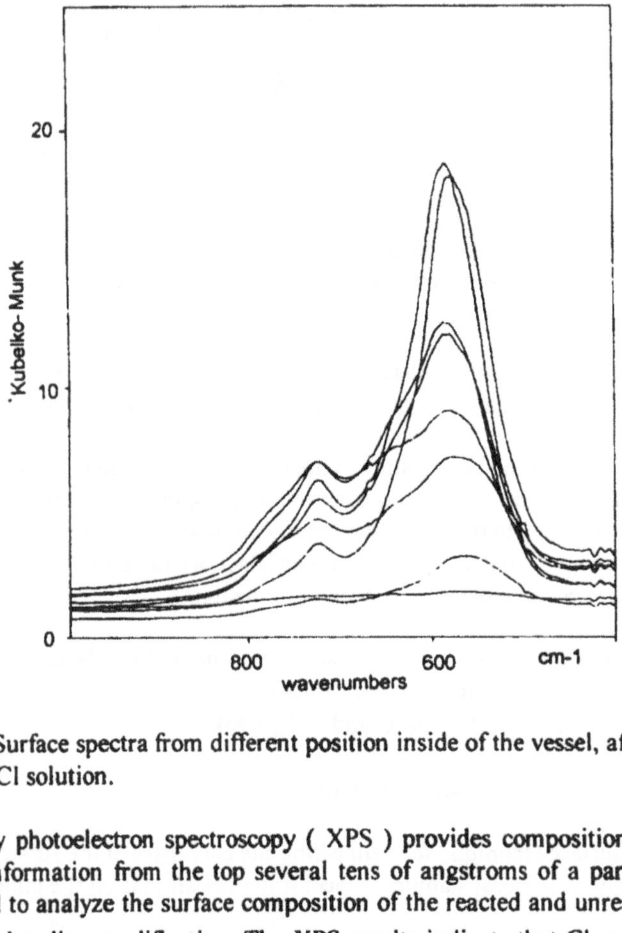


Fig.4. Surface spectra from different position inside of the vessel, after fluorite reacted with HCl solution.

X - ray photoelectron spectroscopy ( XPS ) provides compositional as well as bonding state information from the top several tens of angstroms of a particle. Hence, XPS was applied to analyze the surface composition of the reacted and unreacted fluorite and their surface bonding modification. The XPS results indicate that Cl remains in the $Cl^-$ state at all conditions on the reacted fluorite surface. As one can see in Table 1, spectra for the unreacted fluorite and the material reacted at pH = 3.26 are almost similar. But note that

$F_{1s}$ and $Ca_{2p^{3/2}}$ peak intensities are different between fresh fluorite and reacted fluorite. And also, the binding energies responding to $Cl_{2p}$ are different between the two sort of materials, i.e. between the reacted fluorite and pour $CaCl_2$.

Our combined the XPS study on the mineral - liquid interface and the kinetic experimental results in the flow through packed bed reaction system reveal new informations on the mechanism of controlling the interfacial kinetics.

As reported before, authors investigated that in the flow - reaction system, the output concentrations of F and $Cl^-$ were also changed with flow rate, while fluorite dissolving in HCl - water system [10,14]. As maintain a constant input pH of HCl solution, the output $Cl^-$ were less than input $Cl^-$ in some cases as varying flow rate. It indicates that $Cl^-$ can be also adsorbed on the surface. The concentrations of $Cl$ adsorbed on mineral surface were affected by nature of the input solution, and vary by the different pH of input solutions [11]. The pH values favour decreasing the output $Cl^-$ concentrations and increasing of F.

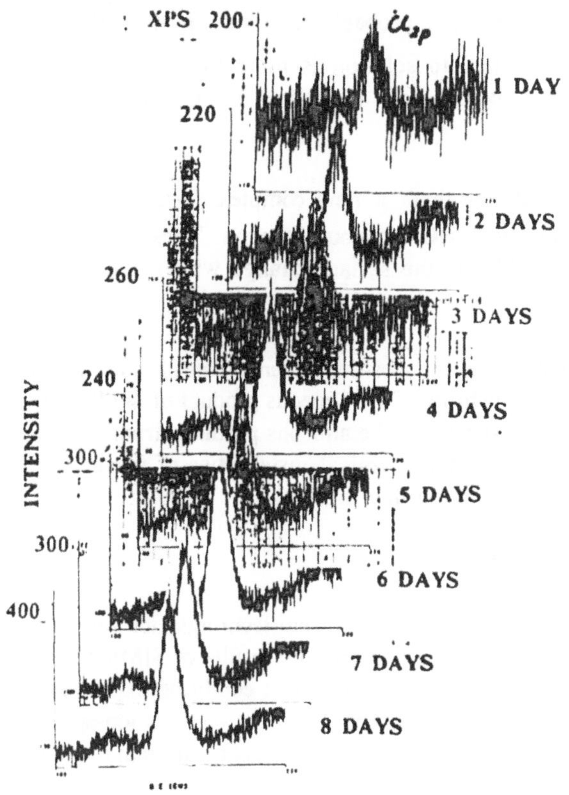


Fig.5 XPS spectra of fluorite after reaction in Fluorite - HCl - H2O flow system at 25°C, pH = 0.5. This diagram displays the XPS spectra of fluorite surface reacted within 8 days.

Fig. 5 shows that XPS spectra of fluorite after reaction in HCl - H2O flow system at 25°C, pH = 0.5. This diagram express that the XPS spectra of fluorite surface vary within 8 days. X -axis shows the binding energy. This plot expresses that the surface modification: $Cl^-$ moves into the surface, and its spectrum intensity is a function of time.

The XPS results arise an evidence for $Cl^-$ absorption on fluorite surface. And there may be an ion exchange reaction on fluorite surface between $F^-$ and $Cl^-$:

$$CaF^+_{fluorite} + Cl^- = CaCl^+_{fluorite} + F^-$$

As operating a long term experiment of fluorite / HCl solution interaction, the XPS show strong $Cl_{2p}$ spectra on surface, which change with time. The surface compositions change with the accumulation reaction time. The XPS spectra suggest that there is a surface complex formed on surface, and probably having the form $Ca(F,Cl)_2$. It proves that $F$ released from fluorite into solution is more than the Ca ion released into solution, deduced from Fig.3, because of formation of $Ca(F,Cl)_2$ on the fluorite surface.

The XPS study also demonstrates that a new complex $Ca(F,Cl)_2$ formed on the fluorite surface after reacted with HCl aqueous solution. These facts suggest that the concentrations of $Ca(F,Cl)_2$ on the surface change with the position and reaction time.

All vibrational variation with time and position, even a complex oscillation of FT - IR spectra indicate the chemical instability of the aqueous product of the fluorite - HCl - H2O system. Raman spectra and UV spectra of the aqueous products from the reaction systems were observed, which also suggest the chemical instabilities of the interface during the reaction in the mineral-aqueous solution flow systems.

*The SIMS study of the reacted surface*

The SIMS, second ion mass spectroscopy, is a new technique for indentify the mineral surface after test the reaction rates. Fig. 6 shows the results of SIMS analysis for fluorite reacted with HCl solution with pH of 0.5, in the same condition expressed in Fig.5. This plot represents the results of SIMS analyses collected between Cs ion-sputtering intervals of the different time, in unit of minute. In fact this is a compositional depth profile.

Fig.6 expresses the comparison of the SIMS results between one of fresh fluorite and the others of reacted samples. It has been found that the $Cl^-$ and proton $H^+$ penetrate into the surface of fluorite after reaction, the $Ca^{2+}$ and $F^-$ removed out of the surface layer. So that, the $H^+$,$Cl^-$, $F^-$ and $Ca^{2+}$ distribution curves in the depth profile are characterized of non-linear feature and the curves are waved in the depth profile. It is an

So that, the $H^+$,$Cl^-$, $F^-$ and $Ca^{2+}$ distribution curves in the depth profile are characterized of non-linear feature and the curves are waved in the depth profile. It is an evidence of the non-linear kinetics that the spatial, non-linear dynamic behavior happens in the solid/liquid interface.

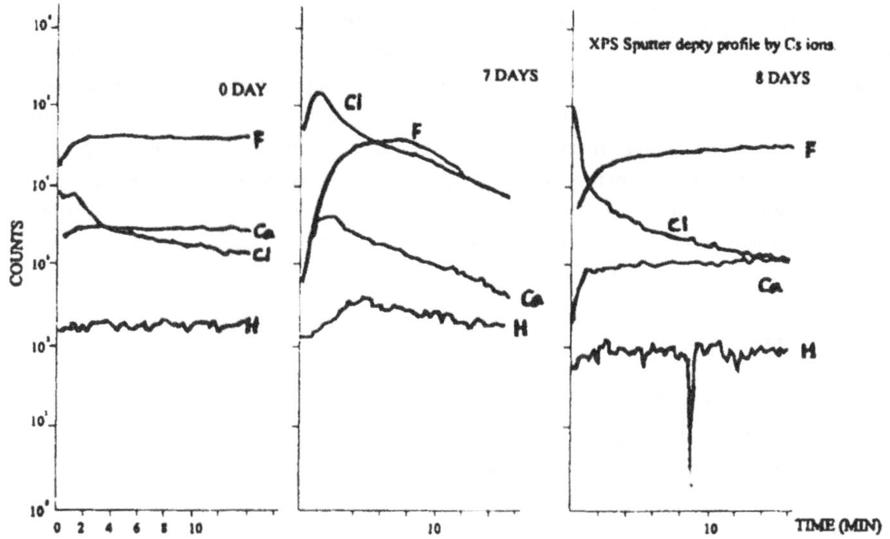


Fig.6 SIMS: Cs ion sputter depth profile results, expressed by time unit in minute. The comparison of the SIMS results of between the fresh fluorite and reacted sample. The Cl and proton H+ penetrate into the surface of fluorite after reaction. And Ca and F removed in the surface layer. Ca, H, Cl and F distribution curves in the depth profile are characterized of a non-linear dynamic process.

## SUMMARY

FT - IR and XPS study on the interface kinetics of fluorite - HCl solution in a flow system indicate that a surface complex $Ca(F,Cl)_2$ formed on fluorite surface, and its concentration on fluorite surface changes with time, input solution, and physico - chemical conditions. $Ca(F,Cl)_2$ is a temporary and transition state compound in some cases, which appears only on fluorite surface.

This phenomenon is discovered that the surface compounds would imply a importance in chemical principle: at first, $Cl^-$ has bigger ionic radius and more strong chemical activity, $Cl^-$ become aqueous species earsier than $F^-$. Why $Cl^-$ can replace $F^-$ on solid surface? The kinetic experiments of fluorite in water and HCl - H2O flow systems have already suggested a new kinetic behavior in the interface of fluorite/aqueous phase : there will be a temporary surface complex on mineral surface that it only can be found by using FT-IR rapid scan technique and by applying the real time and dynamic observation [4,5,6,13,14,15]. It is an important fact obtained in the kinetic experiment that the Cl concentrations of the output solutions from reaction system are changeable, which in some cases are less than input $Cl$ concentrations, during a long term of operating the kinetic experiments in fluorite - HCl - H2O systems. Our combined IR and XPS spectrum study and the expeimental results of the same reaction system suggest that the heterogeneous reaction and the interface kinetics are governed by the detailed structure and chemical bonnding of the minera surface in contact with the fluid. Authors continue this research work with a purpose in answering: what kind of speciation process happened on the interface of fluorite and HCl - H2O slution. Believe that the interface kinetics is taking a important role in mineral dissolution, and also in leading the chemical instability and in conducting the non - linear kinetic processes.

*Acknowledgements:*

This research supported by The Ministry of Geology and Mineral Resources (85-02-215), National Science and Technology Committee and Chinese NSF( 49070179, 49473169). Thank for Ms. Sun Suqin in helping our laboratory work.

Table 1. The binding energy (ev)

| Specimen | $F_{1s}$ | $Ca_{2p^{3/2}}$ | $Cl_{2p}$ | $C_{1s}$ |
|---|---|---|---|---|
| S-11 (after) | 685.2 | 348.3 | 199.9 | 959.5 |
| S-11 (before) | 695.5 | 348.5 | | 960.2 |
| $CaCl_2$ | | 348.6 | 199.3 | 961.5 |
| | | | | 963.7 |

XPS . ES300 type. Analyzed by Liu Zehong, Chemistry Institute of Chinese Academy.S - 11 (before) means fluorite sample before test dissolution rates; S - 11 (after) is the sample after testing the dissolution rates.

## REFERENCE:

1. Buback M., Crerar D. and Vogel Koplitz L., 1987, in: Hydrothermal experimental techniques, eds. Ubern and Barnes, 1987, Wiley.
2. Nakamote K., 1986, Infrared and Raman spectra of inorganic and coordination compounds, Wiley & Sons, 1986 (4th edition).
3. Hochella, Jr. M. F., and White A. F., 1990, ed: Mineral - water interface geochemistry, Rev. Min., Vol. 23, Min. Soc. Am.

4. Hu S. and Zhang R., 1994, Electronic and IR spectroscopy study on mineral (rock) - water reaction kinetics at elevated temperatures and pressures, in book: ed. Zhang T. and Zhang R., Thermodynamics and kinetics of natural processes, Science and technology published house, Beijing.
5. Hu S., Zhang R., and Crerar D., 1992, Chemical kinetic data base on mineral dissolution in hydrothermal systems, Abstract, 29th IGC, Kyoto Japan.
6. Hu S., and Zhang R., 1995, FT-IR and UV spectrum of interface in the mineral /water interaction, Water-rock interaction, Kharaka & Chudaev, eds. Balkema, Rotterdam.
7. White A.F., 1990, Heterogeneous electrochemical reaction associated with oxidation and ferrous oxide and silicata surfaces. In: Mineral - water interface geochemistry, Hochella M.F., and White A.F., eds: Rev. Mineral. vo.23, Min. Soc. Am.
8. Zhang R., Borcsik M., Crerar D., And Hu S., 1990, Kinetic study of mineral dissolutions in hydrothermal - flow system, dissolutions of fluorite and calcite in HCl-H2O systems at 300øc. Abstract, 15th IMA Congress Beijing, China 1990.
9. Zhang R., Posey-Dowty J., Hellmann R., Borcsik M. Crerar D., And Hu S., 1990, Kinetics of mineral - water reactions in hydrothermal flow systems at elevated temperatures and pressures, Science in China (series B) Vol.33, No.9, 1136-1152.( English edition )
10. Zhang R. and Hu S., 1994, Reaction kinetics and non - steady state kinetics of mineral in open flow systems, in book: ed. Zhang T. and Zhang R., Thermodynamics and kinetics of natural processes, Science and technology published house, Beijing.
11. Zhang R., Hu S., Hellmann R., And Crerar D., 1992, book: Chemical kinetics of minerals in hydrothermal systems and mass transfer. Science published House, Beijing ( Chinese with a detail abstract in English)
12. Zhang R., and Hu S.,1992, Chemical kinetic data on mineral aqueous solution at elevated temperatures and pressures. Bulletin CODATA, 1992, 13th CODATA conference, Beijing, CHINA.
13. Zhang R., and Hu S., 1994, Interface kinetics of CaF2 - HCl - H2O interaction in flow system, 1994 Annual meeting of Geological Society of America, Abstracts with programs, Seattle, p96.
14. Zhang R., and Hu s., 1996, Kinetics of fluorite dissolution in open-flow system and surface chemistry, Science in China (D), Vol.26, no.1. p 41-51.
15. Zhang Ronghua and Hu Shumin, Jian Lu, Tong Jianshang et al, 1997. Mineral-fluid reaction in open system and non-linear dynamic process, Science publishing house, Beijing.

# CRUSTAL ABUNDANCE

*Proc. 30th Int'l. Geol. Congr.*, Vol. 19, pp. 65-81
Xie Xuejin (Ed)
VSP 1997

# Chemical Composition of the Continental Crust in North China Platform

YAN MINGCAI and CHI QINGHUA

*Institute of Geophysical and Geochemical Exploration, Ministry of Geology and Mineral Resources*
*Langfang, Hebei 065000, China*

**Abstract**

In an area of some 1.4 million $km^2$ within the North China Platform (NCP), 12,193 various rocks were systematically sampled and combined into 1,207 composite samples which were then analyzed by 15 reliable analytical methods, such as instrumental neutron activation analysis (INAA), X-ray fluorescence analysis (XRF), and atomic absorption spectrophotometry (AAS) etc., with international accepted preliminary national geochemical certified reference materials (CRMs) of identical types for data quality monitoring. The average chemical compositions of igneous rocks, sedimentary rocks and metamorphic rocks as well as high grade granulite facies belts with or without K, Th, U depletion and high grade and medium-low grade amphibolite facies belts in NCP are presented. According to the data of regional geophysics, geology, experimental petrology and geochemistry, a crustal model of 38 km thickness was established and the abundance of 78 elements in the total crust (TC) and exposed crust (EC) were calculated in NCP. The results show that average chemical compositions of both the upper crust (UC) and the middle crust (MC) in NCP are equivalent to granodioritic, that of the lower crust (LC) is equivalent to quartz dioritic, and that of TC is characterized between granodiorite and quartz monzodiorite. The distribution of elements exhibits an obvious vertical zoning: volatile elements like As, Sb, Hg, B, and C are concentrated in the sedimentary cover, U, Th, Rb, Cs, B, and Be decrease remarkably from the sedimentary cover to the lower crust, Nb, Ta, Zr, Hf, W, Sn, and rare earth elements (REE) also tend to decrease gradually, and siderophile elements (SE) and platinum-group elements (PGE) are evidently enriched in LC.

*Keywords: Elemental Abundance, Rock, Total Continental Crust, Exposed Crust, North China Platform*

## INTRODUCTION

Earlier studies of element abundance in the crust were based on the exposed crust model and the two-layer crust model consisting of granitic rock in the upper part and basaltic rock in the lower part to estimate crustal element abundance. Based on extensive studies on GGT and other seismic sections etc., modern geologists have made remarkable advances in understanding the structure and components of the crust. It is now known that the continental crust is quite nonuniform both vertically and laterally. In the study of chemical composition of the crust, therefore, more importance is being attached to measured element abundance in exposed sections of specific crustal units like shield areas [11,10,9], Archean crust [14] and deep-source xenoliths [7], etc. According to available data for regional crust and geophysical and geochemical constraints, different authors have advanced several calculation models and chemical composition table of continental crust[7,9,4,16]. Recently, some authors start to study chemical composition of the whole continental crust by different ways [2,8,16]. But it should be pointed out that the most data for calculation of crustal abundance were collected from the 1960-1980s' literature and were rarely monitored by CRMs at that time.

In this study, numerous rocks were sampled and were used 15 reliable analytical methods with analytical quality monitored by 155 CRMs and 125 duplicate samples. Some newest achievements in Chinese analytical technology were used and, as a result, abundance estimates of some elements difficult to determine were obtained. On the basis of large amount of systematic regional geochemical research performed in the past ten years in China, combining with the regional data of geophysics, geology, and experimental petrology, we present the chemical composition of the various rocks, TC and EC and also UC, MC and LC in NCP.

## GEOLOGIC SETTING

The NCP is an oldest small-sized craton of China, locates at the juncture of important tectonic units, is characterized by obvious mobility. Its main part was formed in Middle to Late Archean, and the various parts were pieced together into a complete body in Early Proterozoic. The exposed basement area is composed of 65% metamorphic rocks and 35% felsic intrusive rocks. Amphibolite facies rocks make up 65% in metamorphic rocks. Granulite facies rocks constituting 18% mainly occur in the north part of the platform and exposed discontinuously for more than 1000 km [12]. Green-schist facies rocks making up 17% are mainly scattered in Proterozoic rift depression area. Since Middle Proterozoic there was deposition of platform cover. During Middle to Late Proterozoic and Early Paleozoic there was marine carbonate deposition, with magnesian materials occupying a major part in the early period. In the Late Paleozoic there was mainly marine facies argillaceous and clastic deposition. After the Triassic, continental deposition took place, and the crust was obviously mobilized, accompanied by large quantity of granitic intrusions.

## SAMPLING AND ANALYSIS

### *Sampling*

Selective sampling was carried out quite uniformly in most stratigraphic-tectonic divisions where bedrock is exposed without Quaternary overburden. The sampling locations along stratigraphic sections and in rock bodies are shown in Fig. 1. Stratigraphic sampling was arranged on the basis of stratigraphic-tectonic divisions and conducted mainly along typical geological sections established during 1:200,000 or 1:50,000 regional geological survey, covering the complete geological epochs and lithologic characters. Igneous rocks were mainly sampled from fairly large and intensively-studied representative bodies of various rock types. The emphasis was placed on the high-grade metamorphic area, and hence rocks were systematically sampled in four granulite facies zones, six high-grade amphibolite facies zones, eight medium to low grade amphibolite facies zones, and five greenschist facies zones, with sampling based on formations and lithologic units. Along more than 300 stratigraphic sections and 200 intrusive bodies, 12,193 rocks were sampled altogether, which were then combined into 1,207 analytical composite samples. The rocks sampled were fresh and unaltered, weighing some 500 g for each.

### *Preparation of Samples*

crushing was performed with a specially-made corundum porcelain pollution-free jaw-breaker, the samples were combined. For stratigraphic unit samples having the same age and lithologic character collected along a certain geological section were combined into a composite sample; for high-grade metamorphic belts, samples from the same lithologic unit were combined into a composite sample; for igneous rocks, samples from a rock body or rock bodies with identical age and lithogic character were combined into a composite sample. A composite sample weighs some 2 kg. After being well mixed, part of a sample was finely pulverized in an agate mill to minus 200 mesh for analysis. For ultratrace elements and elements difficult to determine, we also combined the samples into 29 large analytical composite samples according to lithologic character, metamorphic facies or metamorphic provinces, to reduce the number of analysis.

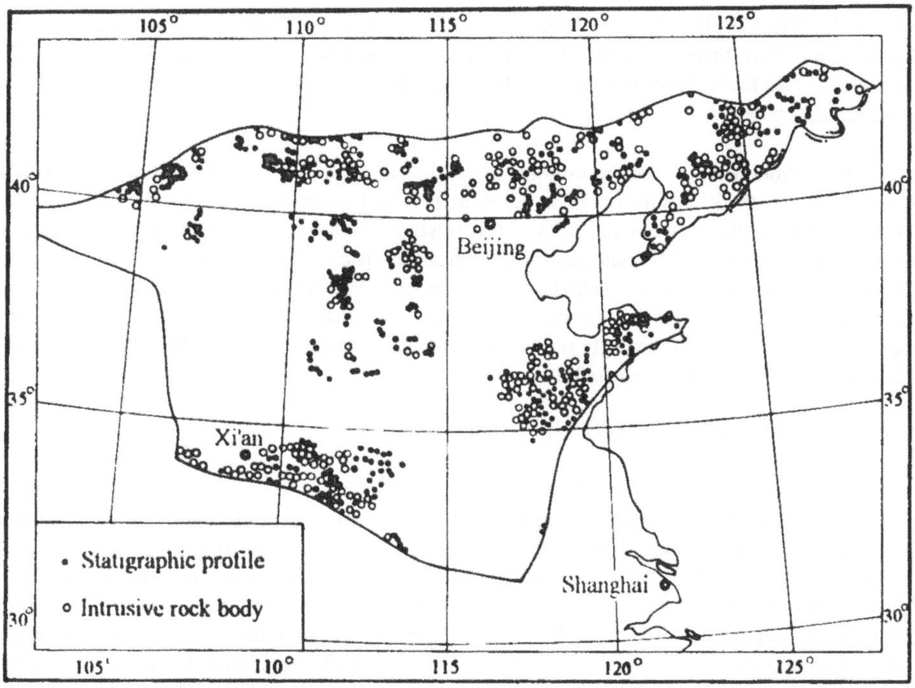


Fig. 1 Sampling sites in North China Platform.

*Analysis and Data Quality Monitoring*

Fifteen analytical methods based on different principles were employed, with major emphasis on INAA and XRF. The analysis was undertaken by experienced laboratories. 62 elements were determined in all composite samples. Seventeen elements, Os, Ir, Pt, Ru, Rh, Pd, Re, Br, I, N, Te, In, Pr, Dy, Ho, Er, and Tm, were determined within the 29 large composite samples. Co, Cr, Cs, Hf, Rb, Sc, Ta, Th, U, La, Ce, Nd, Sm, Eu, Gd, Tb, Yb, and Lu were determined by INAA; Re by neutron activation after chemical enrichment; Ba, Cu, Ga, Mn, Nb, Ni, P, Pb, Rb, Sr, Ti, V, Y, Zn, and Zr by pressed powder disc X-ray fluorescence spectrometry, with 40 high-quality CRMs used to draw working curves. The major elements composition was determined jointly by fused disc X-ray fluorescence spectrometry and chemical methods, with the sum required to be in the range of 99.00%-100.70%; Ag, Be, Cd, In, Tl, and Pd were analyzed by GAAS, with In and Tl pre-enriched by extraction and Pd by fire assay; Cu, Li, Ni, Pb, Na, and K by FAAS; As, Bi, Ge, Hg, Sb, and Se by hydride generation atomic fluorescence photometry; U by laser fluorescence method; Ge, Mo, W, Pt, Rh, and Te by catalytic wave polarography, with Pt and Rh after fire assay enrichment and Te after chemical enrichment; Cl and S by ionic chromatography; F by ion selective electrode method; B, Sn, Au, Pd, and Pt by electric arc emission spectrographic analysis, with Au, Pd, and Pt pre-enriched by fire assay; 15 rare earth elements were analyzed by ICP emission spectrometry or ICP-mass spectrographic analysis after ion exchange enrichment; Br, I, Ir, Os, and Ru by flow-injection spectrophotometric method, with Ir after assay enrichment, and Os and Ru after distillation enrichment; N by volumetric analysis after distillation enrichment.

During the analysis of samples, 10% CRMs (155) [17] and duplicate samples (125) were evenly inserted. The Chinese CRMs inserted are GSR-1 (granite), GSR-2 (andesite), GSR-3 (basalt), GSR-5 (shale), GSR-6 (limestone), GSD-9 (stream sediment), and GSS-8 (loess). These CRMs have been analyzed by 50 institutes and central laboratories, using 21 analytical methods, such as

AAS, COL, GR, ICP, INAA, POL, VOL, and XRF, in which 70-73 elements have been certified. Recent years, more than 30 countries including USA, Canada, Italy, France, UK, and Australia have widely used them. Most analysts admit that the Chinese CRMs are appropriate for monitoring analytical data quality.

The relative errors (RE) are used to estimate the analytical accuracy: $RE(\%)=1/n\sum|C_i-C_r|/C_r\times 100\%$, where $C_i$ is the ith determination of an element in CRM in an analytical batch, and $C_r$ is the certified value of the element in CRM. The relative deviations (RD) of analytical results of replicate samples are used to estimate the analytical precision: $RD(\%)=1/n\sum|C_1-C_2|/((C_1+C_2)/2)\times 100\%$, where $C_1$ and $C_2$ are two analytical results in each pair of the 2n replicate samples, one for basic sample and the other for the coded replicate. The analytical method and detection limit (varying with rock types and sample size) adopted for each element as well as RE of the CRMs and RD of the replicate samples are listed in Table 1.

## CHEMICAL COMPOSITIONS OF ROCKS

*Classification and Nomenclature of Rocks*

The classification of igneous rocks were based on the scheme recommended by the International Union of Geological Sciences (IUGS) subcommission on the systematic of igneous rocks in which rock is named by ANOR-Q'(F') diagram of standard minerals for intrusive rocks and by Total Alkali-$SiO_2$ (TAS) diagram for volcanic rocks with the reference of the field name of rock body. Sedimentary rocks were named by their originals in profile with the reference of their actual chemical composition. The nomenclature of metamorphic rocks are similar to sedimentary rocks; granulites were classified into acidic, intermediate and mafic granulites according to the $SiO_2$ concentration boundaries of 63% and 53%.

*Data Processing*

Abnormal samples having possibly undergone alteration or mineralization were rejected. All data for major elements and REE were used. As for trace elements, the rock type was taken as the statistical unit, and the data outside $X \pm 2s$ (where X was the initial arithmetic mean, and s was standard deviation) were rejected as deviated values. The arithmetic mean was adopted as the optimum estimate.

After taking above two steps, chemical compositions of various types of rocks in NCP are given in Table 2 and Table 3.

## CHEMICAL COMPOSITIONS OF MAIN HIGH-GRADE METAMORPHIC TERRAINS

The NCP is a region where very detailed geological work has been done. Regional geological mapping at 1:50,000 was carried out in most high-grade metamorphic areas, and, as a result, a fairly clear knowledge has been acquired concerning fabrics of metamorphic complexes.

The Qianxi-Qian'an granulite facies belt (Qianxi Group, 8,000 $km^2$ in exposed area) in eastern Hebei and the Xinghe (Inner Mongolia) - Yanggao (Shanxi) - Huai'an (Hebei) granulite facies belt (Xiajining Group, 3600 $km^2$ in exposed area) are planar distribution within large areas, comprising mainly two-pyroxene plagiogneiss, and subordinately, mafic granulite and supracrustal rocks. According to calculations by several investigators, peak metamorphic temperatures and pressures of two-pyroxene rocks are 750-900℃ and 0.7-1.3 GPa respectively [12]. The obvious deficiency of U, Th, K, Rb, Cs and a weak positive anomaly of Eu/Eu* (Table 4) indicate the characteristics of the lower crust.

**Table 1 Analytical methods and data quality evaluation**

| | Method | Low detection limits ($10^{-6}$) | RE(%) n=155 | RD(%) n=125 | | Method | Low detection limits ($10^{-6}$) | RE(%) n=155 | RD(%) n=125 |
|---|---|---|---|---|---|---|---|---|---|
| Li | FAAS | 1 | -7.1 | 15 | Ba | XRF | 20 | 1.4 | 7.7 |
| Be | GAAS | 0.1 | -1.9 | 19 | La | INAA, ICP/MS | 0.2 | -0.32 | 6.9 |
| B | ES | 1 | -2 | 19 | Ce | INAA, ICP/MS | 3, 0.5 | 1.8 | 5.6 |
| N | VOL | 10 | -5 | 15 | Pr | ICP, ICP/MS | 0.2 | (-4.4) | (3.3) |
| F | ISE | 50 | -0.68 | 12 | Nd | INAA, ICP/MS | 3 | 0.44 | 13 |
| P | XRF | 5 | -1.4 | 6.7 | Sm | INAA, ICP/MS | 0.1 | -1.8 | 7.2 |
| S | IC(VOL) | 20(50) | 8.9 | 22 | Eu | INAA, ICP/MS | 0.05 | 0.07 | 6.1 |
| Cl | IC | 20 | 2.3 | 27 | Gd | INAA, ICP/MS | 1, 0.1 | 1.2 | 15 |
| Sc | INAA | 0.1 | 0.95 | 4.4 | Tb | INAA, ICP/MS | 0.1, 0.3 | -1.2 | 14 |
| Ti | XRF | 10 | -0.67 | 4.1 | Dy | ICP, ICP/MS | 0.1 | (-3.0) | (4.0) |
| V | XRF | 5 | 6 | 11 | Ho | ICP, ICP/MS | 0.03 | (4.2) | (6.0) |
| Cr | INAA | 0.5 | -1.4 | 11 | Er | ICP, ICP/MS | 0.05 | (2.5) | (4.5) |
| Mn | XRF | 5 | 3.7 | 4.3 | Tm | ICP, ICP/MS | 0.02 | (-6.3) | (7) |
| Co | INAA | 0.2 | 0.34 | 8.1 | Yb | INAA, ICP/MS | 0.2, 0.1 | -4.8 | 13 |
| Ni | XRF, FAAS | 2, 1 | -2.1 | 14 | Lu | INAA, ICP/MS | 0.02 | -2.0 | 6.0 |
| Cu | XRF, FAAS | 2.1 | -2.1 | 14 | Hf | INAA | 0.2 | -2.9 | 6.5 |
| Zn | XRF | 2 | 0.95 | 4.4 | Ta | INAA | 0.2 | -4.0 | 13 |
| Ga | XRF | 1 | -0.14 | 5.6 | W | POL | 0.2 | -5.1 | 21 |
| Ge | AF, POL | 0.1 | 2.8 | 17 | Re | CNAA | 0.1 | | (20) |
| As | AF | 0.1 | -5.5 | 15 | Os | COL | 0.006 | 5 | (50) |
| Se | AF | 0.01 | -1.3 | 22 | Ir | FA-POL | 0.003 | | |
| Br | COL | 0.2 | (-8.8) | (11) | Pt | FA-ES | 0.1(0.03) | -4.1 | 30 |
| Rb | XRF, INAA | 2 | 1.4 | 5.4 | Au | FA-ES | 0.05 | -4.0 | 24 |
| Sr | XRF | 2 | 0.56 | 1.9 | Hg | AF | 0.002 | -4.3 | 22 |
| Y | XRF, ICP | 3, 0.5 | -6.7 | 3.8 | Tl | GAAS | 0.1 | -0.66 | 24 |
| Zr | XRF | 2 | 2.6 | 2.7 | Pb | XRF, FAAS | 2 | 4.8 | 10 |
| Nb | XRF | 1-2 | 7.1 | 6.4 | Bi | AF | 0.05 | -1.9 | 20 |
| Mo | POL | 0.1-0.2 | -1.2 | 24 | Th | INAA | 0.2 | -2.8 | 8.9 |
| Ru | COL | 0.004 | -10 | (25) | U | INAA, LF | 0.2 | 2.2 | 15 |
| Rh | FA-POL | 0.003 | | | % | | | | |
| Pd | FA-ES | 0.1(0.03) | -1.6 | 17 | $SiO_2$ | XRF, GR | 0.05 | -0.01 | 0.69 |
| Ag | GAAS | 0.02 | 0.87 | 20 | $Al_2O_3$ | XRF, VOL | 0.05 | -1.0 | 2.0 |
| Cd | GAAS | 0.01 | 1.2 | 19 | $TFe_2O_3$ | XRF, VOL | 0.01 | 0.34 | 3.0 |
| In | GAAS | 0.01 | | 25 | FeO | VOL | 0.1 | 0.60 | 8.8 |
| Sn | ES | 0.2 | 5.0 | 22 | MgO | XRF, VOL | 0.05 | 0.6 | 9.6 |
| Sb | AF | 0.05 | 4.6 | 21 | CaO | XRF, VOL | 0.01 | -0.34 | 4.4 |
| Te | POL | 0.01 | | (35) | $Na_2O$ | XRF, FAAS | 0.1 | 2.4 | 8.5 |
| I | COL | 0.1 | (2.2) | (7.3) | $K_2O$ | XRF, FAAS | 0.01 | -0.11 | 3.3 |
| Cs | INAA | 0.3 | -6.8 | 8.7 | $H_2O^+$ | GR | 0.2 | 4.2 | 12 |

Note: For Rh, Ru, Pd, Re, Os, Ir, Pt, Au, and Hg, the units of low detection limits are $10^{-9}$.

FAAS: Flame atomic absorption spectrometry; GAAS: Graphite furnace atomic absorption spectrometry; AF: Atomic fluorescence spectrometry; COL: Colorimetry; ES: Emission spectrography; FA: Fire assay; GR Gravimetry; IC: Ion chromatography; ICP: Inductively coupled plasma spectrometry; ICP/MS ICP-mass spectrography; INAA Instrument neutron activation analysis; ISE: Ion selective electrode; LF Laser fluorescence; CNAA: Neutron activation analysis after chemical enrichment; POL: Polarography; VOL Volumetry; XRF: X-ray fluorescence spectrometry

**Table 2 Average chemical compositions of major igneous and sedimentary rocks in North China Platform**

| Rock | Granite | Grano-diorite | Diorite | Gabbro | Basalt | Felsic-Sandstone | Common Sandstone | Pelite | Lime-stone | Dolomite | Soil |
|---|---|---|---|---|---|---|---|---|---|---|---|
| N | 210 | 67 | 63 | 29 | 34 | 157 | 107 | 43 | 59 | 31 | 155 |
| nN | 2060 | 742 | 726 | 211 | 306 | 2094 | 1329 | 416 | 698 | 590 | |
| $SiO_2$ | 72.07 | 65.68 | 57.88 | 49.45 | 47.92 | 72.66 | 66.91 | 61.15 | 6.58 | 7.89 | 65.0 |
| $Al_2O_3$ | 14.05 | 15.09 | 16.05 | 15.17 | 14.73 | 10.39 | 12.28 | 15.78 | 1.26 | 0.90 | 12.6 |
| $Fe_2O_3$ | 1.20 | 1.93 | 3.05 | 4.18 | 4.98 | 2.58 | 3.07 | 5.25 | 0.37 | 0.42 | 3.4 |
| FeO | 0.92 | 2.68 | 4.08 | 6.57 | 6.47 | 0.76 | 0.93 | 1.08 | 0.30 | 0.31 | 1.2 |
| MgO | 0.60 | 2.08 | 3.53 | 7.29 | 6.90 | 1.18 | 1.52 | 1.95 | 2.97 | 17.10 | 1.8 |
| CaO | 1.30 | 3.20 | 5.66 | 9.08 | 7.87 | 2.99 | 3.96 | 2.26 | 46.85 | 29.92 | 3.2 |
| $Na_2O$ | 3.82 | 3.70 | 4.02 | 2.53 | 3.11 | 1.55 | 2.04 | 1.05 | 0.10 | 0.08 | 1.6 |
| $K_2O$ | 4.27 | 2.92 | 2.22 | 1.14 | 1.54 | 2.40 | 2.67 | 3.78 | 0.39 | 0.38 | 2.5 |
| $H_2O^+$ | 0.90 | 1.28 | 1.45 | 1.71 | 1.64 | 2.46 | 2.90 | 4.32 | 0.74 | 0.69 | 4.2 |
| $CO_2$ | 0.25 | 0.38 | 0.35 | 0.40 | 0.55 | 2.01 | 2.55 | 1.66 | 39.70 | 41.85 | 2.7 |
| Ag | 0.052 | 0.062 | 0.056 | 0.055 | 0.048 | 0.054 | 0.055 | 0.06 | 0.053 | 0.054 | 0.08 |
| As | 0.7 | 1.2 | 1.0 | 1.0 | 0.9 | 3.2 | 3.0 | 5.2 | 2.4 | 2.6 | 10 |
| Au | 0.4 | 0.8 | 0.8 | 0.85 | 0.46 | 0.8 | 0.88 | 1.1 | 0.35 | 0.30 | 1.4 |
| B | 4.5 | 7 | 5.0 | 15 | 4.8 | 33 | 32 | 78 | 18.4 | 19 | 40 |
| Ba | 785 | 910 | 910 | 440 | 530 | 530 | 640 | 565 | 49 | 54 | 500 |
| Be | 2.5 | 1.3 | 1.1 | 0.70 | 0.53 | 1.25 | 1.4 | 2.1 | 0.67 | 0.35 | 1.8 |
| Bi | 0.10 | 0.12 | 0.07 | 0.084 | 0.044 | 0.12 | 0.14 | 0.32 | 0.08 | 0.06 | 0.3 |
| Cd | 0.046 | 0.075 | 0.09 | 0.10 | 0.096 | 0.075 | 0.09 | 0.083 | 0.10 | 0.07 | 0.09 |
| Cl | 60 | 104 | 190 | 180 | 95 | 61 | 64 | 62 | 186 | 227 | 68 |
| Co | 3 | 13 | 23 | 44 | 50 | 7.7 | 10 | 16 | 1.9 | 1.3 | 13 |
| Cr | 6.3 | 54 | 90 | 192 | 180 | 34 | 41 | 68 | 6.3 | 4.8 | 65 |
| Cs | 2.0 | 2.0 | 1.4 | 1.0 | 0.58 | 2.3 | 3.0 | 8.2 | 0.5 | 0.4 | 7 |
| Cu | 4.9 | 23 | 30 | 58 | 50 | 12.6 | 15 | 30 | 4.6 | 3.5 | 24 |
| F | 360 | 570 | 730 | 430 | 520 | 336 | 395 | 755 | 2.63 | 2.60 | 480 |
| Ga | 17.8 | 18.6 | 20.4 | 19.6 | 21.5 | 12.8 | 15.0 | 20.9 | 2.0 | 1.3 | 17.0 |
| Ge | 1.1 | 1.1 | 1.0 | 1.2 | 1.1 | 1.4 | 1.3 | 1.8 | 0.4 | 0.4 | 1.3 |
| Hf | 4.6 | 4.4 | 4.6 | 2.8 | 4.5 | 5.3 | 5.4 | 6.2 | 0.36 | 0.33 | 7.4 |
| Hg | 6.0 | 6 | 6.7 | 10 | 6 | 14 | 13 | 23 | 11.4 | 10 | 40 |
| Ir | (2) | (8) | (25) | (30) | (20) | (15) | (15) | (25) | (8) | (5) | 22 |
| Li | 12 | 16 | 10.7 | 9.7 | 9 | 21 | 24 | 35 | 9.2 | 7.2 | 30 |
| Mn | 282 | 540 | 870 | 1230 | 1300 | 420 | 530 | 520 | 376 | 540 | 600 |
| Mo | 0.5 | 0.48 | 0.49 | 0.30 | 1.2 | 0.55 | 0.48 | 0.76 | 0.47 | 0.47 | 0.8 |
| N | 38 | (34) | (60) | (128) | (96) | (86) | (100) | (320) | (130) | | 640 |
| Nb | 13 | 9.5 | 10 | 9.7 | 34 | 11 | 12 | 17 | (1.5) | (1.5) | 16 |
| Ni | 4.3 | 24 | 38 | 95 | 123 | 14 | 17 | 32 | 4.3 | 3.2 | 26 |
| Os | (12) | (22) | (80) | (50) | (30) | (15) | (20) | (40) | (40) | (40) | 40 |
| P | 310 | 750 | 1290 | 1160 | 2340 | 410 | 490 | 630 | 173 | 100 | 520 |
| Pb | 21 | 16 | 14 | 9.3 | 9.3 | 15 | 16.5 | 23 | 8.4 | 8.2 | 23 |
| Pd | (0.06) | 0.43 | 0.5 | 1.0 | 0.30 | 0.22 | 0.25 | 0.56 | (0.1) | (0.1) | 0.65 |
| Pt | (0.05) | 0.53 | 0.5 | 1.0 | 0.30 | 0.21 | 0.22 | 0.33 | (0.1) | (0.1) | 0.50 |
| Rb | 133 | 88 | 59 | 30 | 25 | 68 | 78 | 147 | 11 | 8 | 100 |
| Rh | (3) | (13) | (110) | (50) | (20) | (13) | (14) | (22) | (4) | (3) | (17) |
| Ru | (6) | (21) | (50) | (50) | (40) | (18) | (20) | (30) | (14) | (10) | (60) |
| S | 70 | 350 | 240 | 380 | 90 | 180 | 180 | 170 | 190 | 110 | 150 |
| Sb | 0.12 | 0.14 | 0.13 | 0.14 | 0.10 | 0.28 | 0.28 | 0.46 | 0.16 | 0.16 | 0.8 |

**Table 2 (continu.)**

| Rock | Granite | Grano-diorite | Diorite | Gabbro | Basalt | Felsic-Sandstone | Common Sandstone | Pelite | Lime-stone | Dolomite | Soil |
|---|---|---|---|---|---|---|---|---|---|---|---|
| Sc | 3.3 | 10 | 17 | 32 | 23 | 6.8 | 8.5 | 14.6 | 1.7 | 0.9 | 11Se |
| | 0.027 | 0.065 | 0.064 | 0.10 | 0.057 | 0.064 | 0.06 | 0.08 | 0.05 | 0.03 | 0.2 |
| Sn | 1.3 | 1.2 | 1.2 | 1.1 | 1.7 | 1.3 | 1.5 | 2.6 | 0.8 | 0.6 | 2.5 |
| Sr | 245 | 440 | 645 | 500 | 640 | 147 | 195 | 145 | 280 | 93 | 170 |
| Ta | 0.91 | 0.51 | 0.52 | 0.40 | 2.2 | 0.63 | 0.71 | 1.1 | 0.11 | 0.08 | 1.1 |
| Th | 13 | 6.6 | 4.1 | 1.7 | 3.3 | 7.9 | 8.5 | 13.4 | 1.3 | 1.0 | 12.5 |
| Ti | 1360 | 2950 | 4680 | 7300 | 11900 | 2560 | 3060 | 4230 | 367 | 290 | 4300 |
| Tl | 0.68 | 0.52 | 0.32 | 0.20 | 0.19 | 0.43 | 0.47 | 0.69 | 0.13 | 0.15 | 0.6 |
| U | 1.7 | 0.87 | 0.76 | 0.55 | 0.84 | 1.74 | 1.9 | 2.5 | 0.77 | 0.6 | 2.7 |
| V | 20 | 72 | 124 | 233 | 180 | 57 | 70 | 97 | 12 | 10 | 82 |
| W | 0.5 | 0.38 | 0.39 | 0.31 | 0.37 | 0.78 | 0.82 | 1.5 | (0.23) | (0.28) | 1.8 |
| Zn | 36 | 65 | 90 | 100 | 122 | 43 | 52 | 83 | 18 | 10 | 68 |
| Zr | 155 | 168 | 177 | 110 | 202 | 190 | 200 | 214 | 18 | 14 | 250 |
| Y | 19 | 15 | 16.5 | 16 | 18 | 15.5 | 17.5 | 25 | 5.9 | 4.3 | 23 |
| La | 40 | 35 | 38 | 19.5 | 32 | 32 | 36 | 49 | 6.3 | 4.3 | 38 |
| Ce | 78 | 68 | 75 | 40 | 61 | 60 | 67 | 90 | 12 | 8.4 | 72 |
| Pr | 7.9 | 7.0 | 8.2 | 4.8 | 6.9 | 6.4 | 7.2 | 9.5 | 1.3 | 0.9 | 8.2 |
| Nd | 29 | 28.5 | 36.5 | 20.5 | 30 | 27 | 30 | 40 | 6.0 | 4.1 | 32 |
| Sm | 4.5 | 4.6 | 6.5 | 4.5 | 6.4 | 4.5 | 5.2 | 7.2 | 1.13 | 0.71 | 5.8 |
| Eu | 0.85 | 1.20 | 1.7 | 1.5 | 2.2 | 0.96 | 1.13 | 1.5 | 0.24 | 0.17 | 1.2 |
| Gd | 3.9 | 4.1 | 5.5 | 4.5 | 5.8 | 3.9 | 4.6 | 6.0 | 1.0 | 0.67 | 5.1 |
| Tb | 0.60 | 0.53 | 0.79 | 0.66 | 0.9 | 0.6 | 0.67 | 0.97 | 0.17 | 0.11 | 0.8 |
| Dy | 3.3 | 2.8 | 4.4 | 3.7 | 5.0 | 3.3 | 3.8 | 5.1 | 0.95 | 0.60 | 4.7 |
| Ho | 0.68 | 0.6 | 0.77 | 0.76 | 0.9 | 0.62 | 0.75 | 1.1 | 0.19 | 0.12 | 1.0 |
| Er | 2.0 | 1.6 | 2.4 | 2.1 | 2.2 | 1.86 | 2.1 | 3.1 | 0.52 | 0.33 | 2.8 |
| Tm | 0.30 | 0.25 | 0.32 | 0.33 | 0.29 | 0.29 | 0.32 | 0.4 | 0.08 | 0.05 | 0.42 |
| Yb | 1.8 | 1.5 | 2.0 | 1.9 | 1.8 | 1.68 | 2.0 | 2.9 | 0.51 | 0.31 | 2.6 |
| Lu | 0.28 | 0.24 | 0.30 | 0.31 | 0.27 | 0.27 | 0.31 | 0.44 | 0.08 | 0.05 | 0.40 |

Data source: this study; Unit: % for major elements; $10^{-12}$ for Ir, Os, Rh, and Ru; $10^{-9}$ for Au, Hg, Pd, and Pt; $10^{-6}$ for others.

The average chemical compositions of the Jianping granulite facies belt (Jianping Group) in western Liaoning and the Jining-Qianlishan granulite facies belt (Shangjining Group) in Inner Mongolia showes no obvious deficiency of U, Th, K, Rb, and Cs (Table 4). The latter belt contains a considerable proportion of sillimanite garnet gneiss, and its estimated metamorphic temperatures and pressures are 700-810℃and 0.55-0.65 GPa respectively, suggesting that it is likely a product either in the upper or the middle part of the lower crust, rather than a typical geological body in LC.

As reported by different authors, metamorphic temperatures and pressures for six high-grade amphibolite facies belts are 500-700℃and 0.4-1.0 GPa respectively whereas those for eight medium-low grade amphibolite facies belts are 500-600℃and 0.4-0.8 GPa respectively. The protoliths are mainly tonalite-trondhjemite-granodiorite (TTG) gneiss series, and subordinately supracrustal rocks and plagioclase amphibolite, showing the characteristics of rocks in the middle and middle-upper crust. Their average chemical compositions of high and medium-low grade amphibolite facies belts are given in Table 4.

## CRUSAL MODEL AND CALCULATION SCHEME

According to the depth of the Mohorovicic discontinuity and data from the crustal section [5,6]

**Table 3 Average chemical compositions of major metamorphic rocks in North China Plaform**

| Rock | Phyllite | Schist | Leptite | Gneiss | Acidic Granulite | Intermediate Granulite | Mafic Granulite | Amphibolite |
|---|---|---|---|---|---|---|---|---|
| **N** | **12** | **30** | **59** | **163** | **19** | **22** | **27** | **40** |
| nN | 88 | 345 | 509 | 1592 | 133 | 144 | 83 | 366 |
| $SiO_2$ | 61.09 | 63.22 | 66.40 | 65.26 | 65.83 | 57.07 | 49.18 | 49.90 |
| $Al_2O_3$ | 16.40 | 16.17 | 14 56 | 14.92 | 15.00 | 16.12 | 13.78 | 13.85 |
| FeO | 3.41 | 3.02 | 2.23 | 2.12 | 1.90 | 3.25 | 5.09 | 4.50 |
| $Fe_2O_3$ | 2.51 | 2.82 | 2.15 | 2.70 | 2.85 | 4.88 | 8.74 | 7.53 |
| MgO | 2.51 | 2.37 | 2.09 | 2.22 | 2.02 | 3.73 | 6.74 | 6.90 |
| CaO | 2.35 | 2.02 | 3.07 | 3.46 | 3.87 | 6.44 | 10.06 | 8.54 |
| $Na_2O$ | 1.20 | 1.96 | 3.48 | 3.68 | 4.01 | 4.06 | 2.48 | 2.60 |
| $K_2O$ | 3.92 | 3.80 | 2.68 | 2.87 | 2.38 | 1.83 | 0.77 | 1.16 |
| $H_2O^+$ | 3.55 | 2.75 | 1.80 | 1.30 | 0.65 | 0.69 | 0.95 | 2.20 |
| $CO_2$ | 1.82 | 0.58 | 0.35 | 0.30 | 0.26 | 0.32 | 0.30 | 0.45 |
| Ag | 0.04 | 0.057 | 0.06 | 0.058 | 0.052 | 0.057 | 0.06 | 0.057 |
| As | 3.4 | 4.4 | 1.8 | 1.0 | 0.6 | 0.9 | 0.9 | 1.2 |
| Au | 0.95 | 1.2 | 0.96 | 0.67 | 1.2 | 1.2 | 1.6 | 0.9 |
| B | 96 | 52 | 13 | 5.2 | 1.3 | 0.9 | 4.7 | 10 |
| Ba | 680 | 745 | 730 | 840 | 780 | 670 | 220 | 350 |
| Be | 2.5 | 2.0 | 1.7 | 1.25 | 0.9 | 0.33 | 0.2 | 0.5 |
| Bi | 0.31 | 0.22 | 0.093 | 0.09 | 0.045 | 0.18 | 0.22 | 0.12 |
| Cd | 0.047 | 0.08 | 0.08 | 0.07 | 0.065 | 0.10 | 0.13 | 0.12 |
| Cl | 44 | 98 | 82 | 115 | 100 | 100 | 110 | 170 |
| Co | 17 | 15 | 12 | 13.5 | 15 | 28 | 57 | 47 |
| Cr | 62 | 70 | 56 | 57 | 60 | 95 | 177 | 165 |
| Cs | 6.4 | 5.1 | 2.1 | 1.6 | 0.37 | 0.5 | 0.7 | 1.3 |
| Cu | 29 | 25 | 23 | 22 | 21 | 36 | 66 | 56 |
| F | 740 | 735 | 500 | 595 | 365 | 620 | 760 | 760 |
| Ga | 22 | 21.8 | 18.1 | 18.7 | 18.4 | 20.6 | 18.6 | 19.5 |
| Ge | 1.7 | 1.6 | 1.1 | 1.0 | 1.0 | 1.2 | 1.3 | 1.4 |
| Hf | 5.0 | 5.1 | 4.7 | 4.6 | 4.3 | 3.9 | 2.6 | 3.3 |
| Hg | 12 | 6.3 | 7.2 | 6.5 | 6 | 7 | 9 | 10 |
| Ir | (25) | (25) | (27) | 22 | (15) | (30) | (35) | (85) |
| Li | 31 | 25 | 15 | 13 | 10 | 8.6 | 10 | 12 |
| Mn | 540 | 520 | 500 | 560 | 590 | 1000 | 1480 | 1650 |
| Mo | 0.43 | 0.55 | 0.47 | 0.50 | 0.50 | 0.41 | 0.53 | 0.21 |
| N | 200 | 115 | (55) | 37 | (10) | (23) | (21) | (20) |
| Nb | 14.5 | 13.7 | 10.4 | 9.7 | 8 | 8.6 | 9 | 11 |
| Ni | 32 | 27 | 23 | 25 | 26 | 42 | 81 | 83 |
| Os | (20) | (40) | (33) | 28 | (14) | (26) | (66) | (130) |
| P | 435 | 540 | 555 | 730 | 740 | 1170 | 880 | 1020 |
| Pb | 16.7 | 17 | 13.4 | 15 | 12 | 9.2 | 9.4 | 14 |
| Pd | 0.46 | 0.4 | 0.56 | 0.44 | 0.37 | 1.5 | 2.8 | 1.6 |
| Pt | 0.42 | 0.4 | 0.56 | 0.46 | 0.36 | 1.5 | 3.2 | 2.0 |
| Rb | 160 | 134 | 80 | 82 | 60 | 31 | 17 | 37 |
| Rh | (30) | (30) | (35) | 35 | (20) | (60) | (140) | (150) |
| Ru | (40) | (40) | (28) | 33 | (20) | (35) | (90) | (230) |
| S | 70 | 360 | 210 | 210 | 330 | 470 | 940 | 340 |
| Sb | 0.23 | 0.25 | 0.18 | 0.13 | 0.10 | 0.09 | 0.11 | 0.15 |
| Sc | 15.4 | 15 | 10 | 10.7 | 12 | 22 | 44 | 38 |

**Table 3 (Continu.)**

| Rock | Phyllite | Schist | Leptite | Gneiss | Acidic Granulite | Intermediate Granulite | Mafic Granulite | Amphibolite |
|---|---|---|---|---|---|---|---|---|
| Se | 0.07 | 0.076 | 0.07 | 0.065 | 0.08 | 0.11 | 0.28 | 0.10 |
| Sn | 3.2 | 2.6 | 1.3 | 1.1 | 0.9 | 0.8 | 1.1 | 1.1 |
| Sr | 106 | 175 | 315 | 400 | 450 | 610 | 210 | 250 |
| Ta | 1.0 | 0.87 | 0.55 | 0.50 | 0.3 | 0.3 | 0.5 | 0.62 |
| Th | 14 | 11.8 | 7.4 | 6.6 | 2.1 | 0.8 | 1.2 | 1.8 |
| Ti | 4060 | 3560 | 2770 | 2950 | 3080 | 5000 | 7040 | 8200 |
| Tl | 0.74 | 0.77 | 0.43 | 0.47 | 0.43 | 0.22 | 0.2 | 0.27 |
| U | 2.4 | 1.9 | 1.2 | 0.87 | 0.34 | 0.3 | 0.3 | 0.58 |
| V | 94 | 92 | 74 | 75 | 80 | 145 | 280 | 250 |
| W | 2.2 | 1.2 | 0.47 | 0.41 | (0.2) | (0.25) | (0.3) | 0.48 |
| Zn | 72 | 76 | 61 | 63 | 60 | 94 | 123 | 125 |
| Zr | 190 | 184 | 167 | 167 | 154 | 150 | 106 | 136 |
| Y | 24.5 | 23 | 17.5 | 16.4 | 13 | 18 | 20 | 18 |
| La | 39 | 43 | 32 | 37 | 28 | 29 | 15 | 17.5 |
| Ce | 76.5 | 82 | 62.5 | 74 | 52 | 58 | 30 | 35 |
| Pr | 8.1 | 9.0 | 6.5 | 7.9 | 6.0 | 7.0 | 3.8 | 4.5 |
| Nd | 35 | 37 | 29 | 32.4 | 24 | 28 | 16.5 | 19.5 |
| Sm | 6.1 | 6.4 | 4.8 | 5.3 | 3.8 | 5.5 | 4.2 | 4.5 |
| Eu | 1.16 | 1.43 | 1.14 | 1.28 | 1.2 | 1.7 | 1.45 | 1.6 |
| Gd | 5.2 | 5.2 | 4.4 | 4.6 | 3.1 | 4.7 | 4.5 | 4.8 |
| Tb | 0.73 | 0.85 | 0.64 | 0.65 | 0.42 | 0.73 | 0.76 | 0.84 |
| Dy | 4.2 | 4.3 | 3.5 | 3.4 | 2.4 | 4.0 | 4.6 | 5.0 |
| Ho | 0.85 | 0.95 | 0.73 | 0.75 | 0.50 | 0.82 | 1.0 | 1.0 |
| Er | 2.8 | 2.7 | 2.0 | 2.0 | 1.4 | 2.2 | 2.9 | 2.9 |
| Tm | 0.42 | 0.44 | 0.30 | 0.30 | 0.20 | 0.31 | 0.44 | 0.42 |
| Yb | 2.54 | 2.54 | 1.8 | 1.8 | 1.3 | 1.9 | 2.75 | 2.6 |
| Lu | 0.40 | 0.41 | 0.28 | 0.28 | 0.18 | 0.28 | 0.44 | 0.40 |

Data source: this study; Unit: % for major elements; $10^{-12}$ for Ir, Os, Rh, and Ru; $10^{-9}$ for Au, Hg, Pd, and Pt; $10^{-6}$ for others.

and other seismic sections [3], the average crustal thickness of the NCP is 38 km. On the basis of geophysical data, combined with geological and geochemical data, the crust is divided into three layers for calculation of element abundance.

*Upper Crust (UC)*

(1) Sedimentary cover (Vp<6.0 km/s) varies greatly in thickness. On the basis of the volume estimated from the integrated stratigraphic thickness of various ages and the paleogeographic area since the Middle Proterozoic together with some consideration of Vp data, it is estimated that the average thickness is 3 km. The value was obtained by weighted average of volumes of various strata which consist of about 36% sandstones, 26% argillaceous rocks, 24% carbonate rocks, and 14% volcanic rocks. (2) Basement rocks (Vp=6.0-6.2 km/s) is 12 km thick, which was obtained by weighted average of exposed areas of various metamorphic rocks as well as intrusive rocks such as granite and diorite in the region (which consist of 60% metamorphic rocks, 35% granite, and 5% diorite).

*Middle Crust (MC)*

MC, being a transitional zone between UC and LC with Vp 6.2-6.5 km/s, has an average thickness of 10 km, which was evaluated from the average of medium- and low-grade amphibolite facies, high-grade amphibolite facies, and granulite facies rocks, attached by two factors within the facies

**Table 4 Average chemical compositions of various granulite facies belts and amphibolite facies belts in NCP**

| Terrain | 1 | 2 | 3 | 4 | Terrain | 1 | 2 | 3 | 4 |
|---|---|---|---|---|---|---|---|---|---|
| N | 81 | 47 | 114 | 145 | P | 760 | 780 | 755 | 550 |
| nN | 298 | 487 | 840 | 1554 | Pb | 10 | 14 | 15 | 14 |
| $SiO_2$ | 60.15 | 62.75 | 61.79 | 63.36 | Pd | 1.3 | 0.77 | 0.74 | 0.43 |
| $Al_2O_3$ | 15.40 | 14.11 | 14.58 | 14.02 | Pt | 1.5 | 0.90 | 0.86 | 0.53 |
| $Fe_2O_3$ | 2.94 | 2.82 | 2.54 | 2.25 | Rb | 32 | 77 | 71 | 77 |
| FeO | 4.30 | 3.70 | 3.10 | 2.57 | Rh | 78 | 69 | 45 | 33 |
| MgO | 3.43 | 3.54 | 5.17 | 2.96 | Ru | 69 | (58) | (30) | (55) |
| CaO | 5.49 | 4.20 | 3.55 | 4.46 | S | 310 | 345 | 230 | 160 |
| $Na_2O$ | 3.82 | 3.02 | 2.50 | 3.26 | Sb | 0.10 | 0.13 | 0.10 | 0.16 |
| $K_2O$ | 1.77 | 2.59 | 1.38 | 2.56 | Sc | 19 | 18 | 16 | 13 |
| Ag | 0.056 | 0.055 | 0.055 | 0.054 | Se | 0.12 | 0.09 | 0.065 | 0.060 |
| As | 1.2 | 1.5 | 0.87 | 1.5 | Sn | 1.0 | 0.98 | 1.1 | 1.3 |
| Au | 1.3 | 0.90 | 0.69 | 0.79 | Sr | 380 | 430 | 390 | 305 |
| B | 10 | 3.4 | 6.7 | 15 | Ta | 0.30 | 0.42 | 0.46 | 0.53 |
| Ba | 780 | 740 | 725 | 740 | Th | 1.1 | 5.0 | 5.6 | 7.2 |
| Be | 0.50 | 0.83 | 1.1 | 1.4 | Ti | 3800 | 3880 | 3590 | 3030 |
| Bi | 0.11 | 0.098 | 0.087 | 0.12 | Tl | 0.23 | 0.47 | 0.38 | 0.45 |
| Cd | 0.080 | 0.088 | 0.075 | 0.069 | U | 0.40 | 0.71 | 0.73 | 1.1 |
| Cl | 98 | 115 | 160 | 110 | V | 120 | 110 | 100 | 81 |
| Co | 25 | 22 | 20 | 15 | W | 0.4 | 0.31 | 0.36 | 0.42 |
| Cr | 100 | 103 | 82 | 64 | Zn | 81 | 72 | 75 | 64 |
| Cs | 0.43 | 1.0 | 1.5 | 1.8 | Zr | 138 | 152 | 160 | 156 |
| Cu | 32 | 28 | 31 | 22 | Y | 12 | 15 | 16 | 15 |
| F | 555 | 610 | 595 | 540 | La | 24 | 35 | 32 | 31 |
| Ga | 19.4 | 18.2 | 19 | 18 | Ce | 46 | 71 | 61 | 63 |
| Ge | 1.1 | 1.2 | 1.1 | 1.1 | Pr | 5.5 | 8.0 | | |
| Hf | 3.6 | 4.1 | 4.2 | 4.2 | Nd | 22 | 32 | 28 | 27 |
| Hg | 8 | 5 | 6 | 6 | Sm | 4.0 | 5.3 | 4.8 | 4.6 |
| In | (50) | | (65) | (65) | Eu | 1.3 | 1.4 | 1.3 | 1.1 |
| Ir | 35 | 36 | 25 | (23) | Gd | 3.5 | 4.6 | 4.0 | 4.3 |
| Li | 10 | 12 | 12 | 13 | Tb | 0.57 | 0.62 | 0.66 | 0.60 |
| Mn | 760 | 720 | 730 | 620 | Dy | 3.0 | 3.6 | | |
| Mo | 0.45 | 0.42 | 0.41 | 0.41 | Ho | 0.57 | 0.67 | | |
| N | 24 | 34 | 58 | 58 | Er | 1.6 | 2.0 | | |
| Nb | 7.8 | 9.2 | 9.0 | 9.7 | Tm | 0.25 | 0.30 | | |
| Ni | 39 | 39 | 36 | 33 | Yb | 1.5 | 1.9 | 2.0 | 1.7 |
| Os | 50 | (35) | (40) | (40) | Lu | 0.24 | 0.28 | 0.28 | 0.29 |

1. Granulite facies belts (with K, Th, U depletion) 2. Granulite facies belts (without K, Th, U depletion) 3. High grade amphibolite facies belts 4. Medium-low grade amphibolite facies belts

Data source: this study; Unit: % for major elements; $10^{-12}$ for Ir, Os, Rh, and Ru; $10^{-9}$ for Au, Hg, In, Pd, and Pt; $10^{-6}$ for others

belt, namely area of the terrain and representativeness of the territory. In a considerably large range from the middle-upper crust (low-grade amphibolite facies) to the middle-lower crust (high-grade amphibolite facies), TTG gneisses are dominant, and chemical composition is fairly consistent.

*Lower Crust (LC)*

LC has an average thickness of 13 km. Its upper part generally has Vp values of 6.4-6.8 km/s (average Vp=6.6 km/s), and is approximately conformable with tonalitic rocks or intermediate granulite facies rocks. The lower part of LC commonly has a gradient layer 3-6 km (average 4 km) in thickness with rapid increase of Vp value (Vp=6.7-7.3 km/s, average Vp =7.0 km/s). According to the data of experimental petrology and the relationships between Vp value and lithologic character, the rocks are inferred to be mainly composed of mafic granulite in the lower part of LC. On the basis of an integrated analysis of the above factors, two granulite facies belts having high temperature-pressure conditions and geochemical characteristics of LC, namely Qianxi-Qian'an (Qianxi Group) and Xinghe-Yanggao-Huai'an (Xiajining Group), represent the upper part of LC (9 km), and the mafic granulites of the Qianxi Group and Xiajining Group represent the lower part of LC (4 km). Thus, element abundance of LC of the NCP was estimated, and TC was finally evaluated according to weighted averages of the thickness of various layers.

## CRUSTAL ABUNDANCE OF CHEMICAL ELEMENTS

Crustal abundance of 78 elements in NCP are listed in Table 5, and data of major elements are shown in Table 6. Of the 78 elements, the abundance of most elements are markedly higher than the corresponding low detection limits. REs of most elements in 155 inserted CRMs are lower than 5%, RDs of most elements in 125 duplicate samples are lower than 15%, and, in addition, there is sufficient number of samples. Therefore, the resultant data are reliable for most elements. For a few difficult elements there are remarkable analytical errors, due to insufficient analytical sensitivity. These values are given in brackets for careful reference.

The locations of chemical composition for each crustal layer in rock classification diagram (ANOR-Q') are shown that average chemical composition of the TC of the NCP is equivalent to from quartz monzodiorite to granodiorite, equivalent to quartz-diorite in LC, and equivalent to granodiorite both MC (more intermediate) and UC (more felsic).

### *Chemical Compositions of the Upper Crust (UC)*

There are significant difference in chemical compositions between sedimentary cover and basement, both of which constitute the UC. As the crustal recycling products, the sedimentary cover is characterized by : (1) Enrichment of $CO_2$, N, and Ca, indicating the accumulation of carbonates and organic matter in the strong biogeochemical environment. (2) Enrichment of volatile trace elements such as Hg, As, and Sb, etc. (3) Enrichment of boron. It can be seen from Table 3 that the average contents of Hg, As, Sb, and B are low in magmatic rocks and high-grade metamorphic rocks, and are significantly concentrated in sediments, sedimentary rocks, and meta-sedimentary rocks. Evidently, it is difficult to proceed chemical equilibrium on the high contents in the sediments with original rocks. The major part of Hg, As, Sb, (and Br, I) are very likely enriched by the circulation of atmosphere. Boron may be enriched mainly by marine sedimentation. Other trace elements in sedimentary cover are close to the basement in contents, mainly inheriting the characteristics of UC.

The basement rocks of UC mostly consists of high-grade metamorphic rocks (dominated by TTG rock series) and granitic rocks, and secondarily greenschist facies rocks. It is characterized by concentration of silicon, alkali metals, lithophile elements of large ion radius, and incompatible elements, with low contents of PGE and Cu, as well as distinct Eu negative anomalies. All these show granitic component and its associated elements are enriched in UC.

### *Chemical Compositions of the Middle Crust (MC)*

Amphibolite facies rocks are the high-grade metamorphic geologic bodies most spread over the

**Table 5 Abundance of chemical elements of the continental crust in North China Platform**

| El. | Unit | TC | UC | MC | LC | EC | El. | Unit | TC | UC | MC | LC | EC |
|---|---|---|---|---|---|---|---|---|---|---|---|---|---|
| H | $10^{-2}$ | 0.15 | 0.17 | 0.16 | 0.11 | 0.20 | Rh | $10^{-9}$ | 0.056 | 0.026 | 0.050 | 0.097 | 0.02 |
| Li | $10^{-6}$ | 13 | 16 | 12 | 9.4 | 17 | Pd | $10^{-9}$ | 1.1 | 0.43 | 0.80 | 2.0 | 0.24 |
| Be | $10^{-6}$ | 1.0 | 1.6 | 1.1 | 0.40 | 1.5 | Ag | $10^{-9}$ | 57 | 55 | 55 | 60 | 53 |
| B | $10^{-6}$ | 10 | 17 | 8.0 | 4 | 21 | Cd | $10^{-9}$ | 80 | 72 | 76 | 96 | 77 |
| C | $10^{-2}$ | 0.40 | 0.84 | (0.16) | (0.07) | 2.1 | In | $10^{-9}$ | (50) | (40) | (50) | (55) | (40) |
| N | $10^{-6}$ | 50 | 82 | 36 | 23 | 150 | Sn | $10^{-6}$ | 1.2 | 1.5 | 1.1 | 0.9 | 1.4 |
| O | $10^{-2}$ | 47.5 | 49.0 | 47.7 | 45.8 | 49.2 | Sb | $10^{-6}$ | 0.15 | 0.18 | 0.14 | 0.11 | 0.21 |
| F | $10^{-6}$ | 570 | 540 | 580 | 595 | 500 | Te | $10^{-9}$ | (8) | (6) | (8) | (10) | (8) |
| Na | $10^{-2}$ | 2.57 | 2.20 | 2.49 | 2.52 | 1.82 | I | $10^{-6}$ | (0.1) | (0.15) | (0.1) | (0.05) | (0.14) |
| Mg | $10^{-2}$ | 2.08 | 1.65 | 1.91 | 2.73 | 1.80 | Cs | $10^{-6}$ | 1.4 | 2.2 | 1.4 | 0.44 | 2.4 |
| Al | $10^{-2}$ | 7.57 | 7.20 | 7.74 | 7.88 | 6.38 | Ba | $10^{-6}$ | 690 | 740 | 740 | 585 | 540 |
| Si | $10^{-2}$ | 28.1 | 28.9 | 29.1 | 26.5 | 26.4 | La | $10^{-6}$ | 29 | 35 | 32 | 19 | 35 |
| P | $10^{-6}$ | 680 | 590 | 715 | 755 | 615 | Ce | $10^{-6}$ | 55 | 67 | 62 | 37 | 67 |
| S | $10^{-6}$ | 300 | 200 | 280 | 440 | 160 | Pr | $10^{-6}$ | 6.2 | 7.2 | 6.7 | 4.5 | 7.1 |
| Cl | $10^{-6}$ | 110 | 100 | 125 | 100 | 105 | Nd | $10^{-6}$ | 25 | 29 | 28 | 19 | 29 |
| K | $10^{-2}$ | 1.93 | 2.54 | 2.00 | 1.20 | 2.22 | Sm | $10^{-6}$ | 4.4 | 4.8 | 4.8 | 3.7 | 4.8 |
| Ca | $10^{-2}$ | 4.66 | 4.40 | 3.39 | 5.25 | 6.35 | Eu | $10^{-6}$ | 1.21 | 1.10 | 1.27 | 1.29 | 1.03 |
| Sc | $10^{-6}$ | 18 | 10.6 | 16 | 27 | 8.6 | Gd | $10^{-6}$ | 3.8 | 4.0 | 4.1 | 3.7 | 3.4 |
| Ti | $10^{-2}$ | 0.36 | 0.30 | 0.35 | 0.46 | 3130 | Tb | $10^{-6}$ | 0.61 | 0.61 | 0.64 | 0.59 | 0.55 |
| V | $10^{-6}$ | 110 | 68 | 100 | 166 | 62 | Dy | $10^{-6}$ | 3.3 | 3.3 | 3.4 | 3.3 | 3.2 |
| Cr | $10^{-6}$ | 84 | 52 | 87 | 120 | 47 | Ho | $10^{-6}$ | 0.67 | 0.67 | 0.67 | 0.66 | 0.65 |
| Mn | $10^{-6}$ | 730 | 575 | 700 | 930 | 610 | Er | $10^{-6}$ | 1.9 | 1.9 | 1.9 | 1.9 | 1.8 |
| Fe | $10^{-2}$ | 4.94 | 3.35 | 4.59 | 6.74 | 2.97 | Tm | $10^{-6}$ | 0.29 | 0.30 | 0.30 | 0.29 | 0.28 |
| Co | $10^{-6}$ | 22 | 12 | 20 | 36 | 12 | Yb | $10^{-6}$ | 1.85 | 1.9 | 1.9 | 1.85 | 1.8 |
| Ni | $10^{-6}$ | 40 | 24 | 39 | 55 | 24 | Lu | $10^{-6}$ | 0.29 | 0.29 | 0.29 | 0.29 | 0.27 |
| Cu | $10^{-6}$ | 30 | 18 | 27 | 45 | 16 | Hf | $10^{-6}$ | 4.0 | 4.5 | 4.2 | 3.1 | 4.6 |
| Zn | $10^{-6}$ | 74 | 60 | 72 | 93 | 58 | Ta | $10^{-6}$ | 0.6 | 0.8 | 0.55 | 0.3 | 0.77 |
| Ga | $10^{-6}$ | 18.2 | 17.4 | 18.5 | 19.0 | 16 | W | $10^{-6}$ | 0.6 | 0.9 | 0.5 | (0.3) | 0.68 |
| Ge | $10^{-6}$ | 1.2 | 1.2 | 1.1 | 1.2 | 1.0 | Re | $10^{-9}$ | (0.4) | (0.5) | (0.4) | (0.35) | (0.4) |
| As | $10^{-6}$ | 1.5 | 1.9 | 1.3 | 1.1 | 2.2 | Os | $10^{-9}$ | 0.052 | 0.053 | 0.036 | 0.062 | 0.05 |
| Se | $10^{-6}$ | 0.11 | 0.06 | 0.08 | 0.18 | 0.054 | Ir | $10^{-9}$ | 0.028 | 0.017 | 0.028 | 0.040 | 0.015 |
| Br | $10^{-6}$ | (0.3) | (0.4) | (0.3) | (0.2) | (0.3) | Pt | $10^{-9}$ | 1.2 | 0.45 | 0.85 | 2.3 | 0.23 |
| Rb | $10^{-6}$ | 63 | 92 | 69 | 26 | 84 | Au | $10^{-9}$ | 1.0 | 0.74 | 0.85 | 1.5 | 0.64 |
| Sr | $10^{-6}$ | 360 | 330 | 370 | 380 | 290 | Hg | $10^{-9}$ | 8 | 8 | 7 | 8 | 9 |
| Y | $10^{-6}$ | 15 | 17 | 15 | 13 | 16 | Tl | $10^{-6}$ | 0.40 | 0.52 | 0.40 | 0.23 | 0.45 |
| Zr | $10^{-6}$ | 146 | 162 | 155 | 122 | 170 | Pb | $10^{-6}$ | 13 | 17 | 13 | 9.6 | 16 |
| Nb | $10^{-6}$ | 10 | 12 | 9.3 | 7.7 | 13 | Bi | $10^{-6}$ | 0.14 | 0.13 | 0.12 | 0.17 | 0.12 |
| Mo | $10^{-6}$ | 0.5 | 0.6 | 0.45 | 0.4 | 0.48 | Th | $10^{-6}$ | 5.0 | 8.6 | 5.6 | 1.5 | 8.4 |
| Ru | $10^{-9}$ | 0.050 | 0.031 | 0.047 | 0.075 | 0.03 | U | $10^{-6}$ | 1.0 | 1.5 | 0.9 | 0.4 | 1.5 |

TC: Total crust UC: Upper crust MC: Middle crust LC: Lower crust EC: Exposed crust

NCP, which mainly consist of TTG series and are the major part of the MC. For two granulite facies belts without significant depletion of U, Th, and K, their basic chemical compositions show no major difference from amphibolite facies rocks (Table 4). They are therefore included in MC for calculation. Compositions of two granulite facies belts with significant depletion U, Th, and K are on the other hand great different from those of MC. Furthermore, contents of most elements in MC are close to the average values of TC.

*Chemical Compositions of the Lower Crust (LC)*

The chemical compositions of the LC of NCP are characterized by significant depletion of the

**Table 6 Average chemical composition of the continental crust in North China Platform (%)**

| | TC | UC | MC | LC | EC | | TC | UC | MC | LC | EC |
|---|---|---|---|---|---|---|---|---|---|---|---|
| $SiO_2$ | 60.14 | 61.77 | 62.24 | 56.64 | 56.50 | CaO | 5.93 | 5.60 | 4.75 | 7.02 | 8.89 |
| $TiO_2$ | 0.62 | 0.48 | 0.58 | 0.80 | 0.52 | $Na_2O$ | 3.46 | 2.96 | 3.36 | 3.40 | 2.45 |
| $Al_2O_3$ | 14.31 | 13.61 | 14.62 | 14.88 | 12.05 | $K_2O$ | 2.33 | 3.06 | 2.41 | 1.44 | 2.67 |
| $Fe_2O_3$ | 2.83 | 2.36 | 2.64 | 3.52 | 2.47 | $H_2O^+$ | 1.30 | 1.55 | 1.40 | 0.95 | 1.78 |
| FeO | 3.81 | 2.18 | 3.53 | 5.90 | 1.60 | $CO_2$ | 1.50 | 3.17 | 0.60 | 0.27 | 6.84 |
| MnO | 0.095 | 0.074 | 0.090 | 0.124 | 0.078 | $P_2O_5$ | 0.160 | 0.138 | 0.164 | 0.182 | 0.141 |
| MgO | 3.50 | 2.73 | 3.18 | 4.53 | 2.99 | Total | 99.88 | 99.68 | 99.56 | 99.66 | 98.98 |

TC: Total crust; UC: Upper crust; MC: Middle crust; LC: Lower crust; EC: Exposed crust

large-ion elements, U, Th, K, Rb, and Cs, and to some extent, Pb and Tl. The incompatible elements Nb, Ta, Li, Be, etc., REE, and volatile elements are also in deficiency. On the other hand, siderophile elements, PGE, Cu, S, etc. are significantly enriched, and Eu showing positive anomalous. By contrast with MC and UC, it indicates that there are substantially changes in LC. Of the changes, the incompatible elements and siderophile elements have had marked differentiation, and most trace elements show deficiency in different degree.

Due to the scarce exposure of the LC rocks, the rock samples are even more difficult to collect and justify. Therefore, chemical compositions for continental LC remain to be the major point in dispute thus far. Modern investigators estimate the chemical compositions of LC mainly by means of Vp and data from granulite facies terrains as well as the medium-high pressure granulite xenoliths, in which ratio of felsic granulite to mafic granulite is carefully assumed. Results based on Vp and most granulite data show that the chemical composition of LC is intermediate which, however, will lead to be basic if calculation is made on the granulite xenoliths. Hence, the estimated values differ greatly, depending on the calculation models used.

Content of $SiO_2$ in TC is 60.14% which well corresponds to the average Vp 6.5km/s obtained from the geoscience transect in the region. Crustal heat flow density estimated from U, Th, and K contents in Table 2 is 30mw/m$^2$. Total ground heat flow density is 60mw/m$^2$ (assuming mantal heat flow density 30mw/m$^2$ ) approaches the average ground heat flow 63-65mw/m$^2$ [1] from the juncture of the plain and mountain areas in which the crust thickness is close to the average crust thickness of NCP.

Vertically, contents of the crustal elements display regular variations. From UC, MC to LC, the contents decrease successively for $CO_2$, S, and large-ion elements as well as incompatible elements, high potential elements and REEs. Ratios of K/Na (1.18, 0.80, and 0.48), Rb/K (3600, 3400 and 2200), and $La_N/Yb_N$ (12.4, 11.4, 7.1), etc. display the same manner, whereas siderophile elements, PGE, Au, Cu, and Zn show a trend of increase from UC to LC. The negative anomalies of Eu/Eu* become weakened from UC to MC, and turn to positive anomaly in LC. Moreover, other elements such as Ag, Cd, Ga, Ge, and F have more or less the same contents in each crustal layer and do not indicate definite variations.

*Chemical Composition of the Exposed Crust*

For the exposed crust except those covered by loose sediments, the weighted average of element contents based on exposed area of all rocks or geologic bodies was adopted. The areas of geological bodies were obtained from 1:500,000 geological map and the weights for various rocks in strata were given by the overall average of standard profiles.

Average chemical composition of regional exposed crust (Tabs. 5 and 6) has important actual significance for the quantitative study and scientific correlation on environment and agriculture. In

NCP, the contents of alkaline earth metal elements are high, in which the average chemical composition is equivalent to granitic-granodioritic.

### *Crustal Abundance of Pt, Pd, Au, and Hg*

Literature about the distribution of PGE in rocks and the crust are scarce. A report [10] on the crustal abundance of Ir in UC is $0.02 \cdot 10^{-9}$. The crustal abundance of $0.1 \cdot 10^{-9}$ for Ir, and $1.0 \cdot 10^{-9}$ for Pd had been suggested by [14]. More recently, the crustal abundance have been given by [16] for Pd, Os, and Ir, and by [15] for PGE. For Au and Hg, although the special monographs for each elemental geochemistry have been published, their crustal abundance are still not given with sufficient reliability. The crustal abundance of Pt, Pd, Au, and Hg in NCP are different from those of the literature except the value of Pd which is close to the one provided by [14,15]. Therefore, the following discussion may be helpful to the readers.

Sample and a part of CRMs analysis for Pt, Pd, Au, and Hg is shown in Table 6. It is noted that the 1207 composite samples used to derive the crustal abundance are combined from original rock samples of dozens of thousand, and are form by weighted average with respect to the crustal layers. Therefore, sample representativeness can be ensured. This is particularly important to Au, Pt, and Pd which are highly inhomogeneous in rocks. The low detection limits of the analytical methods for these elements are 4 to 20 times below the average crustal contents, which ensure a reliable analysis for most samples. Of the analytical methods, emission spectrography (ES) is adhered by relative large errors, which was coped with by the increased determinations. In fact, the analytical accuracy for ES has been justified by the monitored CRMs (The certified values for CRMs were established with multiple analytical methods including ICP/MS, and approved by the state authentication). For Au, a parallel analysis for 300 samples has been made with both neutron activation analysis after chemical enrichment (monitoring by trace gold samples of CRMs, [19]) and emission spectrography pre-enriched by fire assay. The average results are consistent with each other. Therefore, we believe that the abundance values for Pt, Pd, Au, and Hg in NCP are reliable. The inconsistency between ours and those from other authors for Au and Hg may indicate a problem of analytical methods or bias, as insufficient analytical sensitivity would usually cause an overestimate of determinations. The inconsistency for Pt and Pd as compared with those from [15] may imply an error of sample representativeness. Table 2 and Table 3 show the contents of PGE in nine types of main rocks in NCP, contents of Pt and Pd are very sensitive to rock type, causing great difference between different rock types, which, for instance, are higher in amphibolite facies (TTG) than that in granitic rocks, and are higher in granulite facies (particularly mafic granulite) in LC than that in UC rocks. The LC is the main contributor of Pt and Pd to the TC. The PGE values presented by [15] are derived from 17 composite samples of European greywackes. It may be more reasonable to regard the values as the estimates of UC rather than TC. In fact, abundance of Pt and Pd in UC of the NCP (Table 5) are well consistent with those in TC from [15].

### *About Crustal Evolution*

Comparison of chemical compositions made between LC and UC supports the viewpoint of [16] that the uplift of granitic magma may result in fractional distillation of the crust. Based on the viewpoint and the fact that granitic rocks of UC in NCP makes up some 30% of the UC, a mixture can be constituted by adding 70% of average composition of LC (Table 5) and 30% of average composition of the NCP granites (Table 2). The resulting mixture is then compared with the NCP crust in average compositions (Fig. 2), showing that most elements are well close to each other, with exception of As, Sb, Cs, Pt, and Pd. Values of Eu and Eu/Eu* can be balanced too. Obviously, the real processes are quite complicated. For example, the granitic matter in the UC may not all come from LC. And, amphibolite facies rocks may, under a certain condition, partly melt to form the UC granitic matter (e.g., migatization). Moreover, part of residual melts in LC may be carried into the upper mantle. The ratios (30% granitic rocks and 70% LC), therefore, may be regarded as the net effects of the long-term differentiation in both MC and LC.

Based on crustal evolution history of NCP and vertical distribution of chemical compositions, an assumption for chemical evolution of NCP is made as follows. During the middle and late Archean, a great deal of mafic magma differentiated continuously from the mantle, which were accumulated successively in the upper part of the upper mantle or at the bottom of LC, to form a large volume of mafic magmatic reservoir. Then differentiation continued in the mafic reservoir to form TTG rock series (mainly tonalite) which constituted the main body of the crust. Meanwhile, part of the heavy residuals (more mafic and calcic) went back to the upper mantle. As a consequence, the partial melting of the TTG produced high-potassium granitic magma and form rock in the upper part of the crust. The dynamic equilibrium of differentiation lasted the whole late Archean. Finally, the main body of the NCP was formed. Crustal temperature decreased and crustal rigidity enhanced in post-Archean. The Proterozoic basic dikes came into the crust. The partial melting in the deep crust then became weakened, and only the lighter granitic magma with high Si and K could be formed which moved upwards to play the leading role in the consequent crustal chemical evolution of post-Archean. Chemical evolution of granitic rocks in NCP [18] indicates the transformation between Archean and post-Archean. In the meantime, crustal recycle gradually strengthened in post-Archean which caused sedimentary deposits of various thickness to form in different regions. Intrusion and eruption of granitic magma along with the crustal recycle products (including mass exchanges of atmosphere and ocean) constituted the additional UC component, characterized by enrichment of large-ion lithophile elements, incompatible elements and volatile elements as well as significant Eu negative anomalies. By tectonic movement, the post-Archean additional UC, together with Archean UC constituted the complex embedded geologic bodies.

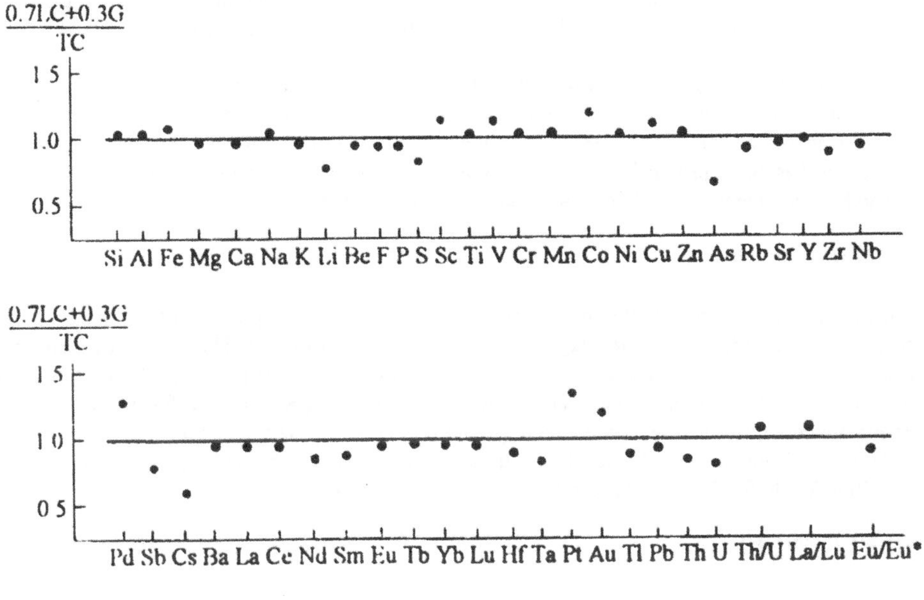


Fig. 2. Ratios of composition of 70% lower crust plus 30% granite in comparison with composition of the total crust in North China Platform.

## CONCLUSIONS

Following the research work done in Russian Platform and Canadian Shield, the study once more provides a complete set of chemical composition data for whole platform which are obtained by systematic sampling and analytical determinations. Crustal abundance of the major elements in this study are close to the values given from literature. The chemical composition of the whole platform may approximately represent the chemical composition of the global continental crust. For this reason, the estimated data of chemical compositions of the NCP would be extensively citable and referable, which, in particular, may contribute important information in studying chemical compositions of the global continental crust.

In this study, the new progresses achieved are: (1) Based on the comprehensive data collected from regional geology, geochemistry, and geophysics, with reference to modern crustal models to provide element abundance estimates for both total crust and each crustal layer. (2) Significantly increase the number of analytical elements. Up to 77 elements are actually analyzed, plus O which is obtained by calculation. (3) Analytical data quality is controlled by the unified monitoring methods, and analytical data for all element come from the same samples. This would avoid the problem of sample representativeness possibly caused by collecting data from different sources, and also avoid introducing uncertainty factors in analytical quality evaluation. Furthermore, owing to the use of same samples and monitored with CRMs, the comparable state among elements is improved, which may facilitate quantitative geochemical research.

The average chemical composition of TC in the NCP is between those of granodiorite and quartz monzodiorite. The average chemical composition of the LC is equivalent to quartz diorite, whereas mafic components are not dominant in LC, which mainly distribute in the lower part of LC. The UC and LC may approximately be imitated and equilibrated with granitic rocks.

As for PGE, Au, Hg, and those with extremely low abundance and large uncertainty, due to the close collaboration of the Chinese analysts, the research work has resulted in significant progresses. With the special and reliable CRMs used in the analysis and quality monitoring, the reliability of the results has been justified, which will consequently make important contribution to further research on continental crustal abundance and basic geochemistry.

## *Acknowledgments*

The present paper represents the main achievements of the research project in basic geochemistry supported by the Ministry of Geology and Mineral Resources of China. During the research work, many geologists and analytical workers gave their support and kind help; for this the authors would like to express their sincere thanks. The authors are especially grateful to Professors Xie Xuejing, Zhang Benren and Lin Cunshan for their guidance, and to the main collaborators Professors Lin Yu'nan, Luo Tingchuan, and Gao Shan. Thanks are also given to Misters Li Guohui, Lu Yinxiu, and Zhou Yunlu for their full cooperation and enthusiastic help.

## REFERENCES

1 Chen M X (1988) Geotherm in North China. Science Press. 218. (in Chinese)

2. Christensen N.I. and Mooney W.D. (1995) Seismic velocity structure and composition of the continental crust: A global view. *J. Geophys. Res.*, Vol. 100, No. B7, 9,761-9,788.

3. Compiler group of National Seismic Bureau " Deep Geophysical Prospecting Results" (1986) *Geophysical Prospecting Results of Crust and Upper Mantle in China*. Seismic Press. 407. (in Chinese)

4. Condie K.C. (1993) Chemical composition and evolution of the upper continental crust: Contrasting results from surface samples and shales. *Chemical Geology*, Vol. 104, 1-37.

5. Feng R (1985) Crustal thickness and density distribution in the upper mantle of China (results of three-divisional gravity inversion). *Acta Seismologica Sinica*, Vol. 7, No. 2, 143-156 (in Chinese)

6. Ma X Y, Liu C Q, and Liu G D (1991) Geoscience Transect from Xiangshui, Jiangsu to Mandula, Inner Mongolia. *Acta Geologica Sinica*, No. 3, 199-215. (in Chinese with an English abstract)

7. Rudnick R.L. and Presper T. (1990) Geochemistry of intermediate- to high-pressure granulites. In: Vielzeuf D. and Vidal P. (eds.), *Granulites and Crustal Evolution*. Kluwer Acad., Norwell, Mass., 523-550.

8. Rudnick R.L. and Fountain D.M. (1995) Nature and composition of the continental crust: a lower crustal perspective. *Rev. Geophys.*, Vol. 33, 267-309.

9. Shaw D.M., Cramer J.J., Higgins M.D., and Truscott M.G. (1986) Composition of the Canadian Precambrian Shield and the continental crust of the earth. In: Dawson J.B., Carswell D.A., Hall J., and Wedepohl K.H.(eds): *The Nature of the Lower Continental Crust*. Geological Society Special Publication. No. 24., 275-282.

10. Shaw D.M., Dostal J., and Keays R.R. (1976) Additional estimates of continental surface Precambrian shield composition in Canada. *Geochim. Cosmochim. Acta*, Vol. 40, 73-84.

11. Shaw D.M., Reilly G.A., Muysson J.R., and Pattenden G.E. (1967) An estimate of the chemical composition of the Canadian Precambrian shield. *Canadian Journal of Earth Sciences*. Vol. 4, 829-853.

12. Shen Q H, Xu H F, and Zhang Z Q et al. (1992) *Early Precambrian Granulites in China*. Geological Publishing House. 237. (in Chinese)

13. Tarney J and Windley B.F. (1977) Chemistry, thermal gradients and evolution of the lower continental crust. *J. Geol. Soc. London*, Vol 134, 153-172.

14 Taylor S.R. and McLennan S.M. (1985) *The Continental Crust: Its Composition and Evolution*. Blackwell Scientific Publications. 312pp.

15. Taylor S.R. and McLennan S.M. (1995) The geochemical evolution of the continental crust. Rev. Geophys., Vol. 33, 241-265.

16. Wedepohl K.H. (1995) The composition of the continental crust. *Geochim. Cosmochim. Acta*, Vol. 59, No. 7, 1,217-1,232.

17. Xie X J, Yan M C, Wang C S et al. (1989) Geochemical standard reference samples GSD 9-12, GSS 1-8 and GSR 1-6. *Geostandards Newsletter*, Vol. 13, No. 1, 83-179.

18. Yan M C and Chi Q H (1997) *Chemical Composition of Crust and Rocks in the eastern part of China*. Science Press, Beijing. (in Chinese) (in press).

19 Yan M C, Wang C S, Cao Q X et al. (1995) Eleven gold geochemical reference samples (GAu 8-18). *Geostandards Newsletter*, Vol 19, No. 2, 125-133.

# GEOCHEMICAL MAPPING

*Proc 30th Int'l. Geol. Congr.*, Vol. 19, pp. 85-87
Xie Xuejin (Ed)
VSP 1997

# ENVIRONMENTAL MONITORING AND GLOBAL CHANGE: THE NEED FOR SYSTEMATIC GEOCHEMICAL BASELINES

*A.G. DARNLEY (Geological Survey of Canada, Ottawa, Canada) and*
*J.A. PLANT (British Geological Survey, Nottingham, UK)*

**Extended Summary**

The purpose of this paper is to draw attention to the urgent need for systematic information on the geochemistry of the surface environment at the global scale, as a basis for the maintenance of the earth's essential life support systems. It outlines the concerns of international bodies, and the steps which have been taken by scientists to prepare an action plan, and invites national organization to participate in implementing the detailed recommendations contained in a report entitled A Global Geochemical Database published by UNESCO in 1995 [1]. Standardized geochemical mapping is required in order to determine the spatial variations in the natural distribution of chemical elements as a basis for understanding land contamination and degradation, and for anticipating and monitoring the effects of global climatic changes.

Concern over the state and sustainability of the global environment was expressed at the UN Conference on the Human Environment in Stockholm in 1972. Although the need for more comprehensive and reliable data describing the natural environment is widely recognized, this has tended to be given a lower priority than analytical and modeling studies. Natural systems are extremely complex, however, and unless all the natural parameters are identified and measured, the results of theoretical studies can be misleading. Predictions concerning future trends in the global environment, which have potentially enormous biological and economic consequences, require sound data. The importance of comprehensive baseline data was clearly stated in the proceedings of the 1991 ASCEND 21 conference, the scientific precursor of the 1992 UN Conference on Environment and Development (UNCED), and geochemical mapping and monitoring were highlighted as needing greater attention in the context of agriculture and land use.

Following the UNCED conference, a Global Environment Monitoring Programme was established but this included only one major activity in the earth science field - global water quality monitoring. Unfortunately, some national earth science organizations were apparently unaware of this programme and were not invited to participate, because leadership was vested in the World Health Organization and the UN Environmental Programme.

A formal connection between geochemical mapping and the UN Global Environment Monitoring Programme was proposed by the UN Committee on Natural Resources in May 1996, under the title of Global Land Monitoring. It was recognized that "a land monitoring programme would be complementary to and would greatly enhance the value of existing programmes" particularly that on water quality monitoring. The resolution relating to land monitoring was forwarded to the UN Economic and Social Council, the senior body concerned with environmental matters, and accepted in July 1996.

The UN Committee on Natural Resources, in the preamble to its Resolution 3/5, noted that "there is currently a huge gap in global environment monitoring programmes, in that current pragrammes do not deal with the natural chemical variability of the land surface or with changes brought about by both natural and orthogenic processes". They also noted that "a comprehensive blue print for such a programme has already been produced by the International Geochemical Mapping Project sponsored by the United nations Educational Scientific and Cultural Organization and the International Union of Geological Sciences, and that there is an urgent need for its implementation.

The reasons for urgency are economic as well as environmental. In recent years, national regulatory agencies and international commissions, stimulated by public concern, have begun to set threshold levels of chemical substances in the environment below which these substances are considered safe. They are also setting 'trigger' or 'action' levels, above which remediation is required. The existence of natural variations in the geochemical background is frequently overlooked or oversimplified. Failure to recognize the extent of these variations has resulted in the setting of unrealistic levels and unattainable standards, causing public alarm and unnecessary legal and economic costs. Reliable quantitative data, collected according to internationally agreed protocols, are needed for educational as well as for scientific purposes, to demonstrate to both legislators and the general public that the natural environment is neither uniform nor "hazard free".

The UNESCO report published in 1995 is the outcome of a six year study undertaken as an International Geological Correlation Programme project, sponsored by UNESCO and IUGS. The principal authors of the report are from eight countries of representing a wide range of experience in different aspects of geochemistry. The report identifies on a country by country basis some of the deficiencies and inconsistencies of available geochemical datasets. The magnitude of the global problem is shown by the fact that no systematic geochemical records exist for 80 percent of the world land surface. The report contains detailed methodologies for international geochemical mapping, including field sampling, sample preparation and analysis, radioelement mapping including airborne radiometric surveys, data management with examples of data applications, and administrative options for implementing the work. The first step in the standardization of geochemical survey data internationally requires the creation of a Global Geochemical Reference Network, entailing the collection, under strictly controlled conditions, of several types of sample media including soil, water, and stream and river sediment. The report was prepared with a variety of earth science organizations in mind, including those which already have experience in conducting geochemical (including radiometric) surveys for environmental, resource and planning purposes, and those which may wish to become involved in such operations in the future. In North America, Europe, Australia and elsewhere, reductions in government funding and questions concerning the need to continue with traditional types of geological mapping have led to uncertainty over the future of geological survey organizations. The provision of high resolution, quantitative, geochemical baseline data, relevant to major environmental issues of public concern, is an activity which many of these organizations are well qualified to undertake and by doing so they would demonstrate their continuing relevance to contemporary society.

In 1996 the IUGS sanctioned the formation of a new Working Group on Global Geochemical Baselines to encourage and facilitate implementation of the recommendations in the UNESCO report and to take forward the results of the International Geological Correlation Programme projects 259 and 360. The working group has been given responsibility for liaison with the various UN Agencies and other regional groups that are becoming involved in the activities. All national earth science organizations are invited to participate in working towards the establishment of a global geochemical database, and are requested to make contact with the Working Group.

1. Darnley, A.G., Bjorklund, A.J., Bolviken, B., Gustavsson, N., Koval, P.V., Plant, J.A., Steenfelt, A., Tauchid, M., and Xie Xuejing, with contributions by Garrett, R.G. and Hall, G.E.M., A global geochemical database for environmental and resource management: recommendations for international geochemical mapping. Ear Science Report 19. UNESCO Publishing, Paris (1995).

*Proc. 30th Int'l. Geol. Congr.*, Vol. 19, pp. 89-109
Xie Xuejin (Ed)
VSP 1997

# Wide-Spaced Floodplain Sediment Sampling Covering the Whole of China: Pilot Survey for International Geochemical Mapping

**CHENG HANGXIN, SHEN XIACHU, YAN GUANGSHENG, GU TIEXIN, LAI ZHIMIN AND XIE XUEJING**
*Institute of Geophysical and Geochemical Exploration, Langfang, Hebei 065000, China*

**Abstract**

In support of IGCP 259 (International Geochemical Mapping) and IGCP 360 (Global Geochemical Baselines), a pioneer project designed to test the feasibility of wide-spaced global geochemical mapping : "Environmental Geochemical Monitoring Networks and Dynamic Geochemical Maps in China" (EGMON), was launched in 1992. The aims of this project are to study the suitability and feasibility of floodplain sediment as a global sampling medium, to develop the methodologies of wide-spaced sampling of such sediment and to study the broad geochemical patterns in China in order to obtain an overall assessment of the country's mineral resource potential and an overview of its environment.

529 floodplain sediment samples with an average density of 1 sample per 15000 $km^2$ were taken all over China. Geochemical patterns revealed by wide-spaced floodplain sediment sampling show great coincidence with RGNR (Regional Geochemistry-National Reconnaissance, China's national geochemical mapping program) and SEBV (Soil Environmental Background Values in China) broad geochemical patterns. Geochemical maps produced by data from wide-spaced floodplain sediment sampling show that there is an important Pt-Pd geochemical province in Southwestern of China. Changes in Hg and P distribution patterns in surface and deep samples are noted; Hg and P pollution is occurring over large areas in China. An area of soil vulnerability is identified in Southeastern China. A chemical time bomb of aluminum similar to the one in central Europe in Southeastern China is predicted preliminarily. Large scale forest dieback may occur within these areas in the future.

***Keywords:** wide-spaced floodplain sampling, international geochemical mapping, Pt and Pd geochemical province, chemical time bomb, aluminum, China*

## INTRODUCTION

The concept of international geochemical mapping (IGM) was developed at the 12th International Geochemical Exploration Symposium held in Orleans in 1987. Later, the International Geochemical Mapping project (phase one, IGCP 259) and the Global

Geochemical Baseline project (phase two, IGCP 360) were accepted in 1988 and in 1993 by the International Union of Geological Sciences and UNESCO, through the International Geological Correlation Program (IGCP)[4,5]. The final objective of the IGM project is to compile surface geochemical maps of the earth as tools in the search for new mineral resources and solutions to environmental problems [19].

After a comprehensive review of methods of regional and national geochemical mapping and an examination of the results obtained, it has been confirmed that wide-spaced geochemical sampling is the only practical way to obtain a controlled, systematic and potentially rapid overview of global geochemistry[5, 6]. The feasibility of wide-spaced global geochemical sampling was still doubtful among many earth scientists before some concrete examples could be presented by large scale pilot studies. In support of IGCP project No. 259 and No. 360, a project designed to test the feasibility of wide-spaced global geochemical mapping, "Environmental Geochemistry Monitoring Networks and Dynamic Geochemical Maps in China" (EGMON) was launched in 1992 with Professor Xie Xuejing as the project leader. The purposes of this project is :

(1) to study the suitability of floodplain sediment as a global sampling medium;
(2) to develop the methodologies of wide-spaced sampling of such sediment;
(3) to study the broad geochemical patterns in China in order to obtain an overall assessment of its mineral resource potential and an overview of its environment.

## SAMPLE COLLECTION AND ANALYSIS

### *Selection of Sample Medium*

Active stream sediments have been widely used as a sampling medium for regional or national geochemical mapping for the past several decades[11], whereas, floodplain sediment and overbank sediment have received special attention in the course of the IGM project[2, 10, 14, 19]. The Glossary of Geology (Bates and Jackson, 1980) defines both as alluvial sediments, but the term overbank sediment is alluvium accumulated adjacent to low-order streams (where stream sediment samples may be collected) and floodplain sediment is alluvium adjoining high-order drainage channels, typically large rivers. The significance of this distinction from a geochemical mapping viewpoint is that overbank samples represent small drainage basins, whereas floodplain samples represent much larger basins [6].

In 1993-1994, a sub-project aiming at systematic study of the representativity of floodplain sediment, by comparing extremely low density floodplain sediment data with the details of stream sediment data derive from China's RGNR project [20,21,23] as a global sampling medium, was carried out in Zhejiang Province [2,19] in parallel with the national wide-spaced floodplain sampling project. This was done to remedy the lack of experimental evidence needed to justify embarking on such huge effort covering the whole of China. Results indicated that floodplain sediments show good suitability as a global sampling medium [2, 19]. Thus , floodplain sediment was chosen as the sampling medium for the national monitoring network.

### *Sample Collection*

Within China, 529 sample stations were chosen to carry out wide-spaced floodplain sediment sampling (Fig.1). Although the surface areas of the super-catchment basins vary considerably, most of the samples in Eastern China were collected at stations located on the plain adjacent to the effluents of rivers draining super-catchment basins ranging from 1000 to 10 000 km$^2$. Two samples were collected at each station using a Louyang shovel, a sampling tool widely used in China designed specifically for collecting deep soil samples. The samples were taken from (a) surface floodplain sediment at a depth of 5-25cm and (b) deep floodplain sediment at a depth of 80-120cm. Sampling stations were at least 500-1000 m from the channel of rivers. At each station, sediment was gathered at three points spaced 500 m apart. Samples from each station were composed of an equal weight of sediment from these three points. The minimum weight of each sample is 2.5 kg.

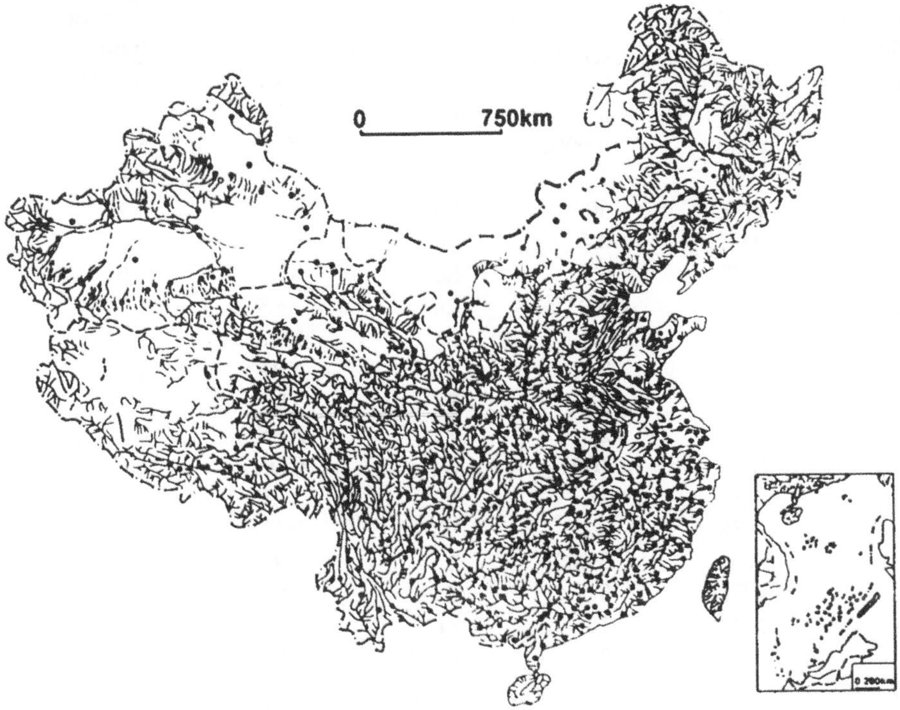


**Figure 1.** Location of stations sampling wide-spaced floodplain sediment

Two photographs were taken at each station. One on the point where a sub-sample was gathered, another of the landscape around the sample station. Pit profiles were kept, ( including sample number, sample type, geographic location, altitude and date of sample as well as stream, environment, landscape and soil characteristics).

*Sample Preparation and Storage*

All samples were dried at room temperature. Later they were passed through a 0.1 mm steel fiber sieve in order to remove the pebble and plant residues.

Each sample was stored for future use in a glass bottle. Bottles were labeled and stored in wooden cases separately for each province. Each sample of 100 g was reduced to -74 μm in an agate ball mill and then divided into four sub-samples for analysis.

*Sample Analysis and Quality Control*

Fifty elements were determined in the present project (Table. 1). Of these, 48 were determined in the central laboratory of IGGE. Pt and Pd were determined by the chemical-spectrometry simultaneous analysis at the central laboratory, Henan Institute of Noble Metals of the Ministry of Geology and Mineral Resources. Requirements for international geochemical mapping [6,17] were followed in all analytical operations.

**Table 1** List of element, analytical methods and detection limits used in EGMON project

| Element determined | Method | Detection (ppm) | Element determined | Method | Detection (ppm) |
|---|---|---|---|---|---|
| Ag | AAS | 0.02 | Pd | CSS | 0.0001 |
| As | AFS | 0.5 | Pt | CSS | 0.0001 |
| Au | AAS | 0.0002 | Rb | XRF | 5 |
| B | AES | 1 | S | VOL | 100 |
| Ba | RXF | 50 | Sb | AFS | 0.2 |
| Be | AES | 0.5 | Sc | XRF | 4 |
| Bi | AFS | 0.1 | Se | AFS | 0.01 |
| Cd | AAS | 0.01 | Sn | AES | 1 |
| Ce | XRF | 3 | Sr | XRF | 10 |
| Co | XRF | 2 | Th | XRF | 3 |
| Cr | XRF | 7 | Ti | XRF | 100 |
| Cu | XRF | 2 | Tl | AES | 0.1 |
| F | SIE | 100 | U | COL | 1 |
| Ga | XRF | 2 | V | XRF | 15 |
| Ge | AES | 0.1 | W | POL | 0.5 |
| Hg | AFS | 0.002 | Y | XRF | 10 |
| I | COL | 0.5 | Zn | XRF | 10 |
| La | XRF | 16 | Zr | XRF | 10 |
| Li | AAS | 3 | $Al_2O_3$ | XRF | 0.1 (%) |
| Mn | XRF | 30 | CaO | XRF | 0.1 (%) |
| Mo | POL | 0.3 | Fe | XRF | 0.1 (%) |
| Nb | XRF | 5 | $K_2O$ | XRF | 0.1 (%) |
| Ni | XRF | 2 | MgO | XRF | 0.1 (%) |
| P | XRF | 100 | $Na_2O$ | XRF | 0.1 (%) |
| Pb | XRF | 5 | $SiO_2$ | XRF | 0.1 (%) |

AAS: Atomic absorption spectrometry; AES: Atomic emission spectrometry; AFS: Atomic flurescence spectrometry; COL: Colorimetry; CSS; Chemical-spectrometry simultaneous; POL: Polarography; SIE: Selective ion electrode; VOL: Volumetric flux method; XRF: X-ray flurescence spectrometry.

The standard reference samples GSD 1-9, GSS1-8 [24, 25], were used for quality control. Analytical detection limits achieved were the same as those recommended for the international geochemical mapping.

## RESULTS AND DISCUSSIONS

### *Comparison of EGMON Data with RGNR and SEBV Data*

At the time of this study, China's national geochemical mapping project, Regional Geochemistry-National Reconnaissance (RGNR) had been underway for 18 years. By 1995, active stream sediment surveys had already been covered 5.17 million $km^2$ of China, with an average sampling density of one sample/$km^2$ in eastern China, one sample/1-50 $km^2$ in the western part of country[20, 21, 23]. Geochemical maps of 39 elements is being compiled on a scale of 1:5 000 000[20]. Simultaneously with the RGNR project, wide-spaced soil sampling for environmental purpose has covered all of China as part of the project entitled " Soil Environmental Background Values" (SEBV,an average density of 1 sample location per 1600$km^2$ in the east part of China and 1 sample location per 6400$km^2$ in remote areas). Atlas with the distribution of 49 elements were published in 1994 [27].

The data obtained from these two projects make it possible to study the suitability and feasibility of using floodplain sediment as a global sampling medium. We have already shown that the data from floodplain sediment display good agreement with the data from RGNR in terms of the geochemical patterns and patterns of pollution and mineralization in Zhejiang province [2, 19].

Figs. 2, 3, 4 compare the Cu patterns detected by RGNR, EGMON and SEBV. The three maps reveal a striking coincidence of broad geochemical patterns of Cu distribution as determined by these three unconnected geochemical investigations, carried out by different organizations at different periods. All the three maps delineate the same two important broad Cu patterns in southeastern and northwestern China. Most known large or giant copper deposits in China occur in the southeastern Cu megaprovince. The northwestern Cu geochemical province was not known before geochemical mapping; by 1995, several large or giant Cu deposits had already been discovered within this area. The search for new copper mineralizations in this Cu geochemical province is still underway.

### *Applications in Mineral Exploration*

As more and more areas are covered by regional and national geochemical mapping projects, increasing attention is being paid to the concept of geochemical province, and some examples of geochemical province have been described in China and northern Fennoscandia[1, 25]. Thus, the strategic role of wide-spaced sampling as a tool in mineral exploration is of growing interests to geological scientists.

The geochemical patterns for Au revealed by wide-spaced floodplain sediment sampling are presented in Fig. 5. Two large areas with Au content of 2.2 ppb, significantly higher than the mean (1.31 ppb in stream sediment, 1.74 ppb in flood plain sediment) for all of China, are distinctly identified in the Southeast and Xiaoxunlin areas. These correspond

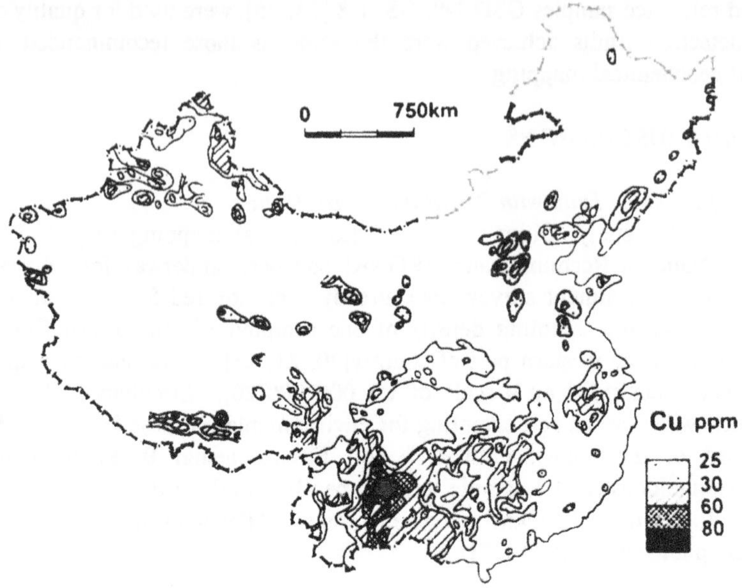


**Figure 2.** Simplified geochemical distribution of Cu, active stream sediment surveys, sampling density 1 sample per 1 to 50 km$^2$ (RGNR project). Modified after Xie[18]

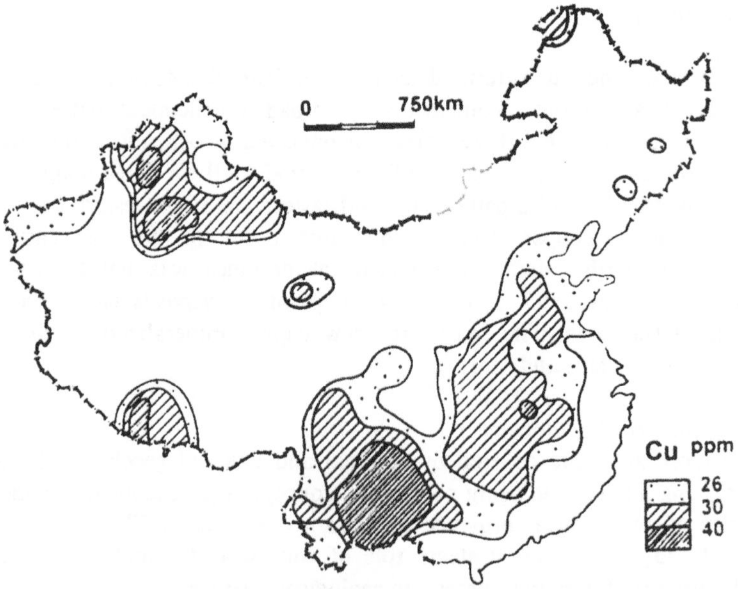


**Figure 3.** Geochemical distribution of Cu in deep floodplain sediment surveys, average sampling density 1 sample per 15000 km$^2$ (EGMON project)

precisely to the three gold mineralization belts (Fig. 6), namely the gold metallogenic province in Shandong, Xiaoxunlin and along the middle and lower Reaches of theYangtze River. Most of China's known large gold deposits are located in these geochemical provinces. Two new gold geochemical provinces were discovered in Southwestern Xinjiang (Northwest China) and Southern Tibet.

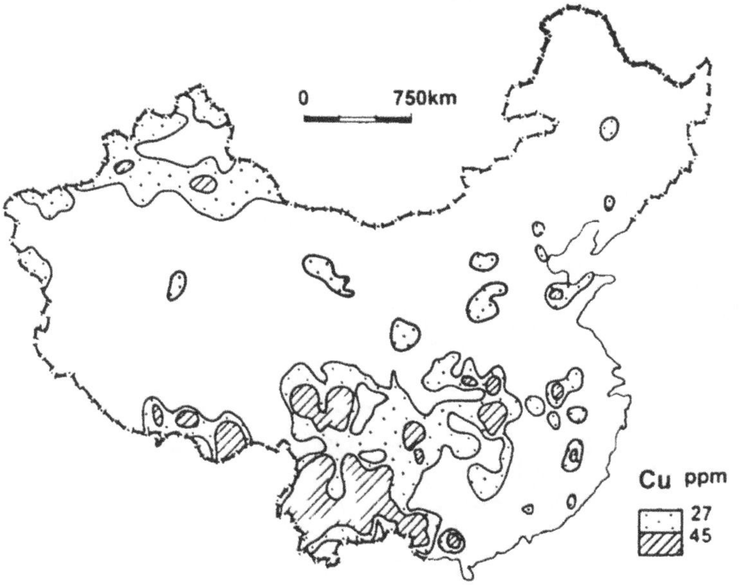


**Figure 4.** Geochemical distribution of Cu in soil surveys, average sampling density 1 sample per 1600 km$^2$ in East China and 1 sample per 1600 km$^2$ in remote areas (SEBV project). Compiled from the maps by Zheng et al. [27]

Determination of platinum group elements (PGE) in floodplain sediment samples benefited from breakthrough of analytical techniques with the detection limits of 0.1-0.2 ppb for Pt, 0.1 ppb for Pd. A remarkable Pt and Pd geochemical province were delineated in Southwestern China. The Pd and Pt contour indicating 1.0 ppb encircles an area of more than 80 000 km$^2$ (Fig. 7). From Fig. 7, it is apparent that the most favorable areas for PGE deposits in China should be in the Southwest. Preliminary study finds that the Pt and Pd geochemical high contents are located in 9 super-catchment basins; high values of P, Cr, Ti, Cu, Ni and Co have been discovered in the same areas [22]. The significance of these remains to be investigated. Thus far there are no known significant Pt and Pd mineralizations in these areas.

Thus we see that the wide-spaced floodplain sediment sampling is a fast and cost-effective way to define geochemical provinces and has strategic significance in the early stages of mineral exploration.

Obvious changes in Hg patterns in surface samples and deep samples have been noted (Fig. 8). It is not surprising that worsening Hg pollution in surface samples was detected in China's gold-mining areas. Mercury is used to concentrate the fine gold particles through amalgamation. Whereas, the scale and pace of Hg pollution caused by industrialization have steadily increased in East of China (Fig. 9). A similar phenomenon is detectable in surface samples for P as a result of widespread use of phosphate fertilizers [22].

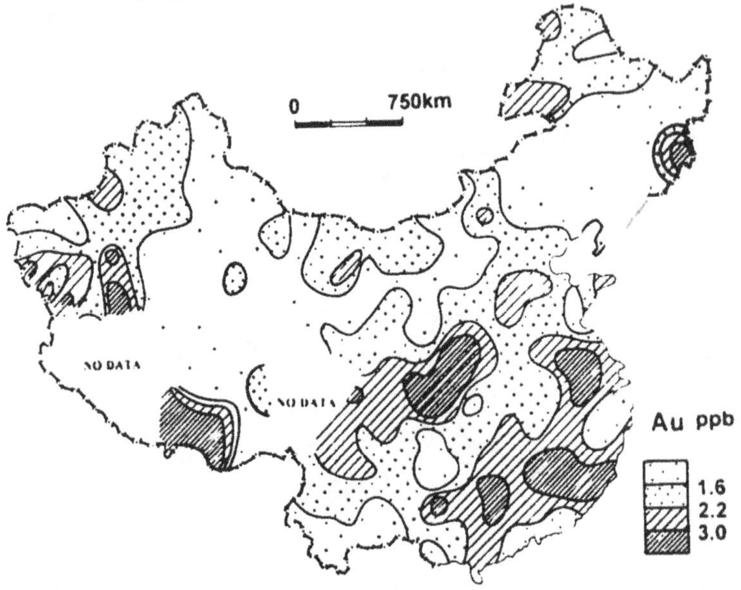


**Figure 5.** Geochemical map of Au in deep floodplain sediment with a depth of 80-120cm

From the geochemical maps we could see that $Al_2O_3$ baseline values in Southeastern China is the highest (Fig. 10), and CaO, MgO, $Na_2O$ are the lowest (Figs. 11,12 and 13). The pH values in Southeastern China are also the lowest (Fig. 14). All these demonstrate the vulnerability of soil in this region to resist acid rain. Fig.14 shows the distribution of acid rain in 1982 and 1992 and soil acidification in China. It can be seen that the areas of acid rain with pH<7 in Southern and Eastern China have increased approximately one million km$^2$ during the past ten years, owing to coal combustion amounting annually to approximately 120 million tons[12].

As a result of acid rain precipitation, S content was greatly increased in surface samples from South and East China (Fig.15 and 16).

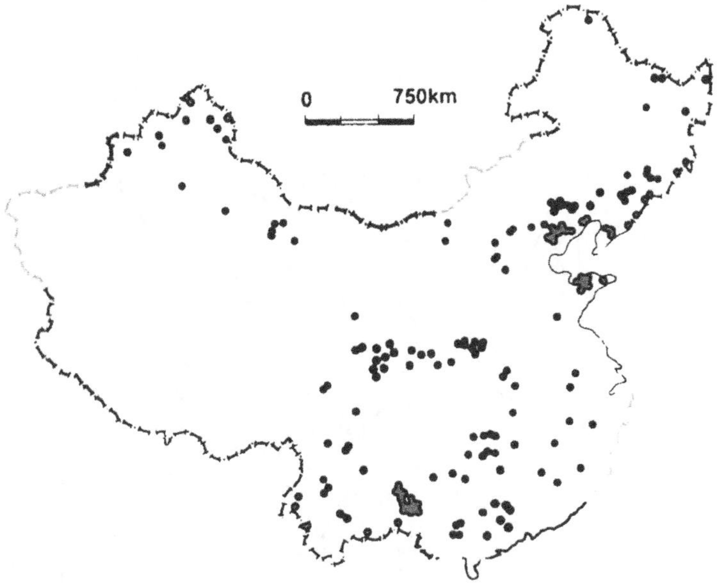


**Figure 6.** Distribution of major gold ore deposits in China. Compiled from the maps by Cun [3]

However, the scale and pace of pH changes in the soil have not matched the increase of acid rain precipitation in soil. Only about one third of the area shows acidification of soil to pH 5.0.

There are only slight differences between the surface and deep CaO patterns demonstrating that the amount of CaO consumed in the course of buffering acid rain in surface soil is still not great enough to change the patterns of CaO in surface samples(Fig.11)

The surface patterns of MgO show more impoverishment than the deep patterns (Fig, 12), $Na_2O$ patterns are more sensitive to acid rain (Fig. 13), but there are also other factors influencing the impoverishment of surface $Na_2O$ patterns such as the larger amount of rainfall in Southeastern China.

Chemical time bomb (CTB) was defined by Stigliani [15] as a chain of events resulting in the delayed and sudden occurrence of harmful effects due to the mobilization of chemicals stored in soils and sediments in response to slow alterations in the environment. In 1988, a CTB of aluminum was recognized in Europe. The example was described as follows : (a) Aluminum present as a natural component of soil is immobile at soil pHs greater than around 4.2. As the pH shifts into the range < 4.2, mobilization is

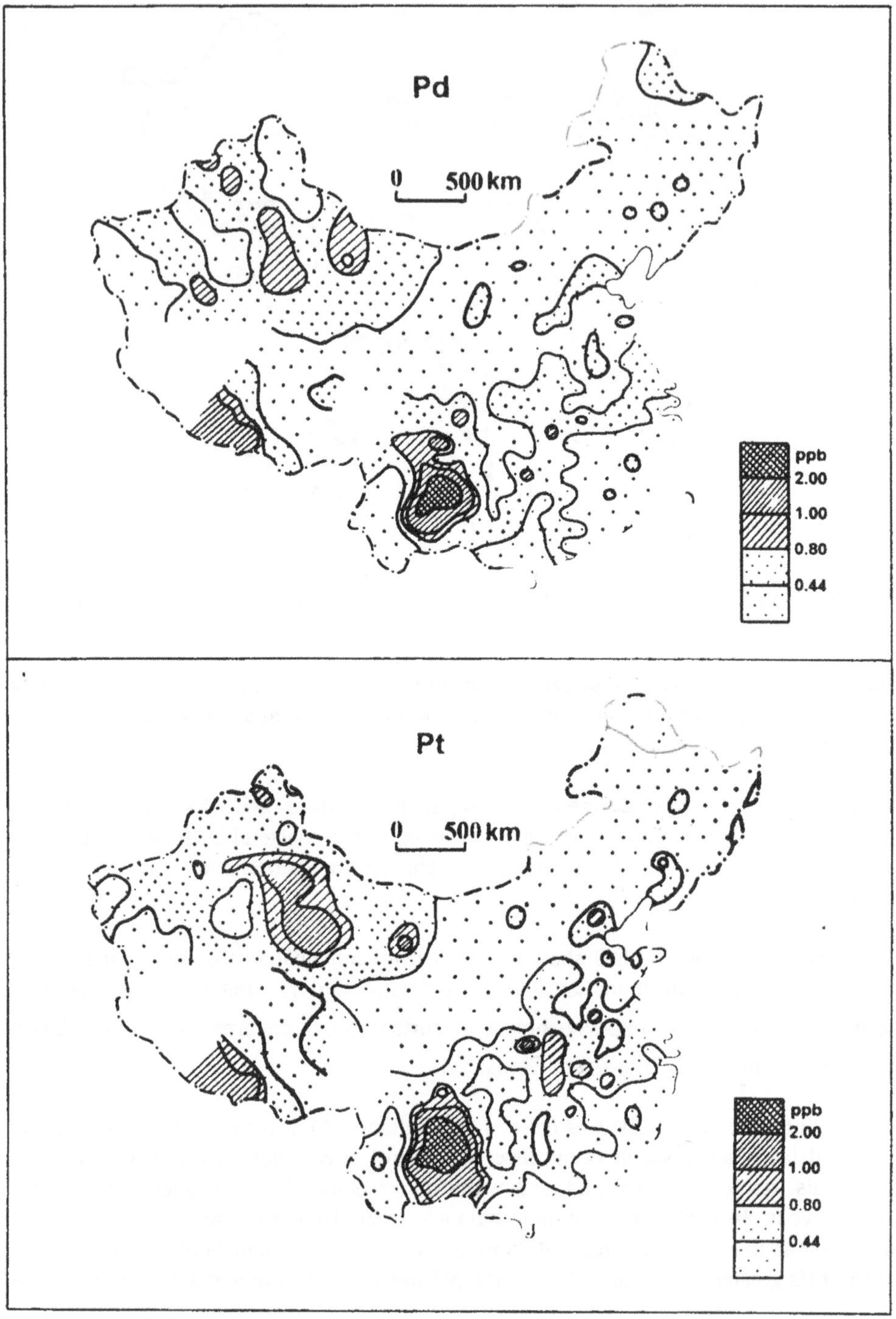


**Figure 7.** Geochemical maps of Pd and Pt in surface floodplain sediment with a depth of 5-25cm

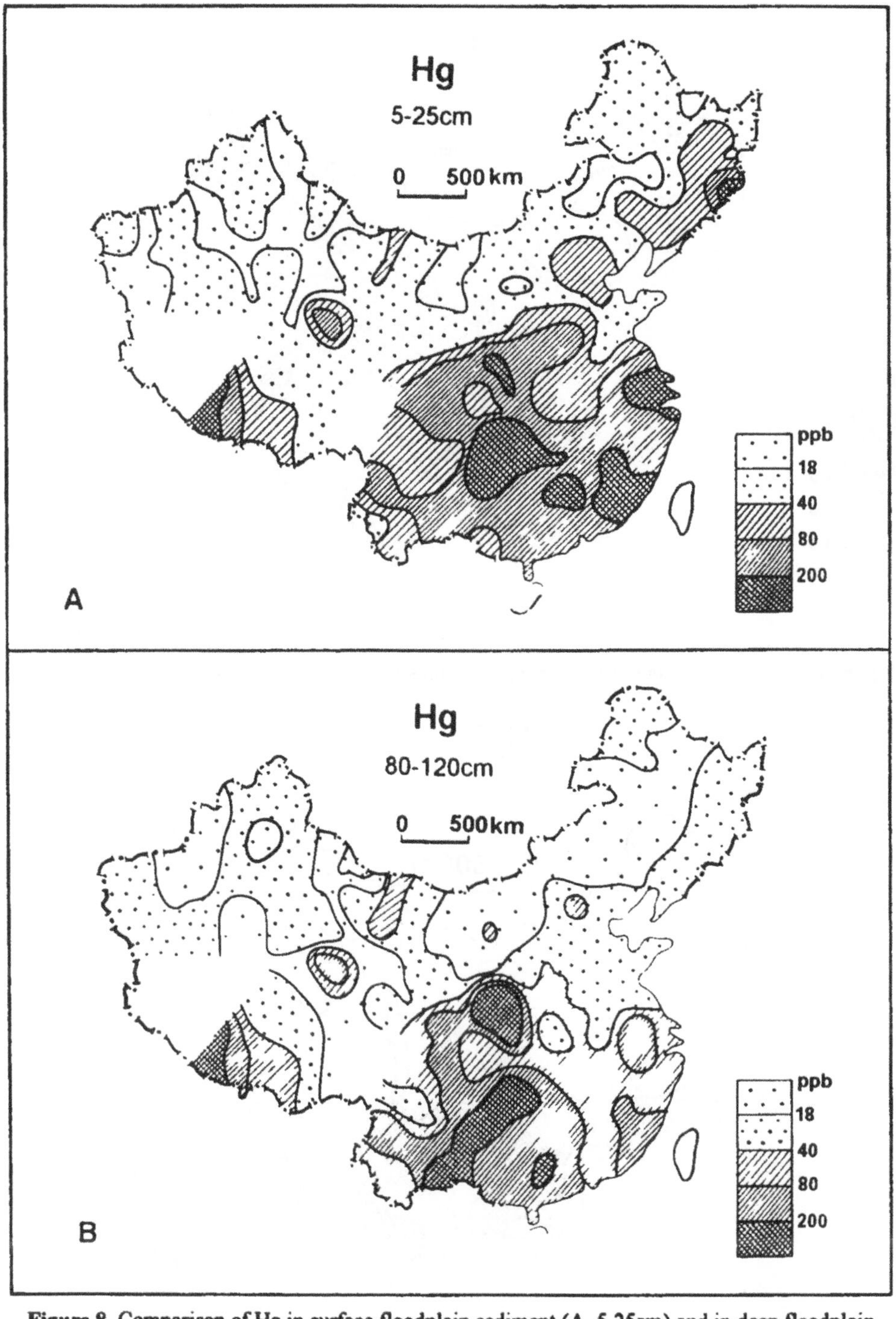


**Figure 8.** Comparison of Hg in surface floodplain sediment (A, 5-25cm) and in deep floodplain sediment (B, 80-120cm)

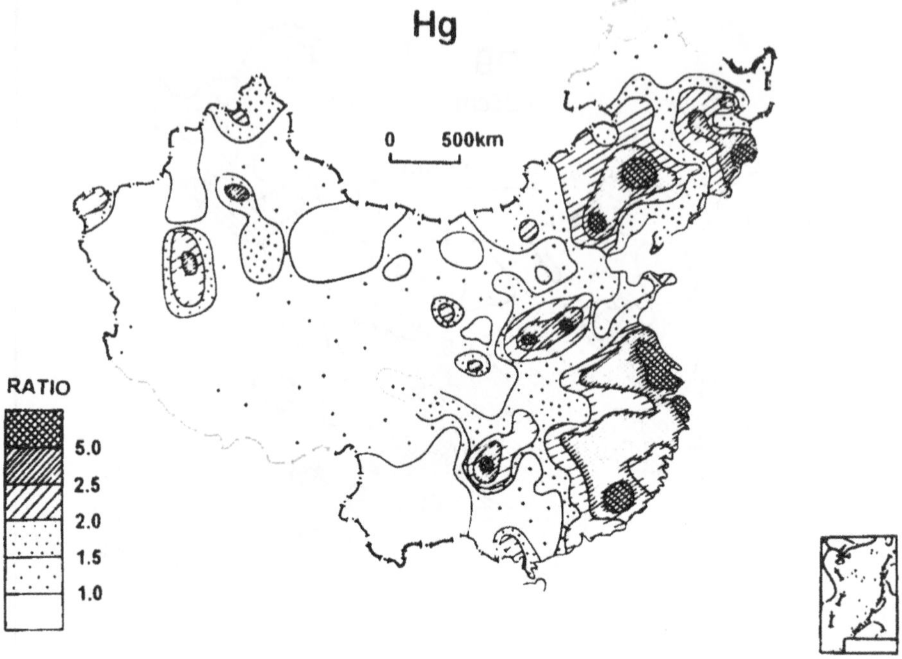


**Figure 9.** The ratio of Hg in surface floodplain sediment to that in deep floodplain sediment

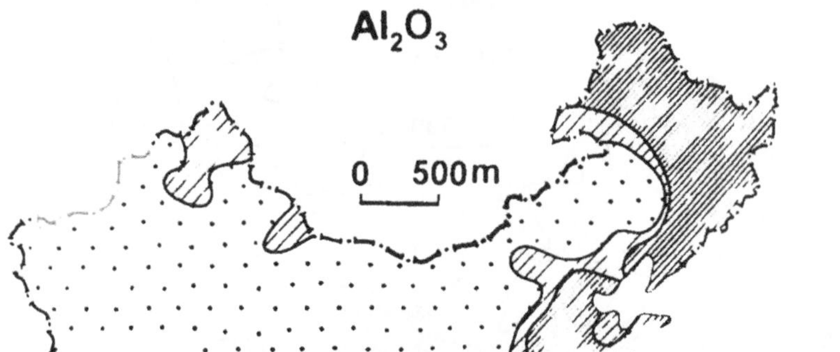

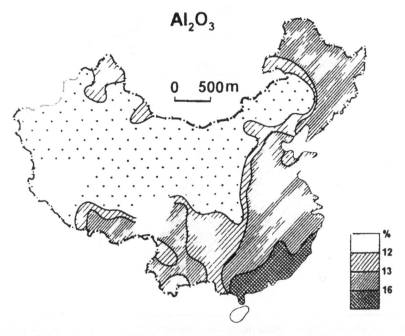


**Figure 10.** Geochemical map of $Al_2O_3$ in surface floodplain sediment with a depth of 5-25cm

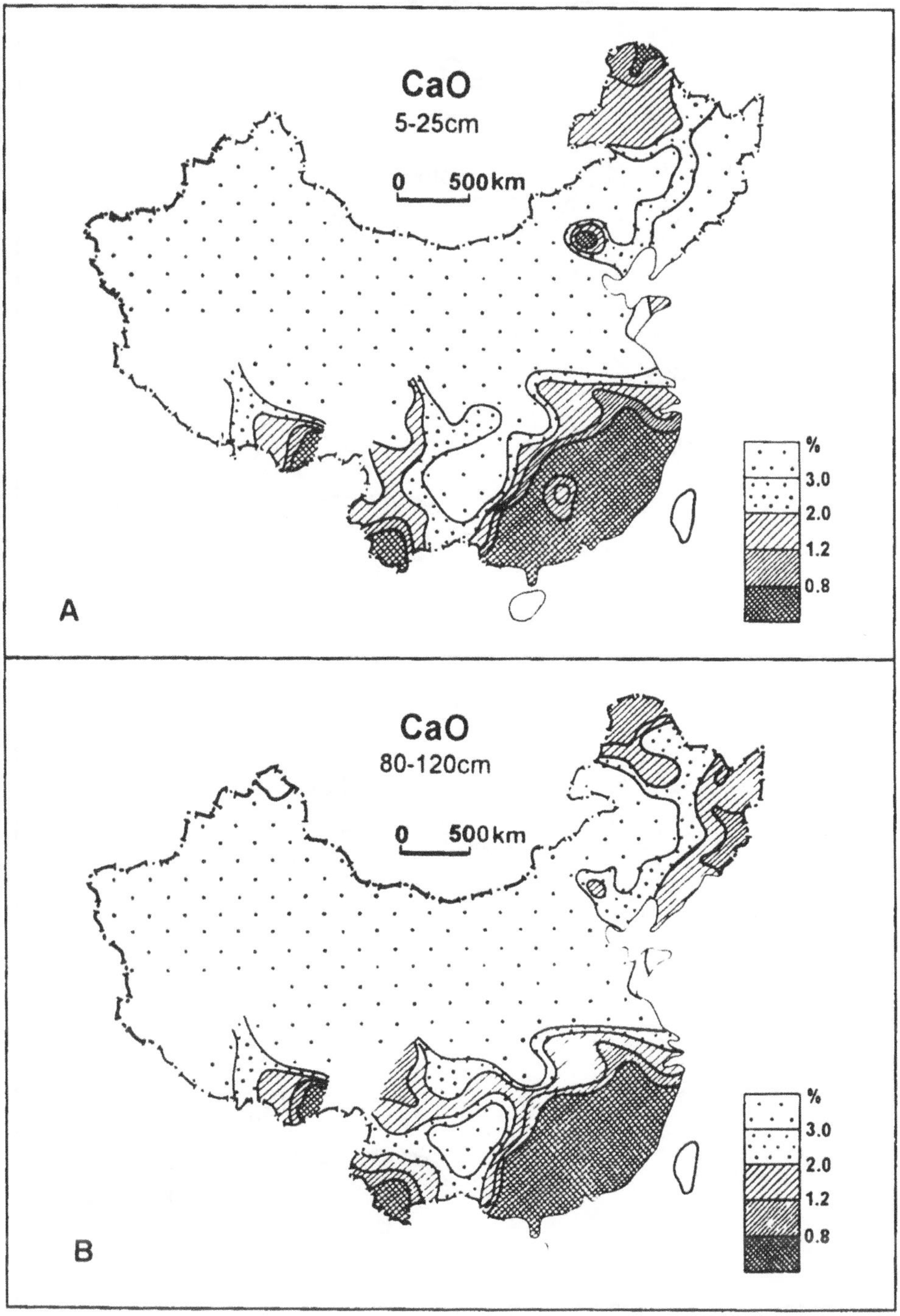


**Figure 11.** Comparison of CaO in surface floodplain sediment (A, 5-25cm) and in deep floodplain sediment (B, 80-120cm)

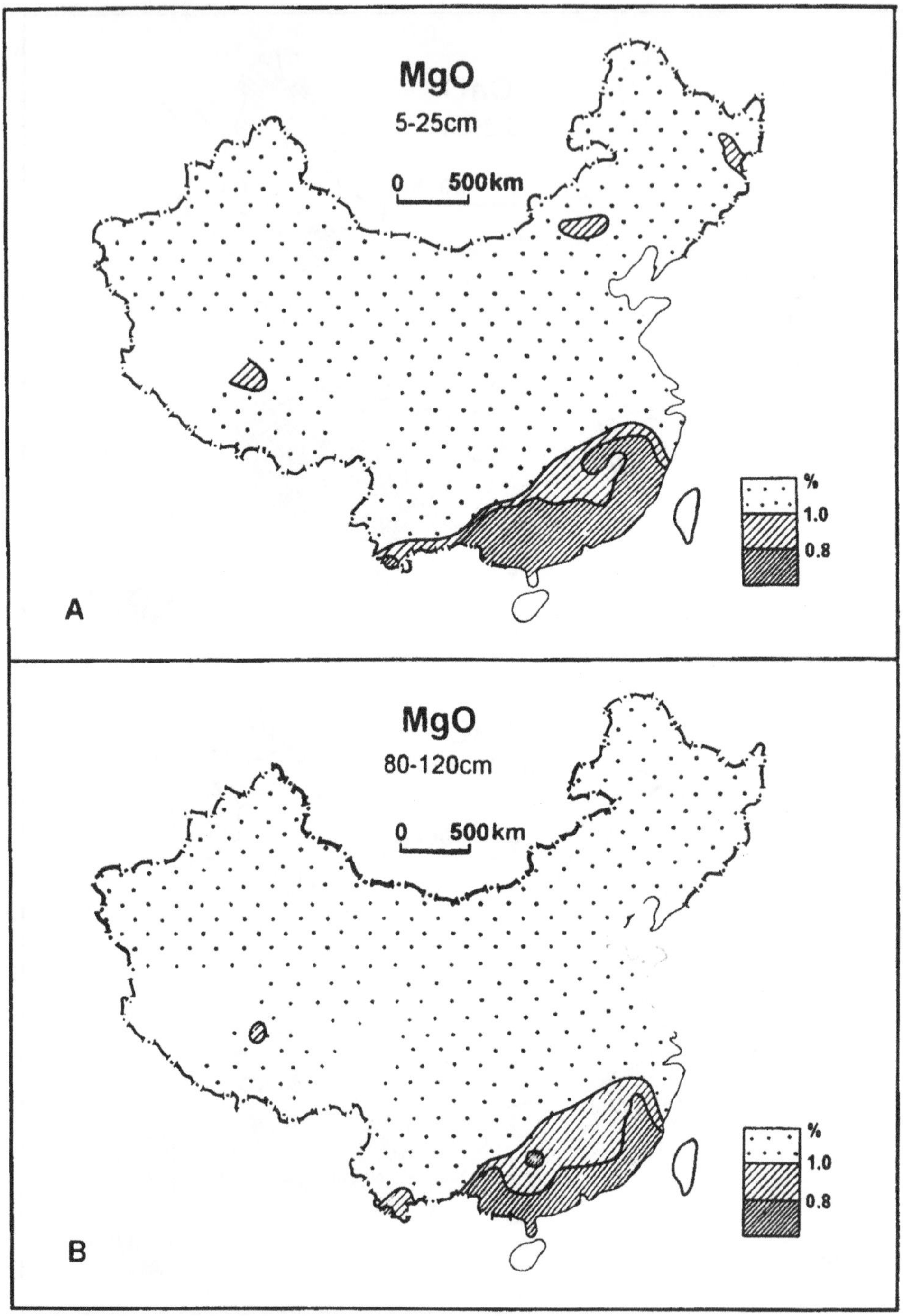


**Figure 12.** Comparison of MgO in surface floodplain sediment (A, 5-25cm) and in deep floodplain sediment (B, 80-120cm)

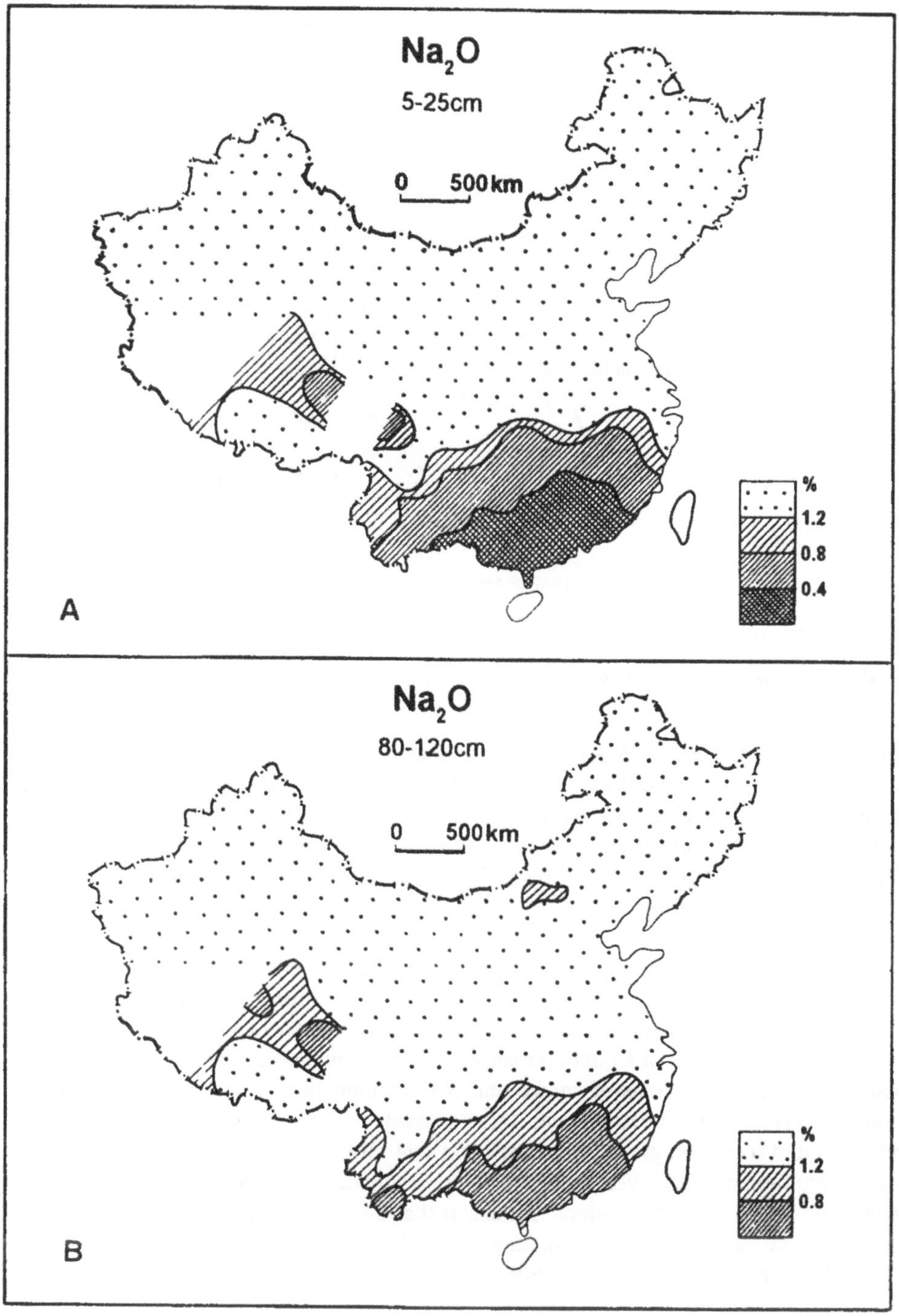


**Figure 13.** Comparison of $Na_2O$ in surface floodplain sediment (A, 5-25cm) and in deep floodplain sediment (B, 80-120cm)

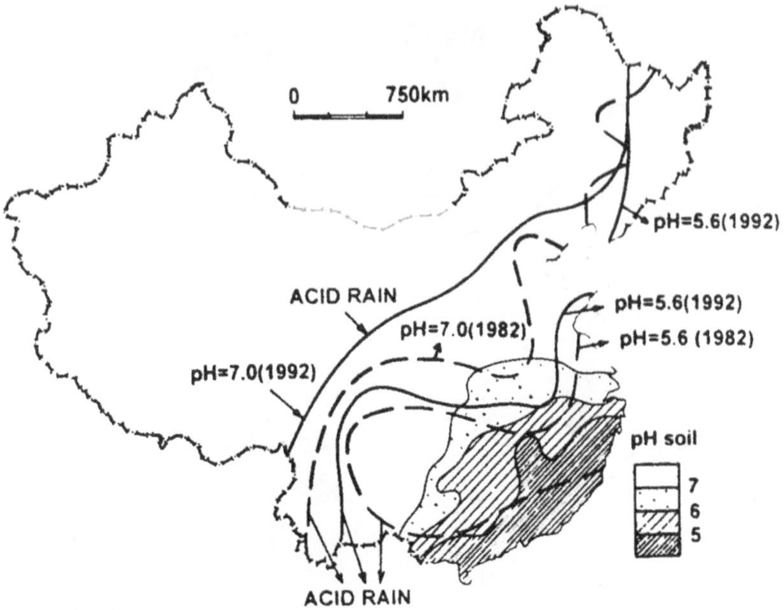


**Figure 14.** Distribution of acid rain between 1992 and 1982, compared with pH in surface floodplain sediment in 1994. Acid rain curves compiled from Wang [16]

greatly enhanced; (b) Acid deposition begins to acidify the forest soils containing the Al. For some time, the pH unaffected by the acid inputs because the soil contains a store of "base cations" that buffers against acidic inputs; (c) After the buffering capacity of the soil is depleted, the soil pH drops rapidly; (d) Around the threshold value of 4.2, the soil begins to leach Al; (e) Adverse effects in the ecosystem (e.g. damage to trees, fish mortalities) begin to be manifested.

A similar aluminum CTB might be predicted by EGMON data in Southeastern China. because first the highest content of aluminum present as an immobile natural component of soil in Eastern and Southeastern China (Fig.10). Second, coal combustion is the main cause of acid rain form in China and smoke particles and sulfur dioxides are its major pollutants. 14 million tons of smoke and 5.99 million tons of sulfur are discharged annually [12] and enter environment; these amounts will not diminish over the next 25 years as rapid economic growth continues, and acid rain will go on acidifying soil with a high aluminum content, even though soil pH is unaffected by acid inputs at present, thanks to the store of base cations in the soil that buffer it against such inputs. Third, content of the base Ca, Mg and Na cations in Southeastern China are the lowest in China showing the vulnerability of soil in these areas are on the increase. Fourth, many

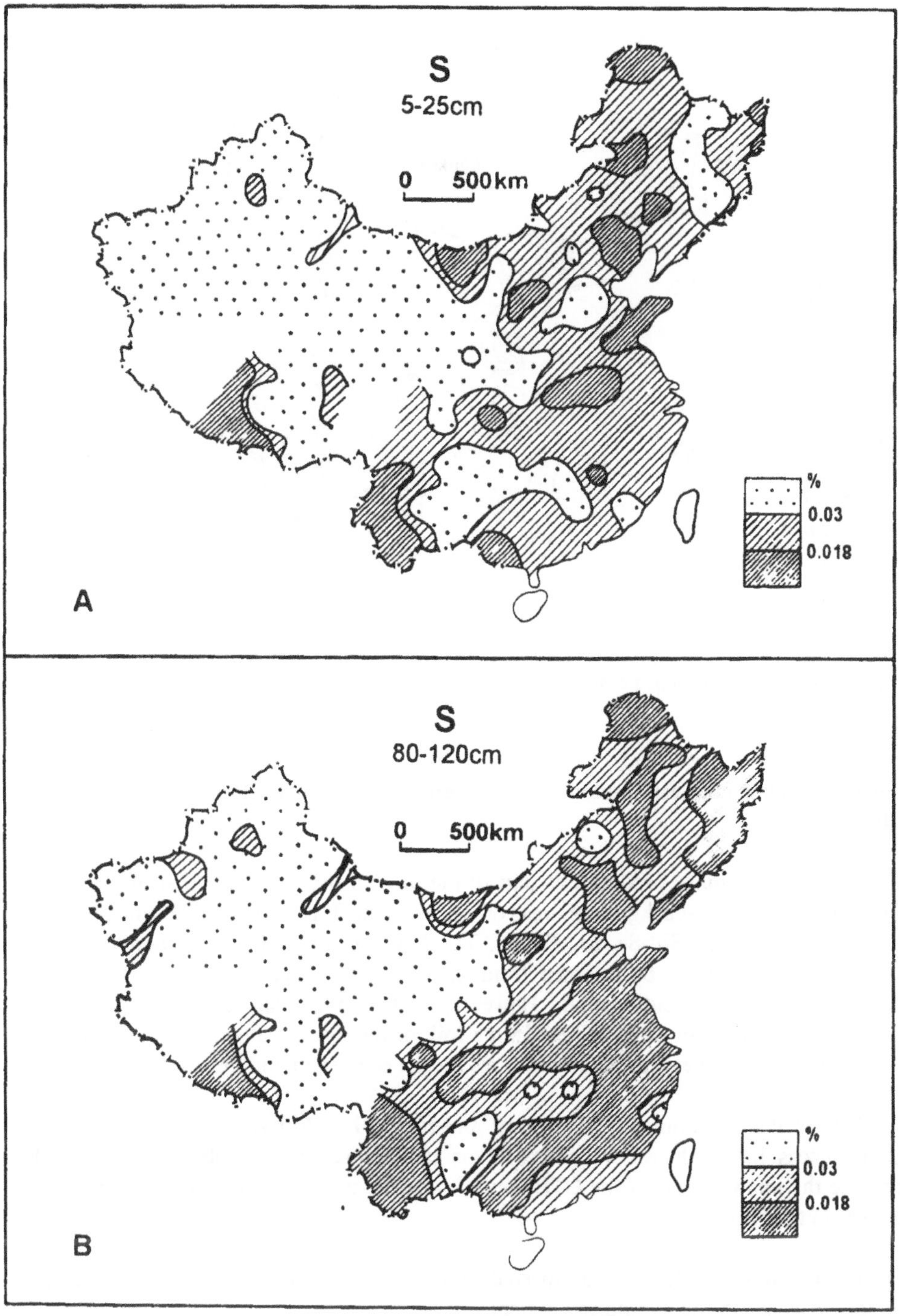


**Figure 15.** Comparison of S in surface floodplain sediment (A, 5-25cm) and in deep floodplain sediment (B, 80-120cm)

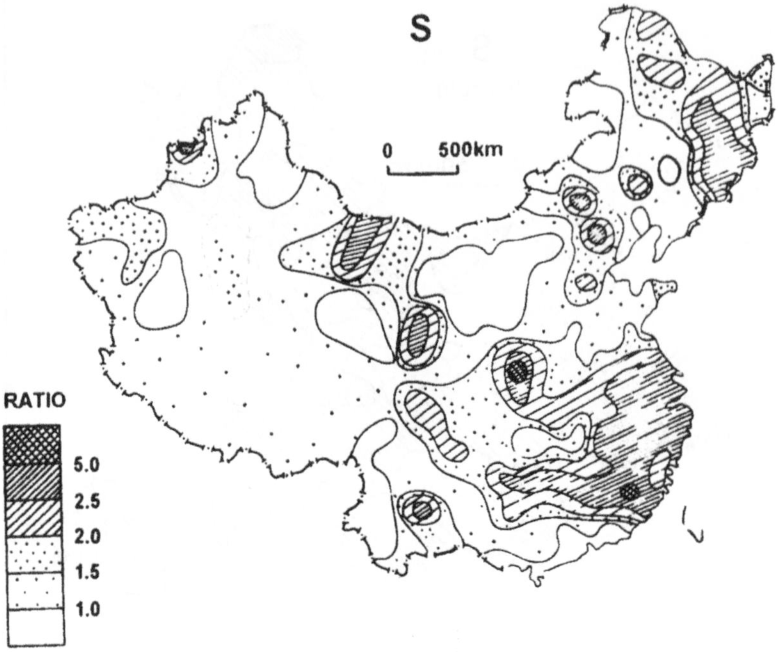


**Figure 16.** The ratio of S in surface floodplain sediment to that in deep floodplain sediment

pines and other conifers limited to China grow widely in Eastern and Southeastern China.

The geochemical maps produced as part of the present project reveal that Southeastern China has large high-background areas of Ag, As, Au, B, Be, Bi, Cd, Ce, Cu, Ga, Ge, I, La, Mo, Nb, Pb, Rb, Sb, Se, Sr, Th, Ti, Tl, U, W, Y, Zn and Zr[22]. Some of those elements may suddenly be released from the soil due to pH changes or the broader impact of continuous acidification. However, we have yet failed to identify the critical state of mobilization of each element in the soil, or the risk entailed in chronic exposure to these elements at very low concentrations for the ecosystem and human health.

We believe that aluminum CTB in local and regional scale have already happened in many places in Southeastern China. It has not aroused much public concerns only because the unawareness of the CTB concept and the more extensive and unamendable danger in the future.

Further research should be done to study the mechanism of the CTB process in this region, to establish the monitoring system.

## CONCLUSIONS

1. Wide-spaced floodplain sediment sampling on the plain adjacent to the effluents of rivers draining super-catchment basins measuring mostly between 1000 and 10 000 $km^2$ can identify geochemical patterns similar to those obtained by China's national (active stream sediment) geochemical project (RGNR) and soil sampling project (SEBV). There is good agreement between data from RGNR and the present project. These facts demonstrate both the suitability and feasibility of using floodplain sediment as a global sampling medium.
2. Wide-spaced floodplain sediment sampling is a fast and cost-effective way to define geochemical provinces and has strategic significance in the early stages of mineral exploration.
3. Geochemical maps produced by data from wide-spaced floodplain sediment sampling reveal two new gold geochemical provinces in Xinjiang and Tibet, as well as three important Pt-Pd geochemical provinces in China.
4. Changes in Hg and P patterns on the surface and in deep samples have been discovered. Hg and P pollution is occurring over large areas in China.
5. An aluminum chemical time bomb similar to the one identified in central Europe is predicted to occur in Southeastern China. Further research should be done to monitor the mechanism of CTB process in this region.

## ACKNOWLEDGMENTS

We would like to express our cordial thanks to Mr. Yan mingcai, Lu Yinxiu and Ren Tianxiang for their advice regarding project design, to Mr. Xiang Yingchuan, Ms. Gao Yanfang and Ms. Shun Hongwei for their assistance with computer work tasks, and to Mr. Chen Fanglun for his help in the analyses of Pt and Pd. We particularly thank Dr. White for his thorough and careful correction of an early draft of the manuscript. Finally many thanks to Ms. Li Xiujue, Ms. Xie Liping, Ms. Wang Xiaohong and Ms. Song Wei for preparing the diagrams.

## REFERENCES

1. Bolviken, B., Kullerud, G. and Loucks, R. R., 1990. Geochemical and metallogenic provinces: a discussion initiated by results from geochemical mapping across northern Fennoscandia. In: A. G. Darnley and R. G. Garrett (Editors). International Geochemical Mapping. J. Geochem. Explore., 39:49-90.
2. Cheng Hangxin, 1994. Representative samples in global wide-spaced sampling: orientation study in China. Inst. Geophys. Geochem. Explor., Master's thesis, 114 pp. (in Chinese with English abstr., unpubl.).
3. Cun Gui, 1992. The atlas of gold mineral resources of China. 62pp (in Chinese, unpubl.).
4. Darnley, A. G., 1990. International geochemical mapping : a new global project. In: A. G. Darnley and R. G. Garrett (Editors), International Geochemical Mapping. J. Geochem. Explor., 39:1-13.
5. Darnley, A. G., 1995. International geochemical mapping--a review. In: Y. T. Maurice and Xie Xuejing (Editors), Geochemical Exploration 1993. J. Geochem. Explor., 55:6-10.

6. Darnley, A. G., Bjorklund, B., Gustavsson, N., Koval, P.V., Plant. J.A., Steenfelt, A., Tauchid, M. and Xie Xuejing, 1995. A global geochemical database for environmental and resource management. Earth Sciences 19. United Nations Educational, Scientific and Cultural Organization, Paris, 122 pp.

7. Dong Deming, Ramsey, M. H. and Thornton, I., 1995. Effect of soil pH on Al availability in soil and its uptake by the soybean plant (Glycine max). In: Y. T. Maurice and Xie Xuejing (Editors), Geochemical Exploration 1993. J. Geochem. Explor., 55:221-230.

8. Garrett, R. G. and Nichol, I., 1967. Regional geochemical reconnaissance in Eastern Sierra Leone. Trans. IMM, Sect.B, Appl. Earth Sci., 76:B97-112.

9. Hauhs, M. and Wright, R. F., 1986. Regional patterns of acid deposition and forest decline along a cross-section through Europe. Water, Air and Soil Pollution, 31:463-474.

10. Ottesen, R.T., Bogen, J., Bolviken, B. and Volden, T., 1989. Overbank sediment: a representative sample medium for regional geochemical mapping. In: S.E. Jenness (Editor). Geochemical Exploration 1987. J. Geochem. Explor., 32:257-277.

11. Plant, J. A., Hale, M. and Ridgway, J., 1988. Development in regional geochemistry for mineral exploration. Trans. Inst. Min. Metall. (Sect. B: Appl. Earth Sci.), 97: B116-140.

12. Qu Geping, 1995. Challenge of environmental problems in China and our response. In: Y. T. Maurice and Xie Xuejing (Editors), Geochemical Exploration 1993. J. Geochem. Explor., 55:1-5.

13. Schulze, E. D., 1989. Air pollution and forest decline in a spruce (Picea abies) forest. Science 244:776-783.

14. Shen Xiachu and Yan Mingcai, 1995. Representativity of wide-spaced lower-layer overbank sediment geochemical sampling. In: Y T. Maurice and Xie Xuejing (Editors), Geochemical Exploration 1993. J. Geochem. Explor., 55:231-248.

15. Stiglinilian, W. M., Doelman, P., Salomons, W., Schulin, R., Smidt, G. R. B. and Van der Zee, S.E.A.T.M., 1991. Chemical time bombs: predicting the unpredictable. Environment, 33: 4-9 and 26-30.

16. Wang Wenxing, 1994. Study of the origin of acid rain formation in China. Journal of China Environmental Science. 14 (5):323-329 ( in Chinese).

17. Xie Xuejing, 1995. Analytical requirements in internationl geochemical mapping. Analyst, 120:1497-1504.

18. Xie Xuejing, 1996. Exploration geochemistry: present status and prospects. Geol. Rev. 42 (4):346-356. (in Chinese).

19. Xie Xuejing and Cheng Hangxin, 1997. The suitability of floodplain sediment as global sampling medium: evidence from China. In: Graham F. Taylor and Richard Davy (Editors), Geochemical Exploration 1995. J. Geochem. Explore., 58:51-62.

20. Xie Xuejing, Mu Xuzan and Ren Tianxiang, 1996. Eighteen years of China's national geochemical mapping. Abstracts of 30th IGC. Beijing, 3:42.

21. Xie Xuejing and Ren Tianxiang, 1991. A decade of regional geochemistry in China-the National Reconnaissance ptoject. Trans Inst. Metall., 100: B57-65.

22. Xie Xuejing, Shen Xiachu, Cheng Hangxin, Yan Guangsheng, Gu Tiexin, Lai Zhimin, Du Paixuan, Yan Mingcai, Lu Yingxiu and Ren Tianxiang, 1996. Environmental geochemistry monitoring network and dynamic geochemistry maps in China. Inst. Geophys. Geochem. Explor. Res Rep. (in Chinese, unpubl.).

23. Xie Xuejing, Sun Huanzhen and Ren Tianxiang, 1989. Regional Geochemistry-National Reconnaissance Project in China. In: Xie Xuejing and S. E. Jenness (Editors), Geochemical Exploration in China. J. Geochem. Explor.,33:1-9.

24. Xie Xuejing, Yan Mingcai, Li Lianzhong and Shen Huijun, 1985. Geochemical reference samples, drainage sediment GSD1-8 from China. Geostand. Newsl., 9:83-159.

25. Xie Xuejing, Yan Mingcai, Li Lianzhong and Shen Huijun, 1985. Usable values for Chinese standard reference samples of stream sediment, soil and rocks GSD 9-12, GSS 1-8 and GSR 1-6. Geostand. Newsl., 9: 227-280.

26. Xie Xuejing and Yin Binchuan, 1993. Geochemical patterns from local to global. In: F. W. Dickson and L.C. Hsu (Editors), Geochemical exploration 1991. J. Geochem. Explor., 47:109-129.

27. Zhen Chunjiang, Li Huiming, Chen Minpei, Wang Wanxing, Zhou Zanao, Jin Shi and Zhang Dongwei, 1994. The atlas of soil environmental background values in the People's Republic of China. China Environmental Science Press, Beijing. 196 pp.

*Proc 30th Int'l. Geol. Congr.*, Vol. 19, pp. 111-125
Xie Xuejin (Ed)


# Geochemical Mapping in Europe.

FIONA FORDYCE[1], JANE PLANT[1], GERARD KLAVER[2], JUAN LOCUTURA[3], REIJO SALMINEN[4] AND KAMIL VRANA[5]

[1] *British Geological Survey, Keyworth, Nottingham, NG12 5GG, UK.*
[2] *Rijks Geologische Dienst, PO Box 157, Haarlem, 2000 AD, The Netherlands.*
[3] *Instituto Technologico Geominero de Espana, Rios Rosas 23, Madrid 28042, Spain.*
[4] *Geological Survey of Finland, PO Box 96, Fin-02151 Espoo, Finland.*
[5] *Geological Survey of the Slovak Republic, Mlynska dol 1, 81704 Bratislava, Slovakia.*

**Abstract**

The Forum of European Geological Surveys (FOREGS) includes Geological Surveys from 34 European countries and is responsible for co-ordinating Geological Survey activities in Europe. The FOREGS Geochemistry Task Group was established in 1994 to supervise European geochemical mapping policy for environmental, legislative, resource-management and scientific purposes. The task group comprises representatives from five countries, charged initially with the compilation of an inventory of geochemical data within FOREGS countries. The preparation of European Geochemical Baseline maps will involve the integration of different national datasets following the recommendations of the International Geological Correlation Programme (IGCP) Project 259 "International Geochemical Mapping". Results of the inventory show that most geochemical surveys in Europe conform to the IGCP 259 recommendations. Stream sediment (26% coverage), surface water (19% coverage), soil/till (21% coverage) and radiometric data (19% coverage) are the most extensive sample types, and the majority of surveys (81%) have been carried out at sampling densities of ≤ 1 sample per 100 $km^2$. Most filtered-water surveys are based on a filter size of 0.45µm, and 83% of stream sediment surveys collect samples sieved at 100-200 µm. The collection of the Global Reference Network (GRN) samples recommended by IGCP 259 to provide internationally standardised geochemical data and the careful use of statistical and map-generation techniques should facilitate the levelling of different national datasets and preparation of a European Geochemical Baseline.

*Keywords: Europe, international geochemical mapping standards, inventory environmental geochemical baselines*

## INTRODUCTION

Public concern about the environment is growing throughout the world and especially in the industrialised countries of North America and Europe. These regions of the world, Europe in particular, have had a long history of mining, industrialisation, intensive agriculture, forestry and urbanisation, leading in some areas to contamination and/or land degradation. These problems are increasingly being extended over the rest of the world as a result of economic growth and population pressure.

In response, national governments and international organisations such as the European Union are attempting to develop policies, legislation and infrastructure such as the European Environment Agency to deal with environmental issues, and attempts are also being made to establish 'safe', 'trigger' and 'action' levels of potentially harmful chemical elements and species (PHES). These measures are often based on inadequate information, as systematic chemical data are available for only 20% of the continental land area of the globe [3, 10].

Geochemical mapping began in Europe and other areas of the world 50 years ago, and for several reasons a variety of methods developed in different countries. Data are often inconsistent between and within countries, and the current situation makes comparison between datasets at an international level difficult. Systematic environmental geochemical baseline data are urgently required to inform policy-makers and provide a sound basis for environmental legislation and resource management.

A standardised World Geochemical Atlas is being prepared by the International Geochemical Mapping Programme lead by Dr A.G. Darnley of Canada and Prof J. A. Plant of the UK. During the first stage of the project entitled 'International Geochemical Mapping' (International Geological Correlation Programme (IGCP) Project 259), standard geochemical mapping, analysis and data management methodologies were developed for national survey organisations. In addition, the project recommended the establishment of a Global Reference Network (GRN) of 5000 sampling sites as an essential first step towards international correlation and the standardisation of present and future national geochemical surveys [3]. The current phase of the project, 'Global Geochemical Baselines' (IGCP Project 360) is concerned with the implementation of the IGCP 259 recommendations and the collection of GRN samples. IGCP 360 will terminate in 1997 and the project will be carried forward by an International Union of Geological Scientists (IUGS)/ International Association of Geochemistry and Cosmochemistry (IAGC) Working Group on Global Geochemical Baselines.

The Forum of European Geological Surveys (FOREGS) is an informal body of 34 Geological Survey directors and is responsible for co-ordinating geological policy on a Europe-wide basis. In 1994, the FOREGS Geochemistry Task Group was established to develop a strategy for the preparation of a European Geochemical Baseline following the recommendations of IGCP 259 and the IUGS/IAGC Working Group. The initial group was chaired by Prof J.A. Plant of the UK, and included representatives from Finland, The Netherlands, Slovakia and Spain. It was charged with compiling an inventory of geochemical data based on the results of a questionnaire completed by Geological Surveys and related organisations throughout the FOREGS community. Results of the inventory and recommendations for the preparation of European Geochemical Maps are detailed in the report of the Task Group [11]. This paper discusses the results of the inventory in relation to the recommendations of IGCP 259/360 and the IUGS/IAGC Working Group.

## THE FOREGS INVENTORY

Information for inclusion in the inventory was collected using a standard form comprising nine sections, each for a particular sample type (Table 1). Detailed information on collection, preparation, analysis and data availability were requested for all sample types, with the exception of rock and biological surveys where information on availability only was required. The form was distributed to 57 Geological Survey and related organisations in FOREGS countries [11]. Mining and exploration companies and universities were generally not included because the surveys which they carry out tend to cover relatively small areas of less than 5000 $km^2$, the lower limit considered relevant for the purpose of the inventory. Completed forms were received from 30 of 34 countries (Fig. 1). Croatia, Iceland, Latvia and Switzerland have not conducted regional geochemical surveys over the minimum area required for the survey.

**Table 1.** Sample types included in the FOREGS geochemical inventory.

| Form Section | Sample Type | Information Required |
|---|---|---|
| A | Drainage Sediment | Full survey procedure |
| B | Lake Sediment | Full survey procedure |
| C | Overbank Sediment | Full survey procedure |
| D | Soil and Regolith | Full survey procedure |
| E | Heavy Mineral | Full survey procedure |
| F | Surface Water | Full survey procedure |
| G | Rock Sample | Information available Yes/No |
| H | Biological Sample | Information available Yes/No |
| I | Radiometric | Full survey procedure |

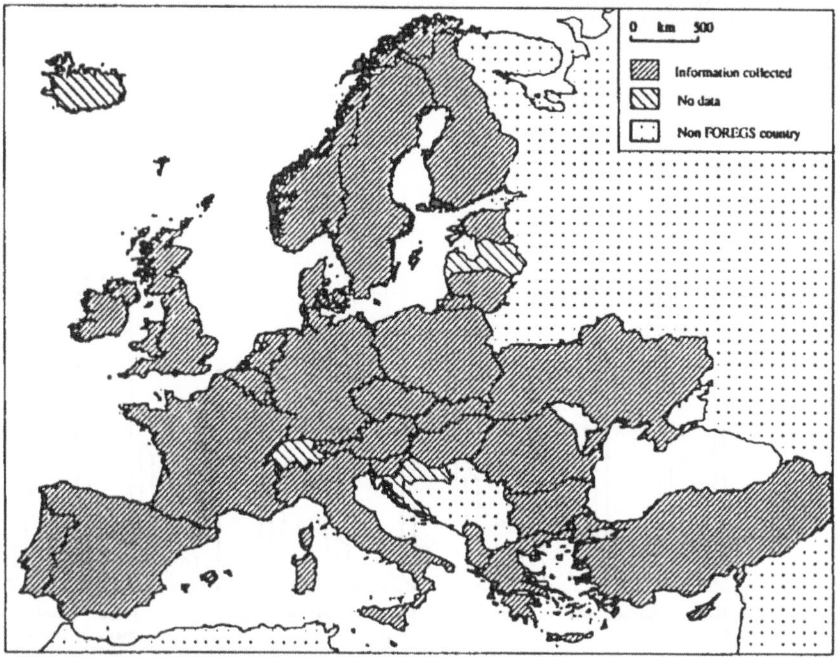


**Figure 1.** Countries included in the FOREGS Geochemical Inventory 1994-1996. Greenland is also included in the inventory but is not shown on the map.

## RECOMMENDATIONS FOR EUROPEAN GEOCHEMICAL MAPS

The FOREGS Geochemistry Task Group recommend that high-resolution national geochemical survey information available for many European countries should be incorporated into geochemical maps of Europe and the globe. The use of existing geochemical data will involve integrating the results of surveys based on different sample- collection, preparation and analytical methods. In order to obtain compatibility between the results of these different surveys, IGCP 259/360 recommends the collection of a Global Reference Network (GRN) of samples, using standard techniques for collection, preparation and analysis [3, 11]. The data for these samples will be directly comparable between different countries and different survey areas, and will be used to level existing national datasets. During the process of data integration, several aspects of data acquisition such as sample type, sampling density and analytical techniques will require careful consideration, and these are discussed in the following sections.

## SAMPLE TYPES

In Europe, as in other areas of the world, a variety of geochemical sample types have been collected by different survey organisations depending on the purpose of the survey and the physical conditions in the survey area. The sample types collected in the FOREGS region are detailed in Figure 2 and Table 2 and include stream sediments, surface waters (including shallow groundwaters), soils, till, overbank sediments, biological samples, heavy-mineral concentrates and rocks. Radiometric data are also included. Figures indicating the percentage of coverage of each sample type are based on the total area of the 34 FOREGS countries which extend to 8 417 427 km².

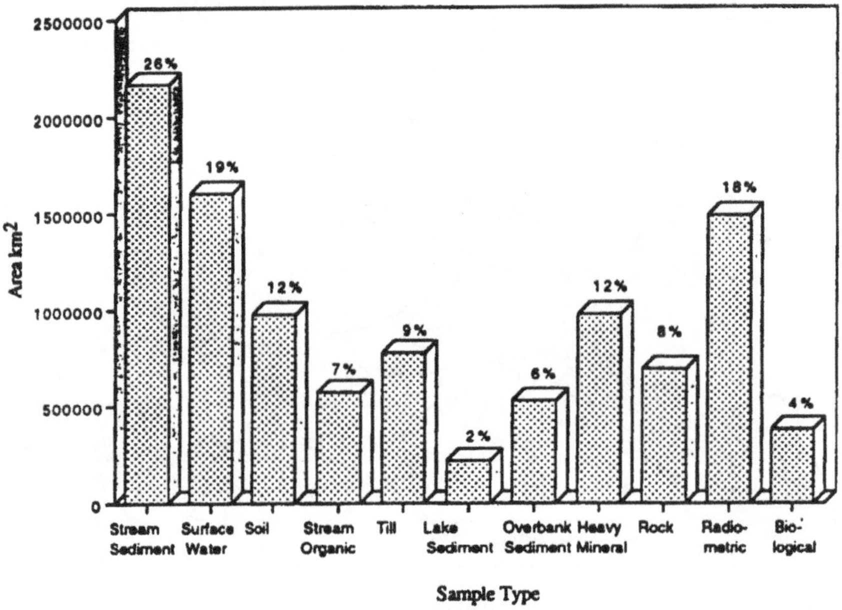


**Figure 2.** Area of FOREGS countries covered by each sample type. (Percentage cover of the total area of FOREGS countries is indicated at the top of each column)

*IGCP 259 recommendations for GRN collection*

The IGCP 259 recommendations for GRN sample collection recognise that a variety of sample types have been collected in different regions of the world. A sampling scheme which includes the collection, where possible, of stream sediment, surface water, residual soil, overbank regolith, floodplain regolith and humus at each GRN site has been devised in order to relate different national datasets based on any one of these sample media (Fig. 3). Other sample types, such as rock and biological samples, are not included in the sampling scheme as surveys of these types generally cover areas that are too small for inclusion in regional and global scale geochemical surveys [3].

Most of the sample types collected in Europe are included in the GRN scheme with the exception of rock, biological, heavy-mineral and stream organic samples. Most FOREGS countries collect more than one geochemical sample type, and only Bulgaria (area 110911 km$^2$, 1% of the FOREGS region) has survey information based on rock geochemistry alone (Table 3). Significant datasets based on rock geochemistry (3% of the FOREGS region) in Cyprus, Czech Republic, Estonia, Finland, Germany, Lithuania, The Netherlands and Slovakia and on biological samples (5% of the FOREGS region) in Finland, Lithuania, Norway, Slovakia and Sweden will not be included in the preparation of European and world geochemical maps. However, these countries conduct geochemical mapping programmes based on stream sediment, soil, till or overbank samples for the same areas (Table 3). Similarly it will not be possible to incorporate datasets based on the collection of stream organic matter in Finland and Sweden (amounting to 6% of the FOREGS region) in the global and European mapping scheme but geochemical data derived from alternative sample types are available for these areas (Table 3). Heavy-mineral data are excluded from the IGCP 259 GRN mapping strategy, since samples of this type have traditionally undergone quantitative rather than qualitative analysis.

**Table 2.** Sample types collected in FOREGS countries

| Stream Sediment | Surface Water | Soil | Stream Organic | Till | Lake Sediment | Overbank Sediment | Heavy Minerals | Rock | Radio metric | Bio logical |
|---|---|---|---|---|---|---|---|---|---|---|
| Albania | Albania | Albania | Finland | Denmark | Czech | Belgium | Czech | Bulgaria | Albania | Finland |
| Austria | Czech | Belgium | Norway | Finland | Finland | Denmark | Denmark | Cyprus | Belgium | Lithuania |
| Belgium | Finland | Cyprus | Sweden | Norway | Lithuania | Greece | Finland | Czech | Czech | Norway |
| Czech | Germany | Czech | | Sweden | | Hungary | Germany | Estonia | Greece | Slovakia |
| Finland | Greenland | Estonia | | | | Luxembourg | Greenland | Finland | Greenland | Sweden |
| France | Norway | France | | | | Netherlands | Norway | Germany | Luxembourg | |
| Germany | Poland | Germany | | | | Norway | Portugal | Lithuania | Poland | |
| Greece | Romania | Lithuania | | | | Slovenia | Spain | Netherlands | Portugal | |
| Greenland | Slovakia | Netherlands | | | | | Sweden | Slovakia | Slovakia | |
| Hungary | Slovenia | Norway | | | | | UK | | Slovenia | |
| Ireland | UK | Poland | | | | | | | Spain | |
| Italy | | Portugal | | | | | | | Sweden | |
| Lithuania | | Slovakia | | | | | | | UK | |
| Norway | | Slovenia | | | | | | | | |
| Poland | | Ukraine | | | | | | | | |
| Portugal | | UK | | | | | | | | |
| Romania | | | | | | | | | | |
| Slovakia | | | | | | | | | | |
| Slovenia | | | | | | | | | | |
| Spain | | | | | | | | | | |
| Sweden | | | | | | | | | | |
| Turkey | | | | | | | | | | |
| UK | | | | | | | | | | |

| Sample Media | Sub-Type / Location | | | Fractions / Amounts |
|---|---|---|---|---|
| Humus | (no subdivision) | | | 10 litres |
| $A_{25}$ surface regolith | Residual soil | Overbank | Floodplain | Bulk / 2 kg: < 2 mm / 2 kg: < 0.18 mm / 5 kg: |
| Deep regolith | C horizon | *Overbank | *Floodplain | < 0.18 mm / 5 kg: |

| | | | |
|---|---|---|---|
| Surface Water | stream | ▲lake | Untreated; acidified; filtered & acidified 100 ml each |
| Drainage Sediment | stream | ▲lake | < 0.18 mm / 5 kg: |

* Collection of deep overbank and floodplain samples is optional
▲ In specific regions only

**Figure 3.** Sample types required for the IGCP 259 Global Reference Network. From Darnley et al. [3].

**Table 3.** FOREGS geochemical datasets based on sample types excluded from the IGCP 259 GRN sampling scheme.

| Country | Sample Type | Area of survey $km^2$ | Alternative sample type for the same area |
|---|---|---|---|
| Bulgaria | Rock | 110911 | None |
| Cyprus | Rock | Unknown | Soil |
| Czech | Rock | 78000 | Stream sediment, overbank sediment, lake sediment, soil |
| Estonia | Rock | 12000 | Soil |
| Finland | Rock | 7000 | Till |
| Finland | Stream organic | 337000 | Till |
| Fin/Nor/Swe | Biological | 250000 | Stream sediment, till |
| Germany | Rock | 34000 | Stream sediment, soil |
| Lithuania | Rock | 64000 | Stream sediment, soil |
| Lithuania | Biological | 64000 | Stream sediment, soil |
| Netherlands | Rock | 10000 | Overbank sediment |
| Norway | Biological | 68600 | Stream sediment, overbank sediment |
| Slovakia | Rock | 49104 | Stream sediment, soil |
| Slovakia | Biological | 49104 | Stream sediment, soil |
| Sweden | Stream organic | 201000 | Till |

*IGCP 259 recommendations for national surveys*

IGCP 259 recommends the preparation of national geochemical surveys based on the collection of stream sediment, stream water, soil and radiometric data wherever possible[3]. In areas where drainage networks are poorly developed, such as the Canadian shield and Scandinavia, lake sediment or till samples should be collected in place of stream sediment samples. The IGCP 259 recommendations for national surveys relate very closely to the situation in Europe, where stream sediment surveys are by far the most extensive and have been carried out in 22 of the 34 countries covering 26% of the FOREGS region (Fig. 2, Table 2). Surface water surveys cover nearly one fifth of the FOREGS region and soils have been collected in 16 of the 34 countries (Figure 2 and Table 2). Radiometric data are available for 19% of the FOREGS region (Table 2) but data have not been collected systematically, a situation which compares unfavourably with other regions of the world such as Australia, North America and the former Soviet Union [3].

In Scandinavian countries, till sampling from depths of 60-200 cm has been carried out in preference to soil sampling. The IGCP 259 report recommends the collection of C horizon soils from the deepest accessible depth and in many areas of Europe the depth of C horizon soil sampling will be similar to that of till sampling in Scandinavia. Studies by Appleton [1] and Flight et al. [4] have shown that it is possible to integrate data for approximately 50% of the elements routinely determined in stream sediments and soils using percentile-percentile plots, despite the different geochemical processes controlling element levels in the two sample media. The geochemical differences between soils and stream sediments are greater than between soil and till therefore it should be possible to integrate geochemical determinations based on till sampling and those derived from C horizon soils to obtain maps which detail element levels in deep regolith samples.

## SAMPLING DENSITY

A wide range of sampling densities have been employed across the FOREGS region, reflecting different survey objectives. Stream sediment survey densities range from 1 sample per <0.5 km$^2$ in France, Greece, Italy, Portugal and Spain for mineral exploration to 1 sample per 2000 km$^2$ in Romania for rapid reconnaissance mapping (Table 4). Most surveys, however, have been carried out in the range of 1 sample per 1 km$^2$ to 1 sample per 5 km$^2$ (Table 4). Surface water surveys range from relatively high densities (< 1 sample per 2.5 km$^2$ in Albania, Germany and the UK) to very low densities in Finland and Romania (1 sample per 290 km$^2$ and 1 sample per 2000 km$^2$ respectively). In general, soil survey sampling densities follow similar trends to those of stream sediments ranging from 1 sample per < 1 km$^2$ in France and Portugal to 1 sample per 3500 km$^2$ in Estonia. Most soil surveys have been conducted in the range 1 sample per 5 km$^2$ to 1 sample per 25 km$^2$.

**Table 4.** Sampling densities employed for stream sediment, surface water and soil surveys in FOREGS countries.

| | Sampling Density (1 sample per x km$^2$) | | | | | | | | |
|---|---|---|---|---|---|---|---|---|---|
| | 0.01-0.30 | 0.4 | 0.5 | 1 | 1.5 | 2 | 2.5 | 3 | 4 |
| Stream Sediments | Portugal<br>Spain | Spain | France<br>Greece<br>Italy | Belgium<br>Turkey | Austria<br>Czech<br>UK | Slovakia<br>Turkey<br>UK | Albania | Italy<br>Norway<br>UK | Germany<br>Hungary<br>Ireland<br>Norway |
| Surface Water | | | | | UK | Germany<br>UK | Albania | Germany<br>Slovakia | |
| Soil | Portugal | | France | Czech<br>Ukraine | | Germany<br>UK | Albania | | |
| | **5** | **6** | **7** | **10-16** | **25** | **30** | **40** | **50** | **60** |
| Stream Sediments | Greenland<br>Poland | | Spain | | Norway<br>Poland | Greenland<br>Norway | Norway | Greenland | Lithuania |
| Surface Water | Poland | Czech | | | Poland | Greenland<br>Norway | | Greenland<br>Norway | |
| Soil | Poland<br>Slovakia | | | Estonia<br>Netherlands | Poland<br>UK | Estonia<br>Slovenia | Norway | Lithuania | |
| | **76-100** | **180** | **200** | **225-290** | **450** | **1800** | **2000** | **2400** | **3500** |
| Stream Sediments | | | Finland<br>Norway<br>Slovenia<br>Sweden | Portugal | | | Romania | | |
| Surface Water | | | Slovenia | | Finland | | | | Romania |
| Soil | Belgium<br>UK | Slovenia | | Portugal | Estonia | Estonia | | Estonia | Estonia |

### *IGCP 259 GRN recommendations*

The IGCP 259 report [3] recommends the collection of a minimum of 5000 samples around the globe from predefined 160x160 km sampling grid squares to form a Global Reference Network (GRN) of samples. The very low sampling density (1 sample per 25600 km$^2$) was chosen to facilitate global coverage in a reasonable time scale and is considerably lower resolution than surveys carried out in Europe. However, the primary purpose of the GRN samples is to provide standard materials for laboratory standardisation and data integration and not for the preparation of geochemical maps. The 5000 samples will each comprise a composite of 5 sub-samples collected from every 160x160 grid-square. The sub-samples should be collected by sub-dividing the square into 4 sub-squares of 80x80 km, one sub-sample should be collected from each sub-square and the fifth sample from any one of the 4 sub-squares chosen at random (Fig.4). The IGCP 259 report recommends that where possible the 5 sub-samples should be analysed in addition to the 1 composite sample for each 160x160 km grid square. In order to gain the greatest amount of information from the GRN samples, the 5 sub-samples will be analysed in Europe thus the GRN sampling density will be increased to 1 sample per 5120 $km^2$.

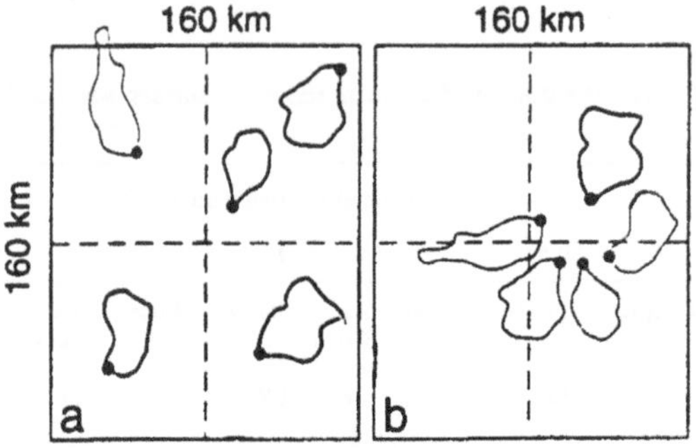


**Figure 4.** Schematic outline of drainage basin sampling pattern for the GRN. Site distribution A is preferable to B. Modified from Darnley et al. [3].

*IGCP 259 recommendations for national surveys*

The wide-spaced GRN samples will provide useful geochemical information in many inaccessible areas of the world where national surveys are not currently active. For most European countries, however, surveys based on much greater sampling densities are available and this data should be incorporated into European and global geochemical maps. The IGCP 259 report recommends a minimum sampling density of 1 sample per 100 $km^2$ for national programmes. Several stream sediment, soil and surface water surveys carried out in Estonia, Finland, Norway, Portugal, Romania, Slovenia and Sweden are based on sampling densities of greater than 1 sample per 100 $km^2$ (Table 5). In Estonia, Finland, Slovenia and Sweden more detailed information is available for stream sediment and soil/till samples for the same areas. However, stream sediment and soil survey programmes in Portugal (0.6% of the FOREGS region), Romania (1.4% of the FOREGS region) surface water surveys in Slovenia (0.2% of the FOREGS region) should be encouraged to provide more detailed information than is currently available if possible (Table 5).

The majority of modern geochemical maps are produced using geographical information systems (GIS) and image-generating software packages which interpolate a regular gridded surface from the randomly distributed point data [2, 7, 8]. The appearance and 'smoothness' of the final maps is dependant on the original sampling density, the interpolating parameters selected and the grid-size of the generated surface. Investigations into methods of combining datasets of differing sampling densities have centred on reducing the density of high resolution surveys prior to interpolation to improve compatibility with low density sampling. Studies by Garrett et al. [6], Ridgway et al. [12] and Fordyce et al. [5] have shown that it is possible to simulate low density survey maps from high density geochemical data using averaging techniques. These methods result in a loss of detailed information from high density surveys as local variations are 'smoothed out' by the averaging process. If the averaging method of data integration was applied to stream sediment datasets in Europe, for example, all datasets would be reduced to the level of the lowest density survey in Romania at 1 sample per 2000 $km^2$ with the loss of much valuable information in other countries.

Alternatively, a surface grid-size greater than the resolution of the raw data can be selected during interpolation. If a grid-size of 5x5 km is applied to raw data collected at a sampling density of 1 per $km^2$, for example, an interpolating function based on inverse distance weighting and selection of nearest neighbour points will select the raw data points with most influence at a given grid point and calculate a 5x5 km grid surface. The calculation of a reduced number of gridded values from more detailed raw data will also result in a loss of resolution of high sampling density information. In order to maintain the resolution of high sampling density surveys (≥ 1 sample per 3 $km^2$), a high density interpolating grid of 1 km or less could be selected. However, application of a detailed grid of this size to the whole of Europe would generate a huge data-file and the level of detail available in low density areas would be falsely presented.

The generation of European Geochemical Maps will therefore require careful consideration. The grid-size to be used in data interpolation should be determined by the scale at which the maps will be presented. Studies [7, 8] have demonstrated that is it possible to combine geochemical surveys of variable density in Finland (1 sample per 4 $km^2$) and Sweden (1 sample per 16 $km^2$) and in Brazil (average sampling density of 1 sample per 1 $km^2$). However, the variation in sample densities across Europe is considerable (1 sample per 0.01 - 1 sample per 3500 $km^2$) and in order to show both high and low density data in a representative manner it may be necessary to interpolate the raw data at two or more grid-sizes. This method of data presentation has been used successfully to combine datasets of differing resolution to demonstrate potentially harmful element concentrations in stream sediments in the UK [1].

**Table 5.** Surveys based on a sampling density < 1 sample per 100 $km^2$ in FOREGS countries.

| Country | Survey Type | Sampling Density 1 per x $km^2$ | Alternative survey for the same area | |
|---|---|---|---|---|
| | | | Survey Type | Sampling Density 1 per x $km^2$ |
| Estonia | Soil | 450-3500 | Soil | 16 |
| Finland | Stream sediment | 200 | Till | 4 |
| Norway | Stream sediment | 200 | Till | 40 |
| Portugal | Soil | 225 | None | |
| Portugal | Stream sediment | 225 | None | |
| Romania | Stream sediment | 2000 | None | |
| Romania | Surface water | 2000 | None | |
| Slovenia | Stream sediment | 200 | Soil | 25 |
| Slovenia | Soil | 180 | Soil | 25 |
| Slovenia | Surface water | 200 | None | |
| Sweden | Stream sediment | 200 | Till | 7, 15 |

## SAMPLE SIZE FRACTIONS

In Europe, the size fractions analysed for the different stream sediment surveys range from < 63 µm (BSI 240 mesh) in the Czech Republic, Romania and Slovenia to < 1000 µm (BSI 16 mesh) in Lithuania (Table 6). Most stream sediment surveys have, however, been based on the collection and analysis of < 177 to < 200 µm (BSI 80 to 76 mesh) fractions (Table 6). All the filtered surface water analyses carried out in the FOREGS region have been based on a filter size of 0.45 µm with the exception of Poland where a hard filter was used (Table 6). The range of grain-size fractions collected for soil surveys is bimodal. Some countries collect < 100 to < 180 μm (BSI 150 to 85 mesh) fractions to integrate with stream sediment surveys, while others follow traditional soil survey practice and use < 1000 or < 2000 μm (BSI 16 or 8 mesh) fractions (Table 6).

**Table 6.** Stream sediment, surface water and soil sample size fractions collected in FOREGS countries.

| | Water<br>Unfiltered<br>None | <br>Filtered<br>0.45 | 63 | 100 | 125 | Size Fraction<br>< μm<br>150 | 177 | 180 | 200 | 1000 | 2000 |
|---|---|---|---|---|---|---|---|---|---|---|---|
| Stream Sediment | | | Czech<br>Romania<br>Slovenia | Greenland<br>Hungary | France<br>Slovakia | Greenland<br>Ireland<br>UK | Austria<br>Greece<br>Spain<br>Turkey | Albania<br>Belgium<br>Finland<br>Norway<br>Portugal<br>Sweden | Germany<br>Poland<br>UK | Lithuania | |
| Surface Water | Albania<br>Czech<br>Finland<br>Germany<br>Greenland<br>Norway<br>Romania<br>Slovakia<br>Slovenia<br>UK | Finland<br>Germany<br>Norway<br>Slovakia<br>Slovenia<br>UK | | | | | | | | | |
| Soil | | | | Cyprus | France<br>Slovakia | UK | | Albania<br>Portugal | | Lithuania<br>Norway<br>Poland<br>Ukraine | Belgium<br>Czech<br>Estonia<br>Germany<br>Nethlnds<br>Slovenia<br>UK |

*IGCP 259 recommendations*

IGCP 259 recommendations for sample size fractions are similar for both GRN and national survey programmes. Collection of < 150 μm size fraction stream sediment samples is preferred whereas < 2000 μm regolith samples should be collected to conform to the International Standard Organisation (ISO) recommendations. The < 150 μm fraction in soils should also be collected whenever possible for comparison with stream sediment data. A filter size of 0.45 μm is recommended for filtered water surveys.

The majority of stream sediment surveys in Europe conform to the IGCP 259 recommendations with the exception of surveys in the Czech Republic, Lithuania, Romania and Slovenia (Table 6). Alternative geochemical data based on the collection of 1000 μm - 2000 μm soils are available in Lithuania and Slovenia but in the Czech Republic (0.6% of the FOREGS region) and Romania (1.4% of the FOREGS region) no additional datasets are available. Similarly, soil and till surveys based on the collection of 63-180 μm size fractions in Albania, Cyprus, Denmark, Finland, France, Norway, Portugal, Slovakia, Sweden and the UK (in total 11% of the FOREGS region) do not conform to the recommendations.

In these cases it may be possible to estimate the concentration in the coarse < 2000 μm fraction from the fine fraction concentrations. Tarvainen [13] has shown that the concentration of certain elements (trace elements in particular) in < 2000 μm Finnish till samples can be estimated from the concentration in the < 63 μm size fraction tills. Linear relationship functions can be calculated between the two datasets if the concentration in both size fractions is known for a proportion of each dataset. Analyses of both fine fraction and < 2000 μm fraction soil and till GRN samples will be required to provide the necessary overlap of information between datasets if these methods of data levelling are to be incorporated into the preparation of European Geochemical Maps .

## ANALYTICAL TECHNIQUES

A range of techniques have been employed to analyse geochemical samples in FOREGS countries, largely reflecting the years during which the survey was conducted. The main analytical methods available include XRF, ICP-AES, ICP-MS, DC-Arc ES, Flame AAS and NAA (Table 7). Surveys in Ireland, Italy and Luxembourg do not have the facilities to analyse regional geochemical samples and geochemical analyses are carried out in commercial and survey laboratories in other countries in these cases.

**Table 7.** Main analytical techniques employed by FOREGS countries

| XRF | DC-Arc ES | ICP-AES | ICP-MS | Flame AAS | NAA |
|---|---|---|---|---|---|
| Albania | Albania | Albania | Finland | Austria | Czech |
| Austria | Austria | Austria | Romania | Belgium | Finland |
| Belgium | Germany | Belgium | Slovenia | Cyprus | Greece |
| Czech | Greenland | Czech | UK | Czech | Greenland |
| Estonia | Lithuania | Finland | | Denmark | Ireland |
| Finland | Ukraine | France | | Estonia | Norway |
| Germany | UK | Germany | | Finland | Sweden |
| Greece | | Greece | | Germany | UK |
| Greenland | | Hungary | | Greece | |
| Lithuania | | Norway | | Greenland | |
| Luxembourg | | Poland | | Hungary | |
| Italy | | Portugal | | Ireland | |
| Netherlands | | Slovakia | | Italy | |
| Norway | | Slovenia | | Luxembourg | |
| Romania | | Spain | | Norway | |
| Sweden | | Sweden | | Spain | |
| UK | | UK | | Sweden | |
| | | | | Turkey | |
| | | | | UK | |

### *IGCP 259 recommendations*

The IGCP 259/360 recommendations for GRN samples are similar to those for national survey programmes. The determination of total concentrations based on XRF, NAA or total acid digestion (HF + $HNO_3$ + $HClO_4$) followed by ICP-AES and ICP-MS analysis are recommended. Determinations based on partial extraction methods are not encouraged as these are less reproducible than total methods [3, 15]. The majority of FOREGS countries employ analytical methods included in the IGCP 259/360 scheme with the exception of Bulgaria, Cyprus, Ireland, Italy, Turkey and Ukraine. It should therefore be possible to standardise geochemical techniques among the majority of countries. The analysis of GRN samples provides an excellent opportunity for standardisation and calibration between laboratories as each suite of GRN samples, for example GRN stream sediments or GRN soils, will be analysed by one laboratory. The GRN samples will also be analysed by each national laboratory allowing comparison between the standard laboratory results and the national laboratory. Standardisation of analytical methods will be enhanced by the inclusion of the international standard reference materials recommended by the IGCP 259/360 analytical committee (Canadian STSD and Chinese GSD and GSS standards) in GRN and national analytical programmes

[3].

Recent work in Brazil [8] has shown that is possible to integrate datasets analysed in different laboratories without inter-laboratory calibration provided the laboratories use the same form of analysis and employ rigorous quality control procedures. However, several European surveys such as those in Albania have been carried out without the analysis of international reference materials or documented quality control procedures (Table 8) and retrospective standardisation of datasets analysed by different laboratories may only be possible if sub-sets of archive material are analysed using the standardised IGCP 259/360 techniques.

The reanalysis of archive material may also be necessary to complete coverage for elements of environmental importance which are not routinely analysed by national sampling programmes. Iodine has been determined on only one water survey, for example, and only Greenland, Norway, Slovakia and the UK have data for Se concentrations in stream sediments, surface waters or soils [11].

**Table 8.** Analysis of international reference standard materials in FOREGS countries.

| International Standards Analysed | | No Response to Inventory |
|---|---|---|
| **Yes** | **No** | **Question** |
| Austria | Albania | Belgium |
| Cyprus | Germany (seds) | Denmark |
| Czech | Greece | Italy |
| Estonia | Poland | Portugal (some surveys) |
| Finland | Spain (some surveys) | Romania |
| France | Ukraine | Turkey |
| Germany (soil) | UK (some surveys) | |
| Greenland | | |
| Hungary | | |
| Ireland | | |
| Lithuania | | |
| Netherlands | | |
| Norway | | |
| Portugal | | |
| Slovakia | | |
| Slovenia | | |
| Spain | | |
| Sweden | | |
| UK | | |

Several elements such Al, Fe, Ca and Ni will be determined by more than one analytical method in the analytical structure proposed by the IGCP 259/360 analytical committee (Fig 4). Overlapping datasets generated by two or more analytical methods are essential when integrating data determined by different techniques. In the UK, for example, determinations of element concentrations in stream sediments by both XRF and DC-Arc ES techniques have been used successfully to 'level' older data based on DC-Arc ES methods and new XRF generated data to produce 'seamless' geochemical maps (T R Lister pers. commun.).

*Total and extractable analytical methods*

Both total and extractable analytical methods have been employed in FOREGS countries (Table 9). Methods of partial extraction vary between surveys making comparisons between total and extractable surveys difficult and it is unlikely that extractable surveys can be used to estimate the total element concentration in various sample types [11]. Investigations into the total and extractable element concentrations in till samples in Finland, for example, proved that it was not possible to estimate the total concentration on the basis of the extractable concentration [9,14]. Significant extractable stream sediment and soil survey datasets from Germany (3% of the FOREGS region ), Greece (0.7% of the FOREGS region) Hungary (0.2% of the FOREGS region) and Poland (4% of the FOREGS region) may be excluded from the preparation of European Geochemical Maps on this basis unless methods of integrating and levelling total and extractable datasets are investigated further (Table 9). Alternatively, the preparation of geochemical maps in these areas may require the reanalysis of sample archive material using standardised IGCP 259/360 analytical techniques.

| | |
|---|---|
| XRF: glass beads:<br>Si, Al, Fe, Ca, Mg, K, Na, Mn, Ti, P, LOI<br>pressed powder:<br>Cu, Pb, Zn, Cr, Ni, Co, Sr, Ba, Rb, Cs, Sc, Y, La, Zr, V, Nb, Th, Ce, S, As | Special elements:<br><br>INAA: (Au to 5 ppb)<br>Ag, As, Au, B, Ba, Br, Ca, Co, Cr, Cs, Fe, Hf, Ir, Mo, Na, Ni, Rb, Sb, Se, Sr, Ta, W, Zn, Sc, La, Ce, Nd, Sm, Eu, Tb, Yb, Lu, U, Th, Dy, Gd |
| DIGESTION:<br>total decomposition, high pressure,<br>HF- $HNO_3$ -$HClO_4$ | Hydride ICP-MS: Hg, Se, Te, Bi<br><br>IC + fusion: I, F, Cl, Br |
| ICP-AES:<br>Al, Ba, Be, Ca, (Cd), Co, Cr, Cu, Fe, K, La, Li, Mg, Mn, Mo, Na, Ni, P, (Pb), Sc, Sr, Ti, V, Y, Zn, (Zr) | DC-Arc ES: B<br><br>Fire-assay: Au, Pt, Pd |
| ICP-MS:<br>Ag, Ba, Cd, Pb, U, (Ta), Th, (Hf), Tl, Ce, Pr, Nd, Pm, Sm, Eu, (Yb) | Leco: C, N, S |

**Figure 4.** The IGCP 259/360 analytical sampling scheme. Modified after Vermeulen [15].

**Table 9.** Surveys based on extractable analytical methods in FOREGS countries.

| Country | Extractable Survey Type | Total Survey Type (for the same area) |
|---|---|---|
| Belgium | Soil | Overbank and stream sediment |
| Czech Republic | Soil | Stream and lake sediment |
| Finland | Till | Till |
| Germany | Soil | None |
| Germany | Stream sediment | None |
| Greece | Stream sediment | None |
| Hungary | Stream sediment | None |
| Norway | Till and stream sediment | Till |
| Poland | Soil and stream sediment | None |
| Poland | Soil and stream sediment | None |
| Portugal | Soil and stream sediment | None |
| Sweden | Till | None |

## CONCLUSIONS

Modern standardised, high-resolution, multi-element international geochemical databases, prepared to the standards agreed by IGCP 259/360 and the FOREGS

Geochemistry Task Group, are urgently required to inform policy-makers and provide a sound basis for environmental legislation and resource management.

The preparation of European Geochemical Maps will incorporate existing geochemical data, and the integration of existing datasets will be greatly aided by collection and analysis of the Global Reference Network (GRN) of samples recommended by IGCP 259/360.

Results of an inventory of geochemical data in FOREGS countries show that most countries in Europe conduct geochemical surveys based on the sample types recommended by IGCP 259/360. Countries that do not currently base their national programmes on stream sediment, stream water, soil and radiometric sampling as recommended by IGCP 259/360 should be encouraged to alter their programmes to include these sample media as soon as possible.

Most surveys carried out in FOREGS countries are based on sampling densities of 1 per 100 $km^2$ or more and conform to the recommendations of IGCP 259/360. For areas where information from lower-density surveys only is available, national programmes should be encouraged to increase the sampling density where possible. Careful consideration should be given to the preparation of European Geochemical Maps. In particular, the parameters involved in computerised map generation should be chosen to reflect the differing resolutions of national datasets (1 sample per 0.05-3500 $km^2$) and maintain detailed survey information wherever possible.

Many geochemical datasets available in Europe conform to the IGCP 259/360 recommended sample size fractions (< 150 μm for stream sediment; < 2000 μm for regolith and < 0.45 μm for filtered water). In areas where information is currently unavailable at the recommended size fraction, linear relationship functions may be applied to estimate element concentrations in the desired size fraction from the existing size fraction, providing data for both size fractions are collected during GRN sampling.

Most FOREGS countries employ analytical methods such as XRF, ICP-AES, ICP-MS and NAA, as recommended by IGCP 259/360 for determination of element concentrations in GRN samples and national survey samples. The collection and analysis of the GRN samples provides an excellent means of standardising laboratory methods across Europe. National surveys should also be encouraged to include the Canadian STSD and Chinese GSD and GSS reference materials recommended by IGCP 259/ 360 in their analytical programmes as soon as possible to aid standardisation between countries. Some reanalysis of sample archive material may be required in areas where no data suitable for inclusion in European Geochemical Baseline maps currently exists.

*Acknowledgements*

The FOREGS Geochemistry Task Group thank the FOREGS Directors, under the Chairmanship of Dr P McArdle (Republic of Ireland), for the opportunity to prepare the FOREGS inventory. The Directors of the Geological Surveys of Finland (Dr V Lappalainen), The Netherlands (Dr C Staudt), Slovakia (RNDr P Grecula), Spain (Dr C Caride) and the UK (Dr P J Cook) are especially thanked for making resources available. We also thank the Survey geochemists throughout the FOREGS countries for supplying the comprehensive data contained in the inventory. Dr H W Haslam is thanked for his constructive comments on this text.

This paper is published with permission of the Director of the British Geological Survey, NERC.

## REFERENCES


1. J.D. Appleton. *Potentially harmful elements from natural sources and mining areas: characteristics, extent and relevance to planning and development in Great Britain.* British Geological Survey Technical Report WP/95/3, Keyworth (1995).
2. British Geological Survey. *Regional Geochemistry of north-east England..* British Geological Survey, Keyworth (1996).
3. A.G. Darnley, A. Bjorklund, B. Bolviken, N. Gustavsson, P.V. Koval, J.A. Plant, A. Steenfelt, M. Tauchid and X. Xie. *A Global Geochemical Database for Environmental and Resource Management. Recommendations for International Geochemical Mapping.* Final Report of IGCP Project 259. Earth Sciences 19, UNESCO, Paris (1995).
4. D.M.A. Flight, R.A. Herd and T.R. Lister. Intercomparison of geochemical baseline data for soils. In: *Abstracts Environmental and Legislative uses of Geochemical Baseline Data for Sustainable Development.* British Geological Survey, Keyworth (1996).
5. F.M. Fordyce, P.M. Green and P.R. Simpson. Simulation of regional geochemical survey maps at variable sample density. *J. Geochem. Explor.* 49 161-175 (1993).
6. R.G. Garrett, R.M.P. Banville and S.W. Adcock. Regional geochemical data compilation and map presentation, Labrador, Canada. *J. Geochem. Explor.* 39 (1/2) 91-117 (1990).
7. N. Gustavsson, E. Lampio, B. Nilsson, G. Norblad, F.Ros and R. Salminen. Geochemical maps of Finland and Sweden. *J. Geochem. Explor.* 51 143-160 (1994).
8. O.A.B. Licht and T. Tarvainen. Multipurpose geochemical maps produced by integration of geochemical exploration datasets in the Parana Shield, Brazil. *J. Geochem. Explor.* 55 3 167-182 (1996).
9. P. Noras. Analytical aspects. In: *Handbook of Exploration Geochemistry. Vol. 5. Regolith Exploration Geochemistry in Arctic and Temperate Terrains.* K. Kauranne, R. Salminen and K. Eriksson (Eds.). pp 185-215. Elsevier, Amsterdam (1992).
10. J.A. Plant, M. Hale and J. Ridgway. Developments in regional geochemistry for mineral exploration. *Trans. Inst. Min. Metall.* 97 B (1988).
11. J.A. Plant, G. Klaver, J. Locutura, R. Salminen, K. Vrana and F.M. Fordyce. *Forum of European Geological Surveys Geochemistry Task Group 1994-1996 Report.* British Geological Survey Technical Report WP/95/14, Keyworth (1996).
12. J. Ridgway, J.D. Appleton and K.B. Greally. Variations in regional geochemical patterns, effects of site-selection and data-processing algorithms. *Trans. Inst. Min. Metall.* 100 B 122-130 (1991).
13. T. Tarvainen. The geochemical correlation between coarse and fine till fractions in southern Finland. *J. Geochem. Explor.* 54 3 187-198 (1995).
14. T. Tarvainen. *Environmental Applications of Geochemical Databases in Finland. Synopsis.* Academic Dissertation Geological Survey of Finland, Espoo (1996).
15. F. Vermeulen. *Minutes of the IGCP 360 Analytical Committee Meeting March 1996.* British Geological Survey, Keyworth. (1996).


The FOREGS Geochemistry Task Group 1994-1996 Report is available from Prof J.A. Plant, British Geological Survey, Keyworth, Nottingham, NG12 5GG, UK. Tel: + 44 (0)115 9363521 Fax: + 44 (0) 115 9363487 E-mail: j.plant@bgs.ac.uk.

The IGCP 259 Final Report 'A Global Geochemical Database' is available from UNESCO, Paris or Dr A.G. Darnley, Geological Survey of Canada, 601 Booth Street Ottawa, Ontario, K1A OE8, Canada. Tel: + 1 613 995 4521 Fax: + 1 613 996 3726 E-mail: darnley@gsc.emr.ca and Prof J.A. Plant, British Geological Survey, Keyworth, Nottingham, NG12 5GG, UK. Tel: + 44 (0)115 9363521 Fax: + 44 (0) 115 9363487 E-mail: j.plant@bgs.ac.uk.

*Proc 30th Int'l. Geol. Congr.*, Vol. 19, pp. 127-140
Xie Xuejin (Ed)
VSP 1997

# Wide-spaced Geochemical Mapping for Giant Ore Deposits in Concealed Terrains

WANG Xueqiu, LIU Dawen, CHENG Zhizhong and XIE Xuejing
Institute of Geophysical and Geochemical Exploration, Langfang, Hebei 102849, China

**Abstract**

Conventional geochemical methods are not suitable for or ineffective in geochemical mapping for concealed deposits hidden bellow the thick cover of transported overburden or thick sequence of post-ore sedimentary or vocanic rocks.
Recently, leaching of mobile forms of metals in overburden (MOMEO) has been developed and widely tested in geochemical mapping by using wide-spaced sampling in diverse concealed terrains. The case histories of pilot study show that large-scale geochemical patterns consistent with or favorable for large metallogenic belts can be delineated using wide-spaced MOMEO geochemical mapping in concealed terrains in different settings. The sequence of cover does not mask the mineralization indications. The methods have potential in strategic exploration for giant ore deposits in concealed terrains in the future.

## INTRODUCTION

China's national geochemical mapping program — Regional Geochemistry-National Reconnaissance (RGNR) project has been carried out 18 years and covered almost all the outcropping or thinly overburden areas approximately 5 millions $km^2$ mainly using stream sediment sampling [1]. Due to the implementation of this project hundreds of gold deposits have been discovered (Fig.1) within outcropping or thinly residual covered terrains [2]. However, no very large gold deposits have been discovered in China yet [Fig. 2]. It would defy geologic probabilities if there were no very large gold deposits in China's diverse terrains of slates, shales, greenstones and igneous rocks [3].
In order to find very large ore deposits in China, it is necessary to follow the lines: ( I ) more detailed survey within extensive and nested geochemical patterns delineated by RGNR project; ( II ) extension of the known deposits by searching for new buried or blind ore bodies; ( III ) delineation of new strategic targets for new large ore deposits in under-explored or unexplored heavily- concealed terrains.
In this paper the effort is made to development of low-cost geochemical methods for the (III). There are large diverse overburden terrains in China (Fig. 3) that still remain to be covered by geochemical mapping. However, conventional geochemical methods are not suitable for or have limited application in searching for deposits hidden under thick cover of transported overburden or thick sequence of post-ore vocanic or sedimentary rocks. Recently, leach of mobile forms of metals in overburden (MOMEO) have been developed

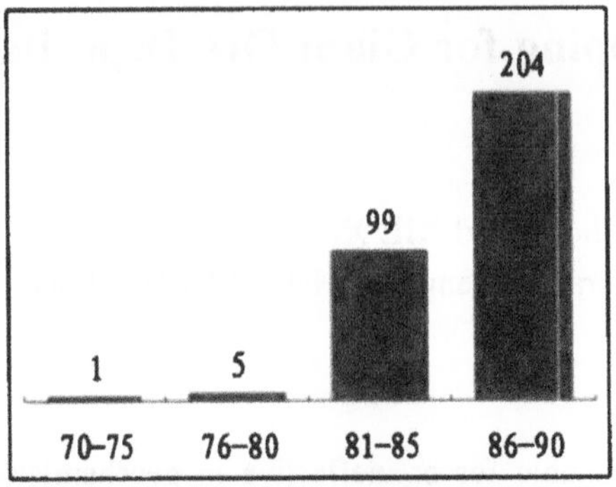


**Figure 1.** Number of new gold deposits discovered by geochemical methods in China during the period 1970-1980. Thanks to the commencement of the RGNR project in 1978, the new discoveries went up by leaps and bounds.

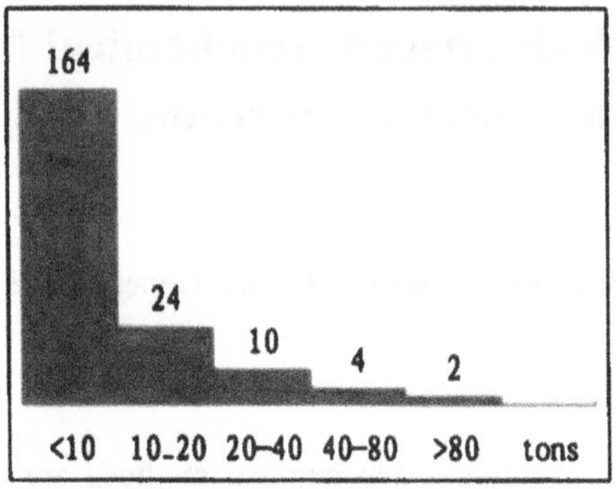


**Figure 2.** Tonnage distribution of new gold deposits. Only 2 deposits with gold reserves more than 80 tons and 4 more than 40 tons have been evaluated. No very large gold deposits have been yet discovered.

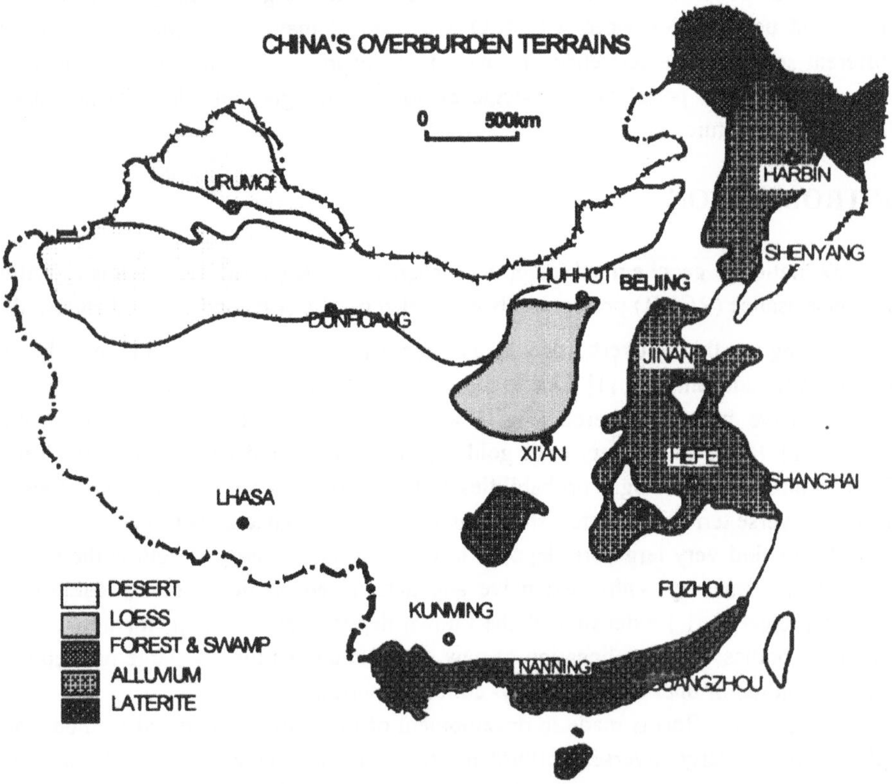


**Figure 3.** Diverse overburden terrains in China still remain to be covered by geochemical mapping.

and widely tested for regional geochemical mapping with wide-spaced sampling in diverse concealed terrains. Several examples are given in this paper.

## GEOCHEMICAL PROBLEMS AND MOMEO METHODS

### Geochemical Problems

In recent years the world-wide mineral exploration activities are concentrating on concealed deposits, particularly on giant ore deposits in concealed terrains. Woodall stated that World-class ore deposits still remain to be discovered by surface prospecting methods but the future belongs to those who can effectively and efficiently prospect concealed terrains [4]. However, development of low-cost and efficient methods is real challenge for regional exploration or location of targets in a large unexplored or under-explored terrains.

Figure 4 illustrated categories of ore deposits (after Erikson, 1973. E and F are added by authors) [5]. A is exposed deposits, the other five types of deposits are concealed either by a premineral or by post-mineral cover such as sedimentary rocks, volcanics and alluvium.

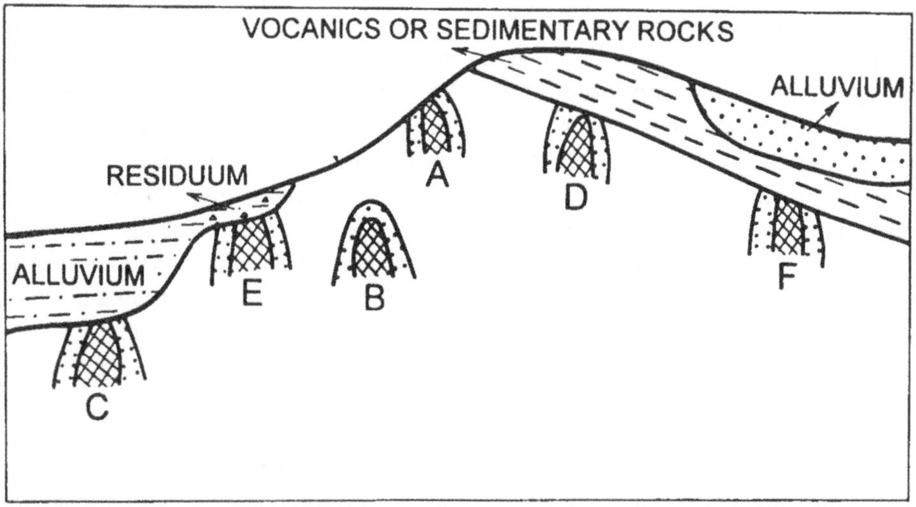


**Figure 4**. Sketch of six categories of ore deposits only one of which (A) is exposed and the other five deposits are concealed. There is an excellent chance of A and E being discovered by stream sediments, and the probability of finding B is greatly reduced by stream sediments. C, D and F are not likely to be detected by geochemical methods as now generally practiced. After Erikson (1973).

The outcropping deposit A is cut by a drainage system, there is an excellent chance of it being discovered during regional reconnaissance geochemical survey by means of the collection and analysis of stream sediments. The subcropping deposit E is covered by thinly veneered residuum. In the case of E, release of metals during physical or chemical weathering can lead to dispersion of mineralization elements in the residuum and the weathering products is easy to be transported to drainage basins and to give rise to a dectatable anomaly. Blind deposit B is concealed by premineral barren rock cover. The probability of finding this deposit is greatly reduced, but the leakage halos can be detected by the analysis of trace elements in stained pebbles or crude pan concentrates collected in the drainages. The buried deposit C is covered by thick transported overburden or deeply weathered cover, geochemical exploration techniques on either a regional or detailed basis, as presently practiced, offer little promise of discovery, there has been some success with the sampling of ground water, vegetation, basal caliche layers and vapors [6]. D and F are concealed by post-ore rocks and alluvium or deeply weathered cover. A thick sequence of variable overlying sediments will completely mask both underlying bedrocks and expressions of concealed deposits and geochemical exploration techniques presently available are not likely to be used. Thus types of C, D and F are not likely to be detected by geochemical methods as now generally practiced. The resource potential of concealed deposits is undoubtedly greater than that of the exposed deposits, and development of methods for their discovery is the real challenge for the present and future [6].

Dispersion of metals from concealed mineralization can produce superimposed anomalies in the overburden at or near surface. It is usually considered that trace elements are transported toward the surface by one or a combination of the following mechanisms: (i) flowing water, (ii) capillary action, (iii) ionic diffusion, (iv) self potential effect, (v) vaporization, (vi) absorption by roots of trees, (vii) as components of gases, and (viii) transportation by gases [7]. For the concealed deposits which are covered with a thick sequence of variable overlying post-mineral rocks and transported overburden, the transportation by gases is probably the most important as demonstrated by many examples [7, 8, 9, 10]. A possible model for formation of superimposed anomalies in areas covered with thick sediments would be as follows: The gasses liberated from the earth's interior would pick up metals from deep-seated ore deposits upward to the surface and trapped in soluble salts, clay minerals, manganese and iron oxide coatings, and in organic matter [8, 9, 10].

A tremendous endowment of materials is necessary for the formation of giant ore deposits, so they are often surrounded by a hierarchy of geochemical patterns from local, regional to provincial and even to domainal scales while minor deposits may be surrounded only with local or small regional geochemical anomalies at most [11]. Xie speculated that there is omnipresent earthgas generated from the earth's interior passing through the earth crust which will not only carry mobile forms of ore's and associated elements from a giant ore deposits but also from a hierarchy of primary geochemical patterns with high abundance of metals in the earth crust to the surface and build up large regional superimposed geochemical patterns at the surface [9]. By using wide-spaced sampling methods and ultra-sensitive analytical methods, such broad hierarchical nested patterns should be recognizable. We call such method the strategic or wide-spaced regional MOMEO method [9, 10].

## METHODS

### Mobile Forms of Metals in Overburden (MOMEO)

Mobile forms of metals in overburden can be present as : (1) ions, complex forms and ultrafine particles in water soluble salts and soil colloidal particles, (2) exchangeable and adsorbed ions and ultrafine particles on clay minerals and onto the surface of iron and manganese oxides, (3) trace constituents bound with organic matter, and (4) ions and ultrafine particles occluded in iron and manganese oxides.

Sequential leaching procedures used to release metals from natural materials and their applications to geochemical exploration have been presented by many authors [12, 13, 14]. The procedure we developed for releasing mobile forms of metals in overburden is different from the partial leach used by other researchers. In our methods, we not only attack the trapped materials such as oxides, clay minerals and organic matter for releasing ions or complex forms from them, but we also cause ultrafine particles of metals on these materials to be released into the solution. Based on our studies, mobile forms of metals occur not only as ions trapped in soluble salts, clay minerals and oxides but also as ultrafine particles (submicron to nanometer) adsorbed on these materials. These ultrafine particles of metals, particularly of gold, are mobile forms and play an important role in the dispersion from mineralization. These ultrafine particles of metals absorbed on their host materials are not dissolved by the solutions and colloidal silica will prevent them from being dissolved into the leach solution used to attack organic matter, oxides and clay minerals. Thus after the leach procedure, a special procedure are needed to treat the leach solution for analyses [15].

The sequential leach procedure used for releasing mobile forms of metals in overburden is as follows: (1) distilled water (ultrapure water) extraction for ions in water-soluble salts, (2) ammonium citrate tribasic solution for metals absorbed on clay minerals, (3) sodium pyrophosphate solution for metals bound with organic matter and (4) ammonium citrate tribasic and Hydroxylamine hydrochloride solution for metals absorbed by oxides.

Determinations were made by AAS for Au, Ag, Cu, Pb, Zn, Fe and Mn. As, Sb, Bi and Hg were determined by AFS ( Atomic Fluorescence Spectrometer ).

### Sampling

Use of low-cost surface geochemical mapping for the search for giant ore deposits in large under-explored or unexplored concealed terrains requires step-by-step reduction of the target size. Firstly, wide-spaced sampling (1 sample/100-1000 $km^2$) is required to outline geochemical provinces associated with mineralized belts or large scale geologic patterns in large areas at low cost and rapid coverage. Then, use of low-density surveys (1 sample/1-100 $km^2$) reduce the size of targets for specific regions likely to contain occurrences of large-scale mineralization. Finally, use of closer spacing detailed survey locate and pinpoint mineralization as exactly as possible prior to drilling.

Orientation studies were performed during the early stage of the projects to find effective sample media in different settings. A-horizon bottom soil samples in the alluvium areas and in grasslands with a developed soil profile and hard caliche layer samples in the desert areas are chosen as the primary media.
In this paper, several examples of wide-spaced geochemical mapping using MOMEO method at sampling density of one sample per 60-800 sq. km for outlining large-scale geochemical patterns associated with metallogenic provinces or mineralization belts in different concealed overburden terrains containing known giant ore deposits or favorable for large or giant ore deposits have been described.

## RESULTS

### Shandong and Anhui, alluvial terrain

In Shandong province a wide-spaced MOMEO survey was carried out which covered over 160 000 $km^2$ at a sampling density of one sample per 800 $km^2$. This study was designed to investigate the usefulness of the MOMEO survey with wide-spaced sampling as a procedure for identifying large-scale regional metal patterns or geochemical provinces associated with the known giant gold deposits. The soil samples were subjected to MOMEO processing and were analyzed for the 7 elements of Au, Ag, Cu, Pb, Zn, Fe and Mn by AAS. The results for Au are presented in Figure 5.
The distribution of MOMEO Au patterns clearly show four large regional anomalies with gold values of more than 2.0 ppb have been delineated in the whole province. Among them, the largest one approximately 6 000 $km^2$ in the northeastern region is not only consistent with the distribution of the known large and giant gold deposits but also extends further in a southern direction into the transported overburden areas. This will facilitate the search for new concealed deposits. The southern anomaly of more than 2.0 ppb has been delineated around the known large gold deposit in the overburden areas. The RGNR project using stream sediment surveys, on the other hand, failed to give response in these two overburden regions. The other two regional anomalies in the NW and SW concealed regions are unknown and still remain to need further verification. The geochemical province delineated by the content contour of 1.2 ppb is distributed along the Tanlu deep faults (Fig. 8). The other elements such as Fe, Mn and Ag revealed similarly high background concentration patterns surrounding gold metallogenic zones and extending into covered areas along the Tanlu deep faults.
After the survey in Shandong province we found that the geochemical province is not enclosed off and extends into Anhui province along the Tanlu deep fault. Thus a project was specifically designed to trace down the extension of Shandong geochemical province along the Tanlu deep faults and try to find new anomalies in alluvium plain. A wide-spaced MOMEO survey in the northern Anhui province which covered over 90 000 $km^2$ at a sampling density of one sample per 400-800 $km^2$ have been conducted. The soil samples were analyzed for 7 elements by AAS.

The results of MOMEO Au and drainage Au are presented in Figure 6. It can be seen that the distribution pattern of MOMEO Au anomalies are not only consistent with the drainage Au anomalies at outcropping areas but also distribute in overburden areas where conventional geochemical methods can not be used to conduct geochemical mapping.

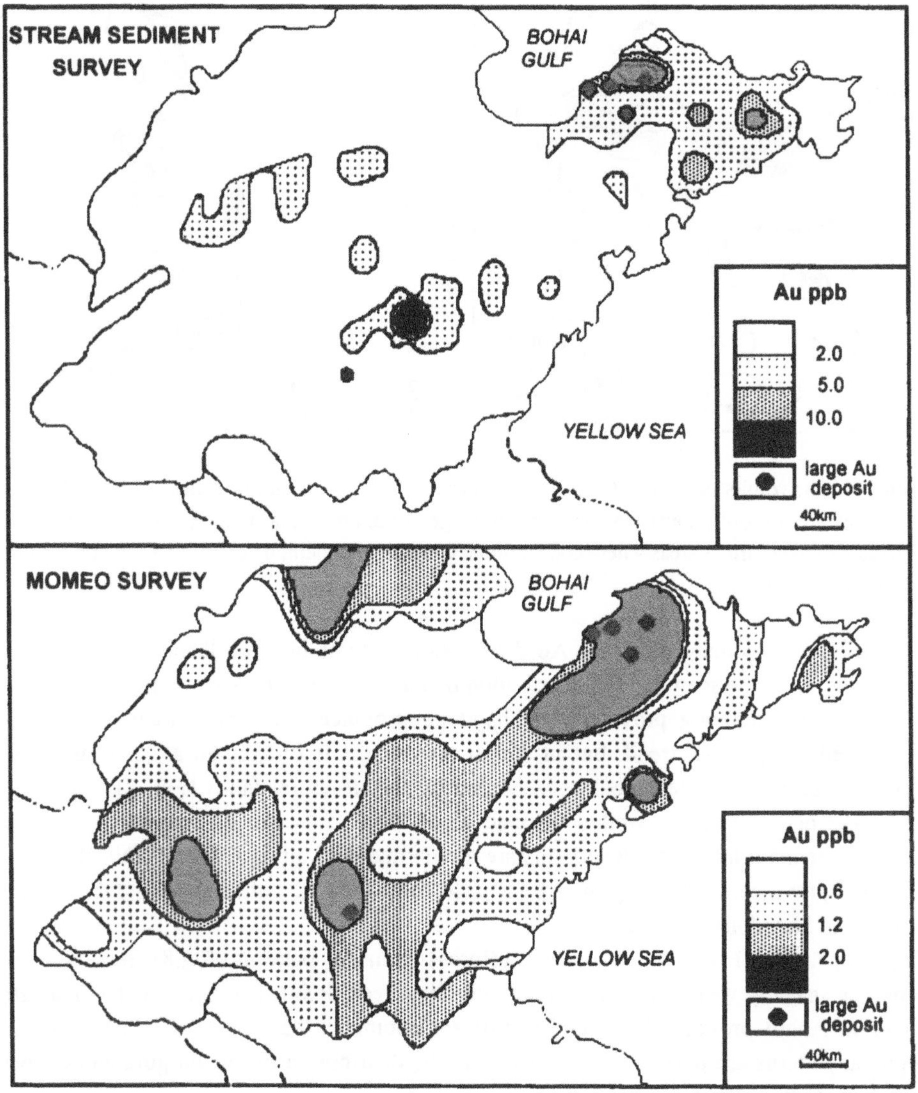


**Figure 5.** Comparison of gold distribution patterns of stream sediment and MOMEO surveys. The distribution of MOMEO Au patterns not only is consistent with the distribution of the known large and giant gold deposits but also extends far into the transported overburden areas, whereas the stream sediment surveys failed to give any response.

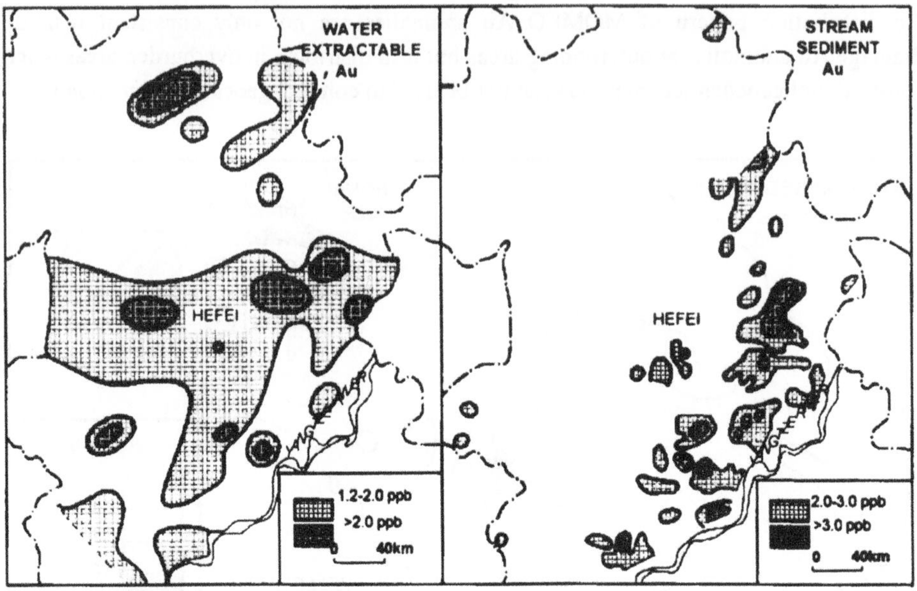


**Figure 6.** The results of MOMEO Au and drainage Au. The distribution pattern of MOMEO Au anomalies are not only consistent with the drainage Au anomalies at outcropping areas but also distribute in overburden areas where conventional geochemical methods can not be used to conduct geochemical mapping.

Figure 7 shows water-extractable Au distribution patterns both in Shandong and Anhui provinces. It can be seen that the distribution pattern of MOMEO Au in Anhui is different from that in Shandong province. In Shandong province the geochemical province is distributed along NNE trend, but in Anhui province it is distributed along nearly EW trend. These patterns are probably controlled by the structural faults (Fig. 8). The main large and deep faults are well developed along NNE direction in Shandong province whereas the main large and deep faults are distributed both along EW and NNE direction in Anhui province. In Anhui province the gold anomalies are not so large and significant as those in Shandong province.

Copper patterns (Fig.9) are also very different from Au patterns (Fig.8). In Shandong province sparsely weak and small anomalies have been delineated, but near the Yangzte River in Anhui province large and significant anomalies have been delineated. These regional patterns are perfectly consistent with the distribution of known gold and copper metallogenic belts both in Shandong and Anhui.

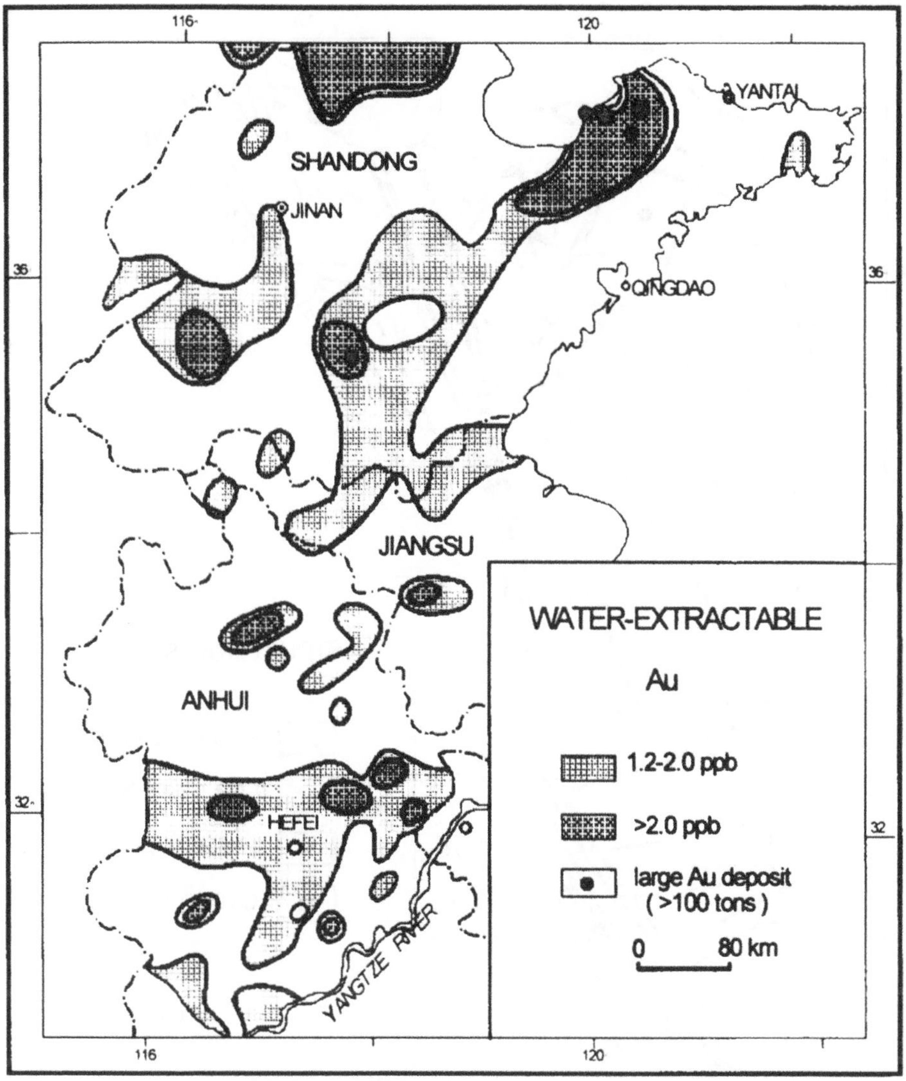


**Figure 7.** Water-extractable Au distribution patterns both in Shandong and Anhui. The distribution pattern of MOMEO Au is different from the pattern in Shandong province. In Shandong province the large regional geochemical anomalies distributed along NNE direction, but in Anhui province they distributed along nearly EW direction. These patterns are probably controlled by the structural faults.

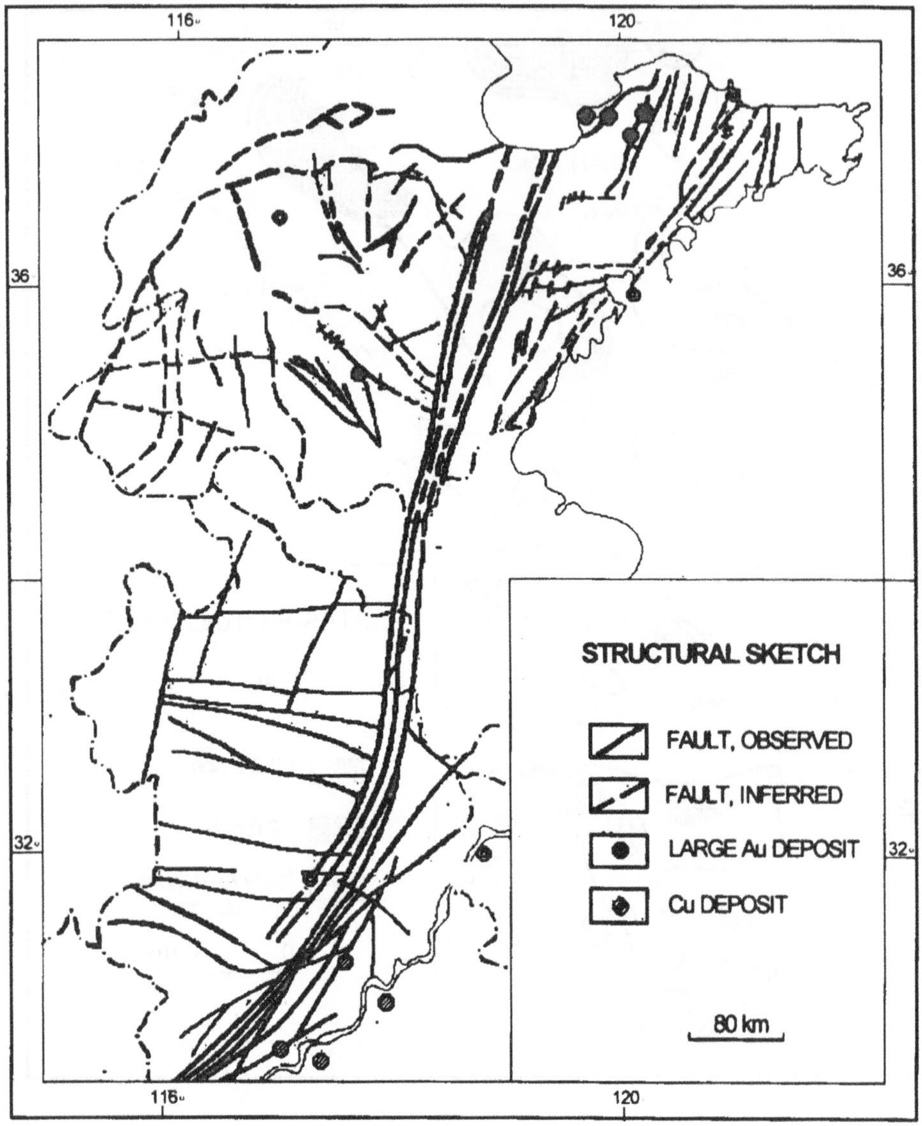


**Figure 8.** Sketch of structures in Shandong and Anhui provinces. The main large and deep faults are well developed along NNE direction in Shandong province whereas the main large and deep faults are distributed both along EW and NNE direction in Anhui province.

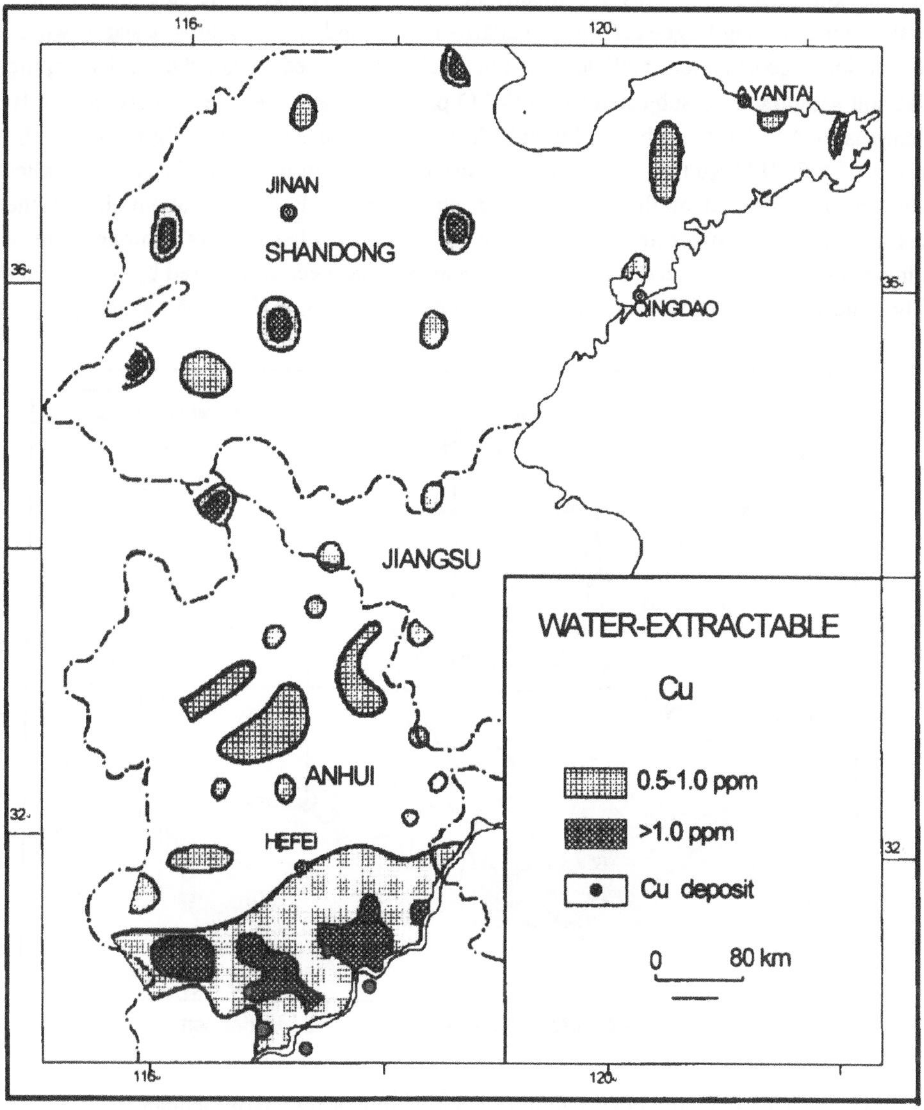


**Figure 9.** Copper patterns in two provinces of Shandong and Anhui. Copper patterns are very different from Au patterns. In Shandong province the Cu anomalies are sparsely weak and small, but in Anhui province Cu anomalies are large and significant.

**Northwestern Sichuan, grasslands**

A wide-spaced MOMEO survey was carried out in swamp grasslands of northwestern Sichuan province which covered over 30 000 $km^2$ at a sampling density of one sample per 300 $km^2$. This study was specifically designed to investigate the suitability of this method for outlining large-scale metal patterns in high and cold swamp grasslands where conventional geochemical methods are not likely to be used in geochemical mapping. The soil samples were subjected to MOMEO processing and then were analyzed for 10 elements of Au, Ag, Cu, Pb, Zn, Fe and Mn by AAS, and As, Sb and Hg by AFS. The results of MOMEO Au (organically bound Au) are presented in figure 10. The anomalies have been delineated at the known large gold deposit. Four new anomalies at the unknown areas are suggested as targets for follow-up detailed surveys. Among them, a mineralization at the southwestern anomaly near Aba has been discovered by preliminary follow-up work and a gold deposit has been found by drilling at Hongyuan anomaly.

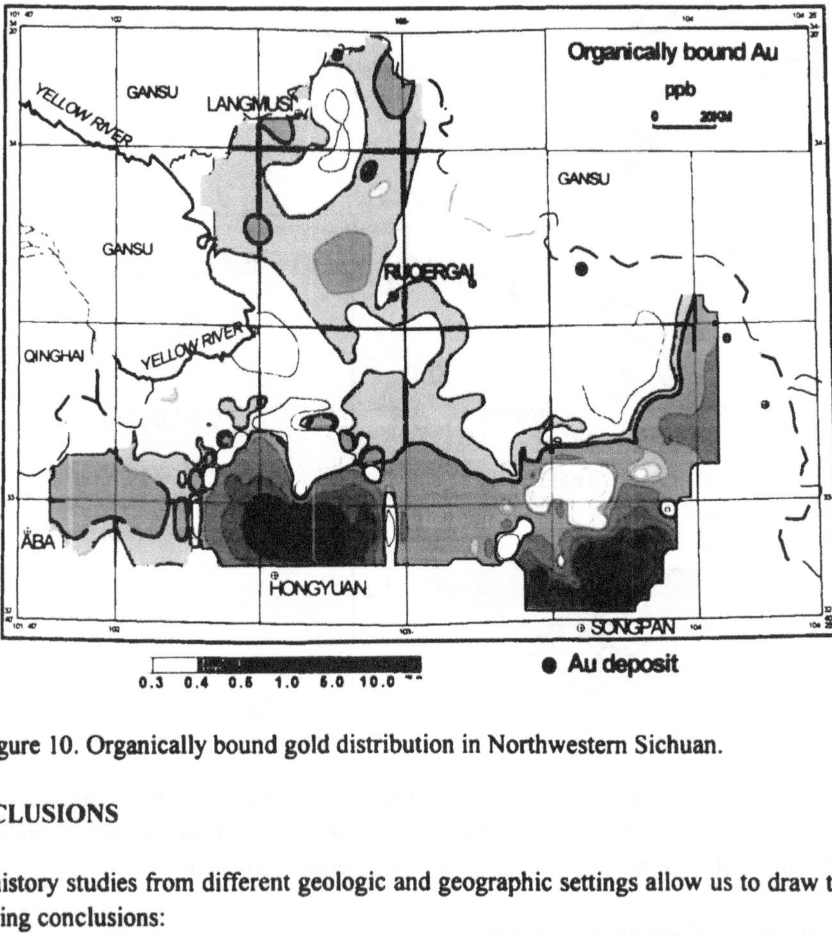


Figure 10. Organically bound gold distribution in Northwestern Sichuan.

## CONCLUSIONS

Case history studies from different geologic and geographic settings allow us to draw the following conclusions:

Giant ore deposits can lead to large-scale expressions of mobile forms of metals in surface overburden. The large-scale geochemical patterns consistent with or favorable for large metallogenic belts can be delineated by using wide-spaced MOMEO geochemical mapping in overburden terrains. The mobile forms of metals in overburden can give a significant response to mineralization covered by sequence of sediments and the cover can not mask the mineralization indicators. This method have potential for wide-spaced geochemical mapping in large overburden terrains.

**Acknowledgments**

We thank the State Science and Technology Commission of China, the Bureau of Geological Survey, the Geological Bureau of Anhui province and the Geological Bureau of Sichuan province for the funding in the support of this research project. Thanks are given to Professor Lu Yinxiu for his laboratory work. Thanks also are due to the many geologists and geochemists for their support and help with us, especially Shao Yue, Sheng Zhonglie, Chu Guozheng, Zhang Qianmin, Zhang Yanying, Ji Zhongming and Li Chengtao.

## REFERENCES

1. Xie Xuejing, Mu Xuzan and Ren Tianxiang. Eighteen years of China's geochemical mapping program. The Abstract of 30th International Geological Congress, Beijing, China, 3, 42 (1996)
2. Ministry of Geology and Mineral Resources, P. R. China. Statistical report on the new discovery of metal deposits by the Ministry during the Seventh Five-Year Plan (1985-1990) in China. (unpublished report, in Chinese) (1991).
3. Wang Xueqiu and Xie Xuejing, 1996. Use of NAMEG and MOMEO methods in the search for concealed deposits. In: SME'96, Phoenix, America.
4. Woodall, R., 1993. The multidisciplinary team approach to successful mineral exploration. Keynote address at the Society of Economic Geologists International Conference.
5. Erikson, R.L.. Presidential address given at the annual general meeting of the Association Geochemists. Vancouver, Newsletter No.9 (1973).
6. Levinson, A.A.. Introduction to exploration geochemistry (second edition). Applied Publishing Ltd., Wilmette, Illinois, U.S.A. (1980).
7. L. Malmqvist and K. Kristiansson. Experimental evidence for an ascending microflow of geogas in the ground. Earth and Planetary Letters, 70, 407-416 (1984).
8. Wang Xueqiu, Xie Xuejing and Ye Shengyong, 1995. Concepts for gold exploration based on the abundance and distribution of ultrafine gold. J. Geochem. Explor. 55 (1-3 ): 93-102.
9. Xie Xuejing. Surfacial and superimposed geochemical exploration for giant ore deposits. Clark, A.H. (editor), Giant ore deposits II, 475-485. Kingston, Canada (1995).
10. Wang Xueqiu, Cheng Zhizhong, Liu Dawen, Xu Li and Xie Xuejing, 1996. Nanoscale metals in earthgas and mobile forms of metals in overburden in wide-spaced regional

exploration for giant ore deposits in overburden terrains. J. Geochem. Explor. (to be published).

11. Xie Xuejing and Yin Bingchuan. Geochemical patterns from local to global. J. Geochem. Explor., 47,109-129 (1993).
12. S.J. Hoffman and W.K. Fletcher. Selective sequential extraction of Cu, Zn, Fe, Mn and Mo from soils and sediments. Watterson, J.R. and Theobald, P.K. (editors), Geochemical Exploration 1978, Rexdale, Ontario, 289-299 (1979).
13. T.T. Chao. Use of partial dissolution techniques in geochemical exploration. J. Geochem. Explor., 20, 101-135 (1984).
14. L.V. Antropova, I.S. Goldberg, N.A. Voroshilov and Ju.S. Ryss. New methods of regional exploration for blind mineralization: application in the USSR. J. Geochem. Explor., 43, 157-166 (1992).

15 Wang Xueqiu, 1997. Leaching of mobile forms of metals in overburden: development and applications. (to be published).

*Proc. 30th Int'l. Geol. Congr.*, Vol. 19, pp. 141-149
Xie Xuejin (Ed)


# New Developments in Regional Stream sediment Geochemical Survey for Platinum-Group Elements

Zhang Hong[1] Chen Fanglun[2] D.de Bruin[3]

1. *Hebei College of Geology, Shijiazhuang, Hebei, 050031, P.R.China*
2. *Henan Central Laboratory of MGMR, Zhengzhou, Henan, 450053, P.R.China*
3. *Council for Geoscience of SA, Private Bag X112, Pretoria, South Africa.*

## Abstract

A new geochemical exploration method is demonstrated for the exploration of platinum-group elements (PGE) in which platinum and palladium are used as direct indicator elements. The Pt, Pd and Au content in 10g geochemical survey samples are determined by emission spectrometry after a pre-concentration procedure and detection limits of 0.2 ppb for Pt and 0.1 ppb for Pd and Au are achieved. In this paper focus is placed on the method and techniques employed and positive results using regional stream sediment surveys for PGE's in the main PGE metallogenic provinces in China and the Bushveld Complex of South Africa are illustrated.

**Keys:** Platinum Group Elements(PGE), Geochemical Stream Sediment Survey, Geochemical mapping.

## INTRODUCTION

A study dealing with the geochemical exploration of the platinum group elements (PGE) has been conducted in China for a number of years. The objective is to develop an adaptable geochemical exploration tool that can be used to explore for PGE deposits under varied surface conditions. The platinum and palladium contents in samples were taken as the direct indicator elements for the search of PGE ores in this survey.

Due to the generally very low abundance of PGE in geochemical exploration samples, an essential part of this study focused on the development of the analytical technique in order to increase the sensitivity for the determination of Pt, Pd and Au. Use was made of a chemical spectrometric analytical method for the simultaneous analysis of ultra-trace level concentrations of the elements Pt, Pd and Au in which detection limits of $0.2\times10^{-9}$for Pt and $0.1\times10^{-9}$ for Pd and Au were achieved. Generally if was found that the analytical precision (RSD) of Pt and Pd to be better than that of Au. No systematic deviations were found between the method used and results from fire assay.

The use of the direct determination of PGE's in geochemical surveys and geochemical mapping of Pt, Pd and Au in China has lead to the delineation of new PGE metallogenic domain.

The regional geochemical exploration of Pt, Pd and Au in the western part of the Bushveld Complex in South Africa have led to the discovery of a previously unknown platinum-mineralized horizon in the Bushveld Complex.

**1. Sampling method**

Stream sediment samples were taken at a sampling density of 1-4 samples/km$^2$. Samples were sieved on site to pass 60 mesh sieve and pulverized in the laboratory to 200 mesh fineness for analysis.

**2. Analytical method**

The AC arc optical emission spectrometry method for the simultaneous determination of ultra-trace levels of Pt, Pd and Au were developed by Chen Fanglun from the Henan Central Laboratory, China[7].

The Pt, Pd and Au contents of geochemical samples are determined in the following way. A 10g sample is digested in a $HCl/H_2O_2$ solution, filtered and the solution is passed through a column of activated charcoal. The addition of $SnCl_2$ to the solution leads to the formation of complex ions of Pt, Pd and Au. In the case of Pt the following complexes are formed:

$$[PtCl_6]^{2-}+SnCl_2=PtCl_2+SnCl_4+2Cl^-$$

$$PtCl_2+5SnCl_2+3Cl^-=[Pt(SnCl_3)_5]^{3-}$$

These complex ions of Pt, Pd and Au are adsorbed by activated charcoal in an acid medium and their recovery rate has been determined to be in excess of 95%. Due to the very low ash content of the processed activated charcoal, the recovered mass obtained after separation and pre-concentration is in the order of 0.3mg. This results in a concentration ratio of 30000 times for Pt, Pd and Au which allows for the direct measurement of the residue by emission spectrograph. A detection limit of 2ng for Pt and 1ng for Pd and Au could be reached. The effective detection limits of 0.2ppb for Pt and 0.1ppb for Pd and Au are achieved in the sample.

The residue after ashing forms a consistent matrix which eliminates matrix effects often presented the cause for inconsistent results by the direct spectrographic analysis on silicate materials. A RSD of 12-32% precision could be achieved at 2-50ppb Pt, Pd and

Au concentration levels.(Table 1). No systematic deviation has been found between results obtained by our method and by fire assay (Table 2).It was however found that samples containing high sulphide, arsenide and organic matter requires roasting at 650 ℃ prior to dissolution to prevent low analytical results.

Table 1 Precision of the method (data obtained by using standard reference materials)

| SRM* | Elements | No of replocate analysis | Average values | RSD% |
|---|---|---|---|---|
| | Pt | | 2.0 | 20.8 |
| GPt2 | Pd | 35 | 2.3 | 17.6 |
| | Au | | 10.9 | 16.0 |
| | Pt | | 5.5 | 14.8 |
| GPt3 | Pd | 33 | 4.9 | 18.5 |
| | Au | | 1.1 | 28.4 |
| | Pt | | 56.2 | 12.8 |
| GPt4 | Pd | 35 | 54.7 | 15.5 |
| | Au | | 4.0 | 32.6 |

* GPt series standard reference samples for precious metals were prepared by Yan Mingcai, Wang Chunshu, Gu Tiexin, Chi Qinghua (unpublished report).

## 3. Nugget effect

Platinum, palladium and gold in rocks, soils and stream sediments consist of discrete particulate plus various microscopic and submicroscopic forms of occurrences. The problem of nugget effect leading to poor reproducibility of gold analysis has long been an annoyance to geochemists dealing with gold exploration.. Very large samples (200-800gm ) should be taken for analysis in order to improve the analytical reproducibility[3]. Xie and Wang[4] found that gold concentration higher than 2ppb will be meaningful for gold exploration and gold with concentration in ppb range will be largely in microscopic or submicroscipic forms. They made suggestion to by-pass the problem of nugget effect by lowering down the detection limit of gold analysis to 0.2ppb, defining favorable regions by a threshold of 2-4ppb, place emphasis on the reproducibility of the location, shape and dimension of the gold anomaly as a whole, attach less importance to individual erratic Au values in either regional of follow-up surveys. Little information on nugget effect in Pt and Pd analysis is available. In our work we are following the approach of Xie and Wang by lowering down the detection limits of Pt and Pd to 0.2 and 0.1 ppb respectively. Only 10gm of sample is taken for analysis. This approach led to very satisfactory results in China and in the Busheld Complex in South Africa.

Table 2 Comparison of the analytical results with fire-Assay method (data obtained by using standard reference materials)

| SRM | Analytical method | Pt | Pd | Au |
|---|---|---|---|---|
| | This method | 0.21 | 0.19 | 0.96 |
| GPt1 | Five Assay | 0.20 | 0.18 | 0.92 |
| | Certified Value | 0.26 | 0.26 | 0.90 |
| | This method | 1.71 | 2.27 | 10.2 |
| GPt2 | Five Assay | 1.78 | 2.38 | 9.6 |
| | Certified Value | 1.6 | 2.3 | 10.0 |
| | This method | .8 | 5.0 | 1.0 |
| GPt3 | Five Assay | 6.0 | 5.6 | 0.92 |
| | Certified Value | 6.4 | 4.6 | 1.1 |
| | This method | 58 | 60 | 4.0 |
| GPt4 | Five Assay | 60 | 58 | 4.5 |
| | Certified Value | 58 | 60 | 4.3 |

## APPLICATIONS

Geochemical surveys for Pt, Pt and Au has been carried out on several known PGE mineralization areas in China in the past years. Some recent examples are given in this paper.

In 1995, two projects of geochemical surveys for Pt, Pd and Au were carried in Hongshila and Gaositai of Hebei Province. Stream sediment samples were collected with a sampling density of 4 samples/km$^2$.

The Hongshila PGE hydrothermal mineralization area can clearly be identified through strong geochemical anomalies of Pt and Pd accompanied by a Au anomaly (Fig 1) while the distribution of Cr, Co, Ni, Cu, V, Mn, Ti, As, Sb, Hg, Ag, Pb shows no indication of mineralization. These findings are consistent with the properties of the Hongshila PGE deposit which is characterized by simple materiel composition and the absence of base metal sulphide and chromate mineralization.

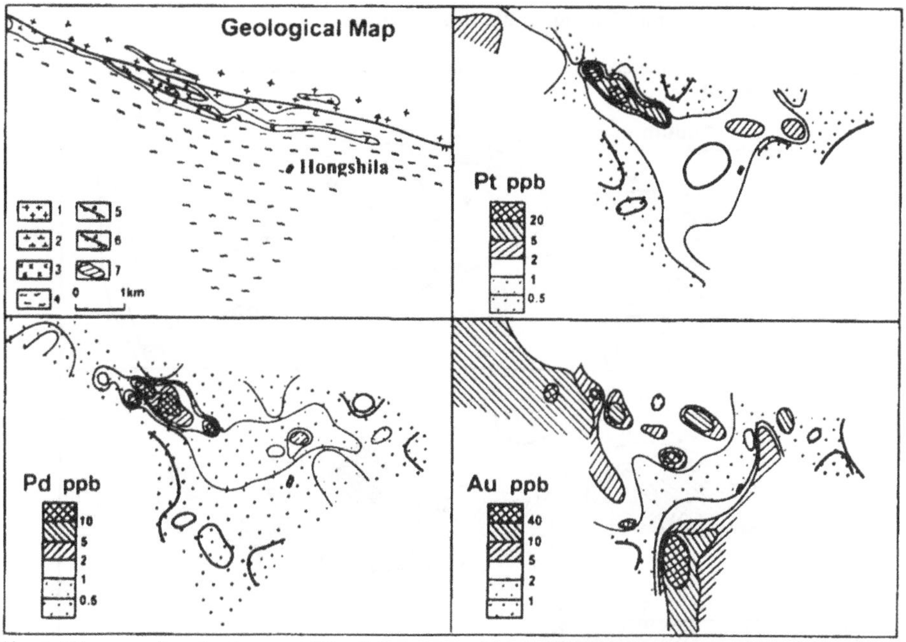


Fig 1. The geochemical map of Pt, Pd and Au, Hongshila PGE ore area, Hebei Province, China

1. Granite; 2. Diorite; 3. Ultra-basic complex; 4. Archeonzoic gneiss; 5. Deep fault zone; 6. Compression fault; 7. PGE ore body.

At the Gaositai PGE bearing chromate deposit, the geochemical anomalies are related to the mineralized chromate and distinct anomalies of Cr, Ni, Co and Pt and weaker anomalies for Ir and Os are developed. The Pd content occurs at background levels. The Pt/Pd ratio is therefore very high (usually in the order of tenth with a maximum in the hundreds). This is obviously different from the anomaly of sulphide-rich PGE mineralization. In the pyroxenite facies in Gaositai Complex a new Pd and Pt anomaly belts have been found (Fig 2).

During 1993-1994 extremely-low density geochemical mapping using floodplain sediment as a sampling medium was carried out in China[5] . Five hundred thirty three samples were collected to cover the whole China's territory each representing about ten to forty thousand $km^2$ drainage area and 48 elements were determined (Cheng et al., this volume ). These samples were sent to us to determine the Pt and Pd contents. Very large Pt and Pd geochemical provinces and a geochemical domain (megaprovince) [6] have been delineated in Sichuan, Yunnan, Guizhou, Guangxi border region , southern Tibet and Gansu, Qinghai, Xinjiang border which are worthy of further study in order to find large

Pt and Pd deposits in China ( Cheng, et al., this volume).

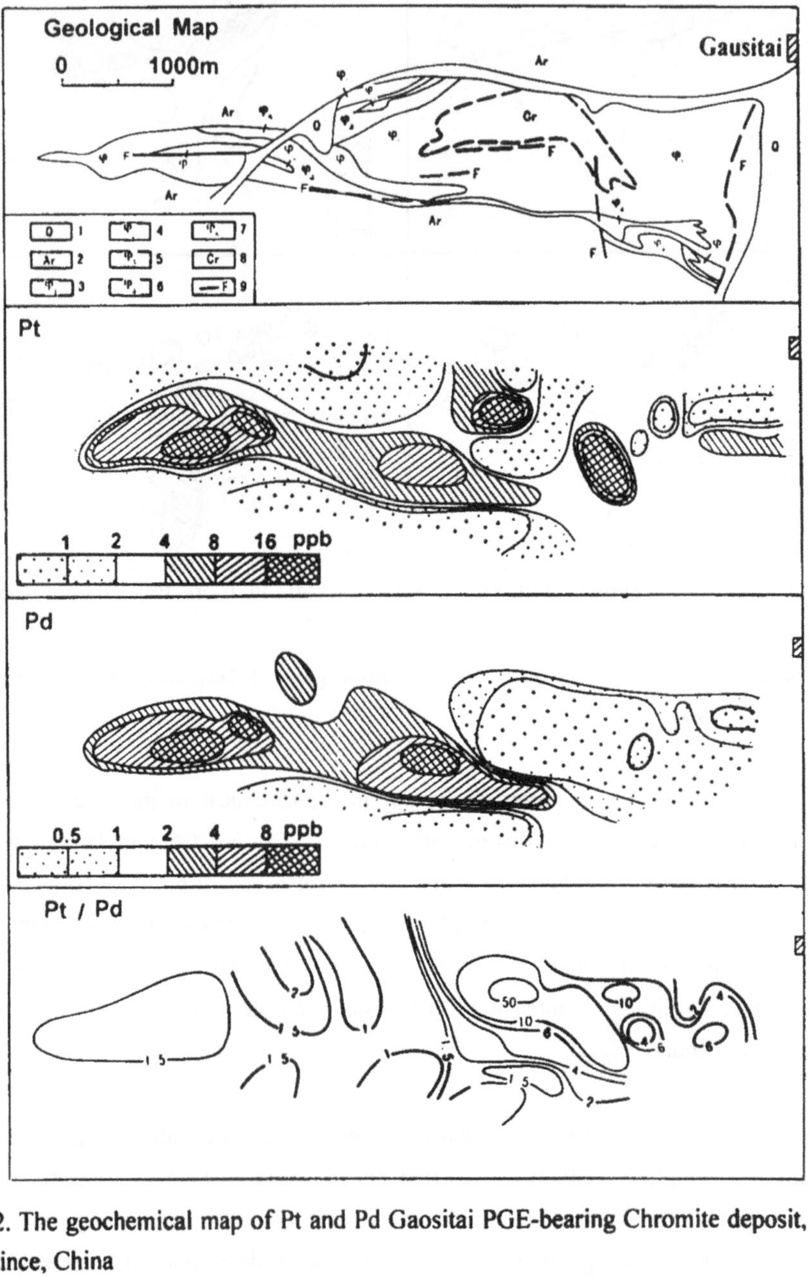


Fig 2. The geochemical map of Pt and Pd Gaositai PGE-bearing Chromite deposit, Hebei Province, China

1. Quaternary; 2. Archeon; 3. Dunite; 4. Pyrolite; 5. Olivine diopsidite; 6. Diopsidite; 7. Pyroxene hornblendite; 8. Chromite mineralization zone; 9. Fault zone.

In 1994 a cooperative study was undertaken by the council for Geoscience of South

Africa , Hebei College of Geology and Henan Central laboratory of MGMR of China. Stream sediment samples were collected at a sampling density of 1 sample per 1km$^2$ in Brits area on the western part of Bushveld Complex. Approximately 1000 samples were collected. The samples were analyzed in South Africa for As, Ba, Co, Cr, Cu, Fe, Mn, Mo, Nb, Ni, Pb, Rb, Sb, Sc, Sn, Sr, Th, Ti, U, V, W, Y, Zr, Zn by x-ray fluorescence spectrometry, and analyzed in Henan Central Laboratory of China for Pt, Pd, Au by this method. Very strong anomalies of Pt, Pd , Cr and Ni were found in Merensky and Chromite horizon in coincidence with the known Pt mineralization (Fig. 3). A new anomalous zone of Pt, Pd associated with Au and Cu, was found along the border of the "Upper zone" and "Main zone", which shows great potential for Pt and Pd mineralization completely unknown in the past. The newly discovered Pt and Pd anomalies and ratio Pt/Pd are similar to the known PGE mineralized belt anomalies. However, there are no accompanied by associated anomalies of Cr, Co and Ni etc. in the newly discovered anomalous zone and the MgO content is lower, the FeO content is higher. These characteristics are obviously different from those of the know ore belts. Pit Botha, Minister of Mineral and Energy Affairs of South Africa said that the find could increase South Africa's platinum group metal reserves by 30%, but industry officials remain cautious over the discovery . They suggest that further information regarding grades and the nature of the ore would need to be determined before any conclusion can be drawn about the viability of the resource[2].

## CONCLUSIONS

This is the first attempt to use Pt and Pd extensively as direct indicator for searching new Pt and Pd deposits. Highly sensitive emission spectrographic method with pre concentration procedure was developed for the determination of Pt , Pd and Au with detection limits of 0.2, 0.1and 0.1ppb respectively. The discovery of new and significant pt and Pd anomalies in Brits area, South Africa presents an illuminating example to show the great potential of this method in searching new Pt and Pd deposits in the future.

## ACKNOWLEDGMENTS

The authors thank Prof. Xie Xuejing, Member of the Chinese Academy of Sciences for his guide to this study and the supply of the samples of floodplain sediment. Dr. Frick, the director of the Council for Geoscience of South Africa is also greatly appreciated for his great support to this cooperation study. Geochemists of Geological Survey of South Africa: Mrs. H.J. Wilhelm, J.H.Elsenbroek and Ms M.Lombard, are thanked for providing the geochemical samples and geological information of the Bushveld Complex.

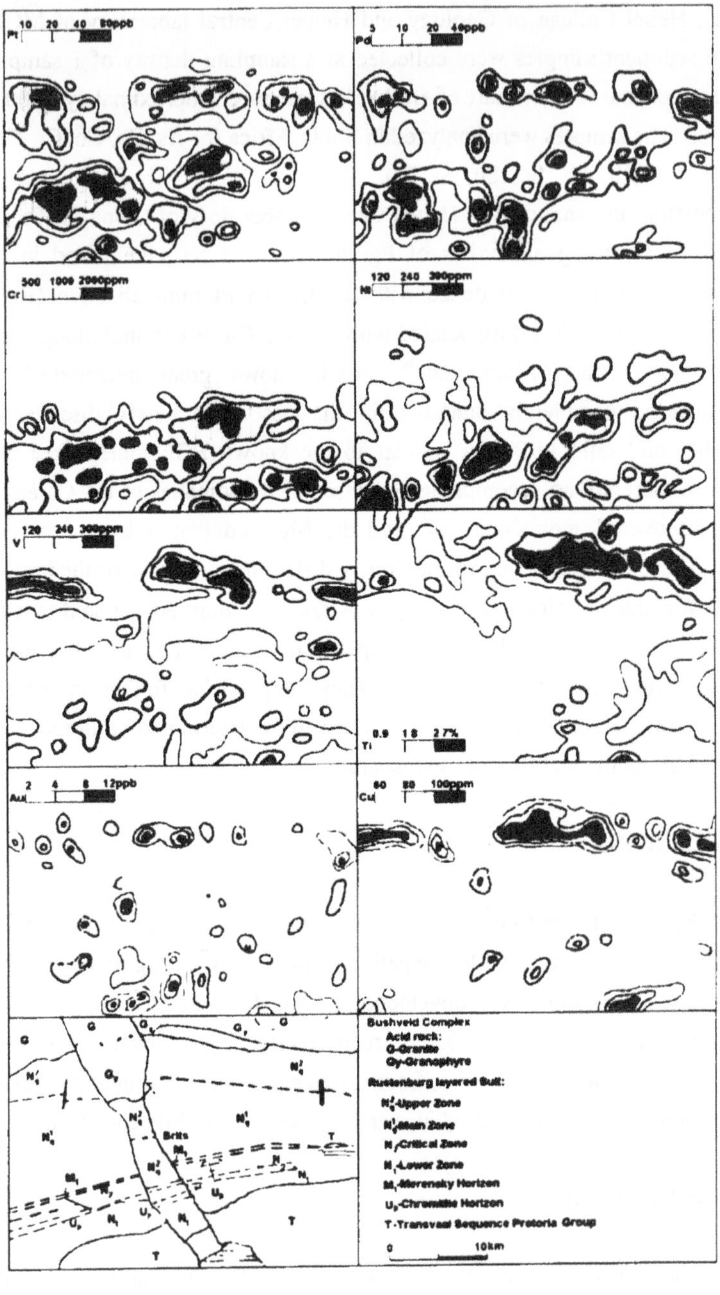


Fig. 3 The geochemical anomaly map, Brits area, South Africe
(Geology based on Eriksson et al.)

## REFERENCES


(1) Eriksson, P.G. Jattingh P.J. and Alterman, W., (1995), An overview the geology of the Transvaal Sequence and Bushveld Complex, South Africa. Min. Dep. 30.

(2) Jones, B., (1995), South Africa platinum find. Mining Journal , vol.324 N.8309, P.13.

(3) I.Nichol. Geochemical exploration of gold, a special problem In: I. Thornton and R.J. Howarth (Eds), Applied Geochemistry in the 1980's. Groham and Trotman, London. 60-85 (1983).

(4) Xie Xuejing and Wang Xueqiu. Geochemical exploration for gold: a new approach to an old problem. J. Geochem. Explor. 40, 25-48 (1989).

(5) Xie Xuejing and Cheng Hangxin. The suitability of floodplain sediment as a global sampling medium: evidence from China. J. Geochem. Explor. 58, 51-62 (1997).

(6) Xie Xuejing and Yin Bingchuan. Geochemical patterns from local to global. J. Geochem. Explor. 47, 109-129 (1993).

(7) Zhang Hong, Chen FangLun, 1996, Platinum Group Element: Analytical method. Geochemistry of PGE deposit and Geochemical Exploration. Geological Publishing House. Beijing (in Chinese).

# ENVIRONMENTAL GEOCHEMISTRY

*Proc. 30th Int'l. Geol. Congr.*, Vol. 19, pp. 153-163
Xie Xuejin (Ed)
VSP 1997

# Comparative Risk Assessment of Lead Exposure in Birmingham and Shanghai

YONG WANG, IAIN THORNTON, MARGARET E. FARAGO
*Environmental Geochemistry Research Group, Centre for Environmental Technology, Royal School of Mines, Imperial College of Science, Technology and Medicine, London SW7 2BP, UK*

**Abstract**

Since the mid 1970s, environmental lead concentrations have declined significantly in most developed countries as a result of successive reductions in lead emissions. Nevertheless, low-level exposure to lead continues to threaten health of vulnerable groups, such as young children. In some developing countries, rapid economic growth and traffic increase have given rise to concern about risk associated with lead in the urban environment. However, until recently, there has been little quantitative information on exposure to lead in these areas. Birmingham, UK and Shanghai, China were chosen to represent different environmental features and provide differences in risks from exposure to lead. Surveys were implemented in and around houses within these cities between 1995 and 1996. In Birmingham, concentrations of lead measured in 85 houses showed a considerable fall over the period from 1984 to 1996. In Shanghai, whilst concentrations of lead in soils and roadside dusts remained relatively lower, elevated concentrations were found in the dusts collected from 65 houses. It was thought that the internal enrichment of lead was mainly due to the use of lead-containing paint in the decoration of floor spaces. The US-EPA Integrated Exposure Uptake Biokinetic (IEUBK 0.99d) lead exposure model for children was validated, updated, and applied to predict blood lead levels and provide comparative assessment of risks. As predicted, health risk has declined significantly in Birmingham over the period 1984-1996, while higher level of lead was predicted in Shanghai's young children. The results suggested that the reduction of environmental lead in the developing world is of equal importance to that in developed countries.

*Keywords: Birmingham, Shanghai, Risk assessment, Blood lead, Lead concentrations, Model*

## INTRODUCTION

Lead is one of the earliest metal products and one of the major toxic contaminants in modern environment [6, 22]. Human activities, such as mining and smelting and the use of lead piping, solder, and cable sheathing, have increasingly given rise to concern about human exposure to environmental lead, in particular that of young children [7, 12, 17, 19, 20, 25, 30]. In light of new evidence of potential effects at levels previously believed to be safe, US-EPA and the Centres for Disease Control (CDC) lowered the level of concern for childhood blood lead from 20 μg/dl to 10 μg/dl in 1986 and 1991 respectively [23].

It's increasingly reported that concentrations of blood lead have declined significantly in most developed counties over the last 20 years [3, 10, 13, 23]. In Britain the annual decrease was estimated as 5% over the period 1976-1984 and a further downward trend followed over the period 1984-1987 [13]. In a very recent health survey of England, blood lead concentrations was estimated to have fallen by a factor of 3.6-5.0 in children since 1984-1987 [10]. In the US, a 37% decline has been reported between 1976 and 1980. In the third national Health and Nutritional Examination Survey between 1988 and 1991, it was shown that mean blood lead concentrations of children aged 1 to 5 years declined 77% (13.7 μg/dl to 3.2 μg/dl ) for non-Hispanic white children and 72% (20.2 to 5.6 μg/dl) for non-Hispanic black children [3].

Despite the success of the reduction measures implemented since the mid 1970s, some developed countries have identified fractions of population that continue to exceed national concern levels for lead. In the US, 8.9% children aged 1 to 5 years, or approximately 1.7 million, were estimated to have blood lead levels of 10 μg/dl or greater [3]. In parallel, concentrations of lead in blood in the UK have been found to be significantly higher in men, older people, heavy smokers and drinkers than in other groups, and in certain parts of the UK such as north west England [14]. Exposure to lead at low levels that may adversely affect the health of young children remains a problem, particularly for those who are from low-income families, urban areas, and minorities [3]. Therefore, future efforts in the identification of the vulnerable groups are needed to further reduce lead exposure.

Moreover, lead exposure is likely to increase in the developing countries with rapid economic growth and accelerated traffic. For example, Asian demand for lead increased by a factor of 6 during the last two decades, Asia has now become the third largest and fastest growing lead-consuming region [12]. In China, seventeen preliminary investigations reported elevated blood lead levels in children [18, 28]. It is vital to establish data for lead exposure and identify population at risk in many cities, such as Shanghai, China.

The observed decline in blood lead levels in developed countries has been ascribed to a wide range of measures taken to phase out lead emissions into the environment. The major influence is attributed to the changeover to unleaded petrol. This interpretation, however, lacks data concerning the background of changes in environmental lead which contributes to blood lead via various routes.

Risk assessment models have been increasingly used to predict the likelihood of various unwanted events and adverse effects. Despite numerous methodological uncertainties remaining, risk assessment can provide reasonably accurate predictions of human exposure to lead and relatively accurate estimates of health risk. A program was initiated and designed to provide up-to-date information about differences in the exposure of children to, health risks associated with, lead in the cities of Birmingham in the UK and Shanghai in China. It aims to examine the benefits of the policies for the reduction in emissions of lead from petrol and other sources in Birmingham and provide developing countries with valuable experience in lead reduction. In this study, the US-EPA IEUBK lead model for children ( 0.99d ) [37] was used to compare risks of exposure in both Birmingham and Shanghai.

## SITE DESCRIPTION

*Shanghai*
As the largest city in China, with population of some 13 millions and an area of 6,340.5 $km^2$ , Shanghai is located at 31° 14' north latitude and 121° 29' east longitude. Bordering on Jiangsu and Zhejiang provinces on the west, Shanghai faces the East Sea on the east with the Changjiang River to the north and the Hangzhou Bay to the south.

Shanghai also comprises one of three provincial-level municipal governments in China, consisting of 13 urban districts, 6 suburban counties and the Pudong New Area. Situated on a central location along China's coastal line, Shanghai is endowed with a favourable geographical location. It is the first Chinese port to be opened to the western trade and was well-known as the "Orient Paris" in the 1930's. Since the early 1990, Shanghai has been undergoing rapid development to take a leading role in economic development of the Changjiang River plain while establishing itself as an international center of economics, finance and trade. In 1994, Shanghai 's GNP reached US$ 24 billion, up 14.3% from the previous year [27].

In parallel, increasing consumption of leaded petrol and lead-containing paint for house decoration have given rise to concern about the potential hazard of lead to health, particularly of children in the city area [26, 28]. In spite of a few surveys focusing on geochemical "hotspots" such as battery factories, little quantitative information is available about the degree and extent of various lead emissions and risks of exposure to lead.

*Birmingham*
Birmingham is the UK's second largest city, with the highest concentration of people and economic activity outside London. It is the regional capital of the West Midlands, covering an area of 265 $km^2$. The population is nearly 1 million, and 6 million people live within 50 miles of the inner city area. By the 18$^{th}$ century, Birmingham was internationally known as the "workshop of the world". Currently, however, Birmingham is becoming more of a leisure and conference centre; it is well known as a major European city.

Birmingham is one of the first cities in the UK in which elevated lead concentrations in the environment received scientific and public attentions. Since the 1970s, many surveys have been undertaken to monitor lead concentrations in the environment within the city [1, 2, 8, 9, 11, 33]. Though these studies showed some evidences of a decline in lead concentrations over a period from 1975 to 1982, data of environmental lead reported prior to 1984 are not strictly comparable, as the sample collection procedures and analytical techniques have varied. Furthermore, lack of up-to-date information has made it difficult to establish a quantitative decline in risk of exposure and subsequently evaluate the benefits of the reduction policies.

## METHODOLOGY

*Sampling*
Sampling was designed to establish comparable data sets in the two cities. Sites were selected within the cities on a grid basis to provide representative samples. Differences

in dust lead levels between wipe-based methods and vacuum-based methods were characterised by recent studies [15]. As the vacuum-based method has been frequently used, an adapted vacuum cleaner, (refereed to as the IC cleaner) was employed in this study. All the samples were taken by the same person using the same procedure to reduce possible biases. The exposed floor space was sampled and the dust was passed into a pre-weighed cellulose filter thimble fitted in the receptor tube of the cleaner to collect dust before entering the collection bag [40]. This method allowed dust loading and hence dust lead loading to be calculated as the amount of lead per square meter. In parallel, roadside dusts were sampled by sweeping a composite of nine subsamples at regular intervals from the gutter with a plastic dustpan and brush. All roadside samples were collected after 5 rain-free days. Soil samples (0-5 cm) were taken from exposed surfaces in gardens using a stainless steel trowel.

A comprehensive questionnaire was administered in each household to collect information (e.g. age of property, distance from road, presence of paint) which could be used in the incorporate with the interpretation of environmental measurements.

Shanghai:
Since the early 1990s, the inner city of Shanghai has expanded to an area of 2057 $km^2$ as a result of rapid urbanisation, particularly in east Shanghai (Pudong), but this present survey was limited to the central city with a sampling area of 298 $km^2$ , and did not include 4 newly authorised districts. 65 households were sampled on a square grid basis with one random sample in each grid square; this resulted in a approximately one sample per 4 $km^2$ or 100,000 population. 15 roads were chosen to represent a broad range of traffic densities. Of 85 urban soil samples, 31 were taken from parks, 29 from public gardens, and 25 from private gardens of houses in which dust samples were taken.

Only one sample of house dust was collected in each home using the IC cleaner, as few householders had their own cleaners in Shanghai. One composite soil sample and one dust sample were taken at the same intervals along beside each road.

Birmingham:
In order to provide a comparison with results from the survey carried out in 1984/85 [8], soil and dust samples were taken in and around 85 of the previously sampled houses within the city in June and July 1996. Of the houses, 24 (18.2%) were the same as those sampled in 1984/1985, 44 (51.8%) were the neighbouring houses, and 17 (20.0%) were alternative locations selected. Thirteen of the original (1984/1985) householders were still living in the same houses, although it has been reported that the residents change their houses on average every 18 months (personal communication with The Birmingham City Council).

(1) House dusts: (A) One sample was taken from the householder's own vacuum cleaner (Own cleaner), representing a composite sample for the householder. (B) One sample was collected from a measured floor space in the room where children spend most time within the home using the IC cleaner.

(2) Road dust: Composite samples of 25 subsamples were collected immediately outside each of these locations.
(3) Soil. One composite topsoil sample (0-5 cm) of 25 subsamples was collected from exposed areas of soil in the front and back gardens of each house.

*Analysis*
To compare data sets previously reported in Birmingham and Shanghai, an inter-laboratory evaluation of analytical measurements was implemented before the surveys. 53 samples of Shanghai topsoil were analysed by AAS in Shanghai and ICP-AES in the UK. Lead concentrations determined by both methods were significantly correlated. Data based on the analysis previously employed in Shanghai and Britain are thus compatible and can be used to compare the exposure.

The methods of sample preparation and analysis employed were the same as in 1984/85 Birmingham Program [8]. The procedures have been described elsewhere [31, 40]. In brief, all samples were dried at 80°C over night. The dust samples were then passed through a 1 mm sieve, principally to remove carpet fibres and other large particles waste. A 0.1 g subsample was taken for analysis. The choice of the 1 mm dust fraction is based on the result of previous study [35]. Soil samples were disaggregated in a ceramic pestle and mortar to pass through 2 mm sieve. The sieved samples were then ground in an agate mill to pass through a 80 mesh sieve and a 0.25g sample was taken obtained for analysis. The prepared samples were first digested in 4:1 nitric-perchloric acid, then taken to dryness over 24 hours (3 hours at 50°C, 3 hours at 150°C, 18 hours at 190°C, and 0.1 hour at 195°C), leached in 2 ml 6m hydrochloric acid and finally made up to 10 ml with de-ionised water prior to the analysis for Pb and other elements by ICP-AES. Rigorous data quality control was effected by the insertion of duplicates (10%) to assess within-batch precision, reference materials (5%) to determine bias, and reagent blanks (10%) to check contamination during analysis. Analytical error was within ± 10% for both soils and dusts.

*Risk Assessment*
Monitoring procedures of blood lead are complicated and expensive. For ethical and other reasons, it is impossible to re-measure blood lead of the young children who were sampled in Birmingham in 1984/85. As an alternative, exposure and blood lead concentrations were predicted using the US-EPA IEUBK lead exposure model for children. The model was initially known as the UBK model by the EPA Office of Air Quality Planning and Standard (OAQPS) and New York University in 1985 [37]. The basis of upgraded IEUBK model is the construction of a detailed and thorough exposure scenario for children aged 6 months ~7 years that can be adjusted to match the exposure of any child. The model simulates lead uptake, distribution within the body and elimination of lead from the body, using standard age-weighted exposure parameters for consumption of food, drinking water, soil, and dust and inhalation of air, combined with site-specific concentrations of lead in these media.

The model was validated and updated by 70 complete Birmingham-specific data sets of environmental lead and the blood lead value measured in 1984/1985. The geometric mean value of the predicted and the observed blood lead matched closely, being 11.5

μg/g and 11.6 μg/dl for 70 children aged 22-26 months respectively. Though the model generally provides slightly lower risk estimates using urban-specific lead values, it is considered to be reliable to evaluate health risks in children from exposure to lead when the EPA default values of exposure parameters are replaced by the site-specific data.

## RESULTS AND DISCUSSION

### *Comparison of lead concentrations*

The concentrations of lead in soils are summarised in Table 1. Compared with a background value of 26 μg/g Pb in Shanghai soils [39], elevated concentrations were found in roadside soils and garden soils, with the geometric means of 143 μg/g and 55 μg/g Pb, respectively. These values are significantly lower than those in Britain soils (1981) and in Birmingham (1996). A slight decrease of lead concentration was noted in the soil samples collected from gardens in Birmingham city over the period 1984-1996. The geometric mean fell from 313 μg/g Pb in 1984 to 278 μg/g in 1996.

**Table 1.** Comparison of lead concentrations in urban garden soils (μg/g)

| Area | Year | N | GM | Min | Max | Reference |
|---|---|---|---|---|---|---|
| Birmingham | 1981 | 160 | 279 | 48 | 1310 | [5, 32] |
| | 1984 | 86 | 313 | 56 | 1650 | [8] |
| | 1996 | 84 | 278 | 82 | 1660 | This Study |
| Shanghai | 1995 | 85 | 55 | 13 | 2500 | This Study |
| Britain | 1981 | 4126 | 266 | 13 | 14100 | [34] |

Concentrations of lead in dusts are summarised in Table 2. In inner city Shanghai, a considerably lower geometric mean value for lead (196 μg/g ) was found in roadside dusts compared with 622 μg/g Pb in house dusts, reflecting elevated lead concentrations within the homes. The internal "enrichment" has been attributed to the use of lead-containing paint which is still commonly used in the decoration of houses in Shanghai, particularly for the decoration of floors. In contrast, a successive decline in

**Table 2.** Comparison of concentrations of lead in urban dusts (μg/g)

| City | Year | N | GM | Min | Max | Reference |
|---|---|---|---|---|---|---|
| **House dust** | | | | | | |
| Birmingham | 1981 | 159 | 419 | 58 | 22000 | [32] |
| | 1984 | 96 | 311 | 62 | 5600 | [8] |
| | 1996 | 85 | 133 | 30 | 848 | This Study |
| Shanghai | 1995 | 65 | 622 | 92 | 12934 | This Study |
| **Roadside dust** | | | | | | |
| Birmingham | 1979 | 12 | 555 | 260 | 370 | [12] |
| | 1984 | 97 | 527 | 80 | 2100 | [8] |
| | 1996 | 82 | 262 | 23 | 9500 | This Study |
| Shanghai | 1995 | 15 | 196 | 125 | 287 | This Study |

lead concentrations was shown in house and roadside dusts in Birmingham over the period. 1984-1996. This fall is clearly demonstrated in the changes of lead levels in 24 households sampled both in 1984 and 1996 (Figure 1). With the exception of lead in soils, lead levels have fallen by more than 50% over this period.

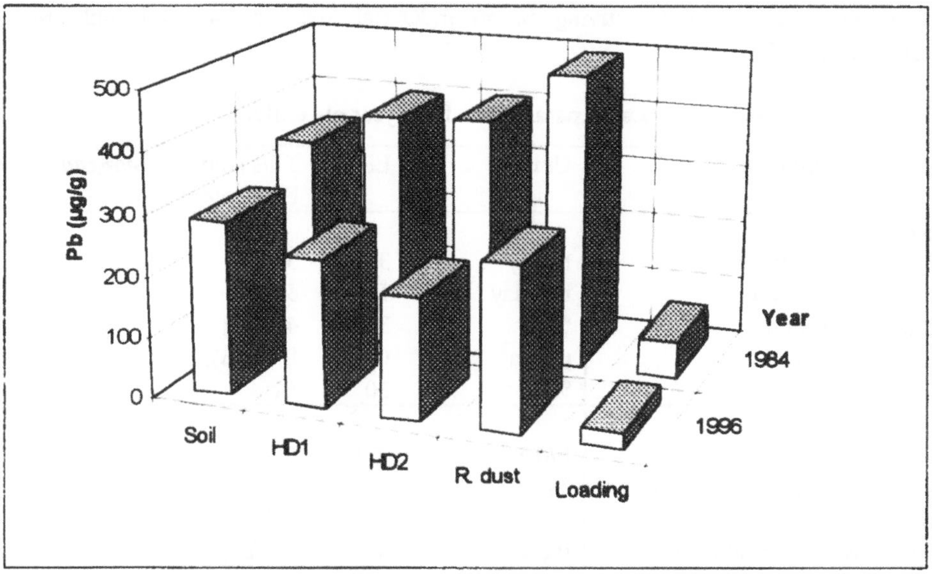


**Figure 1.** Comparison of geometric means of lead concentrations in 24 sites sampled in both 1984 and 1996, Birmingham. Symbols: HD1 - house dust from householder's own cleaner; HD2 - house dust by the IC cleaner; R. dust - roadside dust; Loading - lead loading in dust ( $\mu g/m^2$ ).

*Comparison of Risk*

The updated IEUBK model was used as a risk assessment tool to predict the likely blood lead distribution for children exposed to lead in both cities and to estimate the probability of children's blood lead concentrations exceeding a health-based level of concern. 2 year old children, the most vulnerable group of the population, were selected to characterise risks. Thus risk indices were comparable spatially and temporally between the two cities. Input assumptions of exposure parameters and site-specific data of current environmental lead are briefly listed in Table 3.

The predictions of the geometric means of total lead uptake and blood lead concentrations are shown in Table 4. The results indicated a considerable reduction in risk for two-year old children in Birmingham over the period 1984-1996. The risk to have blood lead value over 10 µg/dl was reduced from 59.8% in 1984 to 28.3% in 1996. On the other hand, assuming the same values for uptake parameters as in Birmingham, a geometric mean of 21.9 µg/dl blood lead with a risk of 95% exceeding the concern level (10 µg/dl) was calculated for Shanghai.

In parallel, based on the detailed measurements of environmental dietary and blood lead, a geometric mean of 11.6 µg/dl blood lead was found in 70 two-year old children in Birmingham. Davies (1987) calculated a slightly higher total Pb uptake of 36 μg/day, with 3% from inhalation and 97% from ingestion. In parallel, a fall of blood lead from 10.1 μg/dl in 1984 to 7.9 μg/dl in 1987 was reported for 6-7 year old children living near busy major roads in Birmingham [13]. A preliminary survey in Shanghai also reported geometric mean values of 20.6 μg/dl and 15.9 μg/dl Pb in blood samples of primary school children living in an industrial area and a residential area, respectively [18].

**Table 3.** Input values of exposure parameters for 2 year old children

| Exposure Parameter | Unit | EPA | This study (S-*Shanghai;* B-*Birmingham*) |
|---|---|---|---|
| **INHALATION** | | | |
| Time spent outdoor | h /day | 3 | 3 |
| Respired air volume | $m^3$ /day | 5 | 5 |
| Lung absorption | % | 32 | 40 [41] |
| Outdoor air lead level | μg/ $m^3$ | 0.1 | 0.43 (B); 0.52 (S) [26] |
| Indoor air lead level | % | 30 | 61 |
| **DIGESTION** | | | |
| Drinking water intake | l /day | 0.5 | 0.5 |
| Gastrointestinal absorption from water and diet | % | 50 | 50 |
| Drinking water lead level | μg /l | 4 | 19 (B); 4.2 (S) [26] |
| Dietary intake | μg /day | 5.78 | 12.5 (B); 137 (S) [24] |
| Soil & dust ingestion amount | mg /day | 135 | 135 |
| Ingestion weighing factor | % soil /% dust | 45/55 | 45/55 |
| Gastrointestinal absorption from soil and dust | % | 30 | 50 |
| Soil lead level | μg /g | 200 | 278 (B); 55 (S) |
| House dust lead level | μg /g | 154 | 133 (B); 622 (S) |
| **MATERNAL SOURCE** | μg $dl^{-1}$ | 2.5 | 2.5 |
| **OTHER SOURCES** | μg $day^{-1}$ | 0 | 0 |

**Table 4.** Total Pb uptake and blood lead concentrations predicted for 2-year old children

| City | Year | Total Pb uptake (μg/day) | Blood Pb level (μg/dl) | Risk1 (%) | Risk2 (%) |
|---|---|---|---|---|---|
| Shanghai | 1995 | 58.4 | 21.9 | 95.0 | 35.7 |
| Birmingham | 1984 | 30 | 11.4 | 59.8 | 3.1 |
| Birmingham | 1996 | 20.8 | 8.1 | 28.3 | 0.4 |

Risk1 - percentage of children with blood lead concentrations exceeding 10 μg/dl (US-EPA guideline); Risk2 - percentage of children with blood lead concentrations exceeding 25 μg/dl (EU guideline).

*Discussion*

In the UK, the maximum permissible level of lead in petrol was reduced from 0.84 g/l in 1972 to 0.4 g/l in 1981, and a further reduction to 0.15 g/l at the beginning of 1986. All new vehicles were required to operate on unleaded petrol (<0.013 g/l) from 1990. In the UK, unleaded petrol has become widely available, and a 5p per litre difference in duty between leaded and unleaded petrol has encouraged sales. This change has contributed to a reduction in emissions of lead from petrol driven vehicles, particularly to the 60% fall in 1986 [13]. Assuming that 67.5% of total Pb intake is derived from direct and indirect atmospheric deposition and 70% of total lead emission from vehicle exhaust [13], emission of lead from petrol contributed to approximately 47% of total Pb intake. Thus the decline in risk from lead exposure was principally due to the reduction of lead in petrol. In Shanghai, however, leaded petrol has been and is still commonly used in motor vehicles; and lead-containing paint is also frequently used for interior decoration in households. Thus the risk is likely to increase before policies are applied for the reduction of lead.

This study has shown that whilst lead concentrations in house dusts remains high in Shanghai, concentrations of lead have fallen considerably in the urban environment of Birmingham over the last decade. Consequently, in Birmingham, health risks associated with lead have fallen substantially due to the removal of lead from petrol and other sources, while the risk is likely to increase in Shanghai city as increased traffic densities still use leaded petrol and add further lead to already large amounts in floor dust from the use of lead-rich paint. This suggests that there is an urgent need for the reduction in lead in such an area with rapid economic growth. Lessons learnt from developed countries will be beneficial to this new challenge in the developing countries.

The IEUBK model provides regional and temporal differences in risks of children exposure to lead. In spite of uncertainties derived from biokinetic and exposure assumptions, the accuracy of the model has been improved as a result of validation with complete observed data sets and the input of house-specific information. Risk characterisation can be further improved with the assistance of GIS techniques by providing more detailed information on the distribution of risk within a city.

*Acknowledgements*

We are grateful to the Sino-British Friendship Scholarship Scheme (SBFSS) and the British Council for financial support and the Environmental Services Department, Birmingham City Council for sampling assistance. We especially appreciate the co-operation of the participating householders both in Birmingham and Shanghai.

## REFERENCES

1. A. Archer and R. S. Barratt. Lead levels in Birmingham dust, *The Science of the Total Environment* 6, 275-286 (1976).
2. A. Archer and R. S. Barratt. Lead levels in the environment: Monitoring for lead, *Journal of Royal Society of Health* 96, 173-176 (1976).
3. D. J. Brody, J. L. Pirkle, R. A. Kramer, K. M. Flegal, T. D. Matte, E. W. Gunter and D. C. Paschal. Blood lead levels in the US population, phase 1 of the third national health and nutrition examination survey (NHANES III, 1988-1991). *Journal of the American Medical Association* 272, 277-283 (1994).
4. H. T. Chon, K. W. Kim and J. Y. Kim. Metal contamination of soils and dusts in Seoul Metropolitan City, Korea, *Environmental Geochemistry and Health* 19, 139-146 (1995).

5. E. B. Culbard, I. Thornton, J. Watt, M. Wheatley, S. Moorcroft and M. Thompson. Metal contamination in British urban dusts and soils, *Journal of Environmental Quality* **17**(2), 226-234 (1988).
6. B. E. Davies. Trace metals in soil, *Chemistry in Britain* **24**, 149-154 (1988).
7. D. J. A. Davies, I. Thornton, J. Watt, E. B. Culbard, P. G. Harvey, H. T. Delves, J. C. Sherlock, G.A. Smart, J. F. A. Thomas and M. J. Quinn. Lead intake and blood lead in two year old UK urban children, *The Science of the Total Environment* **90**, 13-29 (1988).
8. D. J. A. Davies, J. Watt and I. Thornton. Lead levels in Birmingham dusts and soils, *The Science of the Total Environment* **67** (2-3), 177-185 (1987).
9. D. J. A. Davies. In: I. Thornton and E. Culbard, (Eds), *Lead in the Home Environment, Science Reviews Limited, Northwood, UK.* 189-196 (1987).
10. H. J. Delves, S. J. Diaper, Soiggyd, S. Oppert, P. Prescoott-Clarke, W. Dong, H. Colhoun and D. Gompertz. Blood lead concentrations in United Kingdom have fallen substantially since 1984, *British Medical Journal*, **313**, 884 (1996).
11. Department of the Environment, *Lead Pollution in Birmingham, Pollution Paper No. 14.* HMSO, London, UK, (1978).
12. Department of the Environment, *Blood lead concentrations in pre-school children in Birmingham, Pollution Report No. 15.* HMSO, London, UK, (1982).
13 Department of the Environment, *UK blood lead monitoring program 1984-1987, Pollution Report No. 28.* HMSO, London, UK, (1990).
14. Department of the Environment, *The UK Environment*. HMSO, London, UK, (1992).
15. M. R. Farfel, J. J. Chisolm and C. A. Rohde. The long-term effectiveness of residential lead paint abatement, *Environmental Research* **66**(2), 217-221 (1994).
16. M. A. Francer, V. Pan, J. H. Hanko and B. Makima. Home condition and lead levels-a case study from the homes of pre-schoolers in MT Pleasant, Michigan, *Environmental Research* **29**(9), 1879-1886 (1994).
17. A. K. Goodman, H. Shultz, S. Klitzman, M. Kimmelblatt and W. Spadaro. Preventing lead poisoning in New York City: Priorities for lead abatement in housing. *Bulletin of New York Academy of Medicine* **70**(3), 236-250 (1993).
18. L. Jiang. *Effects of lead absorption on children development* (Unpublished PhD thesis). The First Shanghai Medical College, Shanghai, China, (1988).
19. H. W. Miekle. Lead dust contaminated U.S.A. communities: Comparison of Louisiana and Minnesota. *Applied Geochemistry* **6**, 1-5 (1993).
20. H. W. Miekle, J. L. C. Anderson, K. J. Berry, P. W. Miekle, R. L. Chaney and M. Leech. Lead concentrations in inner-city soils as a factor in the child lead problem, *American Journal of Public Health* **73**, 1366-1369 (1983).
21. P. Mushak. Defining lead as the premiere environmental health issue for children in America: criteria and their quantitative applications, *Environmental Research* **59**, 281-309 (1992).
22. J. O. Nrigu. *Lead and lead poisoning in antiquity*. John Wiley & Sons Ltd, New York, USA, (1983).
23. OECD. *Risk reduction monograph No. 1: lead--Background and national experience with reducing risk.* Environment Directorate, OECD, Paris, France, (1993).
24. J. H. Pan (Ed.). *Impact of land contamination on agricultural environment and products.* Shanghai Academy of Agriculture, Shanghai, China, (1992).
25. Royal Commission on Environmental Pollution (RCEP). *Ninth Report, Lead in the Environment*. HMSO, London, UK, (1983).
26. Shanghai Environmental Protection Bureau. *Shanghai Environmental Quality Report*. Shanghai, China, (1994).
27. Shanghai Statistics Bureau. *Annual Outline of Shanghai.* Shanghai Municipal Government, Shanghai, China, (1994).

28. X. M. Shen, J. F. Rosen, D. Guo and S. M. Wu. Childhood lead poisoning in China, *The Science of the Total Environment* **181**, 101-109 (1996).
29. Y. C. Song and Z. Y. Gu. Monitoring air pollution by concentrations of heavy metals in urban trees, *Urban Environment and Ecology* **1**, 34-38 (1988).
30. P. M. Sutton, M. Athanasoullis, P. Flessel, G. Guirguis, M. Haan, R. Schlag and L. R. Goldman. Lead levels in the household environment of children in 3 high-risk communities in California. *Environment Research* **68**(1), 45-57 (1995).
31. M. Thompson and J. N. Walsh. *A hand book of inductively coupled plasma atomic emission spectrometry, 2nd* edition. Blackie, London, England, (1988).
32. I. Thornton, D. J. A. Davies, J. Watt and M. J. Quinn. Lead-exposure in young-children from dust and soil in the United Kingdom, *Environmental Health Perspectives* **89**(11), 55-60 (1990).
33. I. Thornton and C. Elisabeth. Lead in the home Environment, *Science Reviews Limited.* Northwood, UK, (1988).
34. I. Thornton, J. Watt, D. J. A. Davies, A. Hunt, J. Cotter-Howells and D. L. Johnson. Lead contamination of UK dusts and soils and implication for childhood exposure: an overview of the work of the Environmental Geochemistry Research Group, imperial College, London, England 1981-1992, *Environmental Geochemistry and Health* **16**(3/4), 113-122 (1994).
35. I. Thornton, S. John, S. Moorcroft and J. Watt. In: *Trace Substances in Environmental Health XIV*. Hemphill, D. D. (Ed.), University of Missouri, Columbia, 27-37 (1980).
36. N. F. Tom et al. Heavy metal pollution in roadside urban parks and gardens in Hong Kong, *The Science of The Total Environment* **59**, 325-328 (1987).
37. US Environmental Protection Agency (US EPA). *Guidance manual for the integrated exposure uptake biokinetic model for lead in children, EPA/540/R-93/08*. Office of Emergency and Remedial Response, Washington, D. C., USA, 1994.
38. Y. Von-Schirnding, D. Bradshaw, R. Fuggle and M. Stokol. Blood lead levels in South African inner-city children, *Environmental Health Perspectives* **94**, 125-130 (1991).
39. Y. Wang, H. L. Lu and Y. G. Wan (Eds). *Metal background values in Shanghai soils*. China Environmental Science Press, Beijing, China, (1991).
40. J. Watt, S. Moorcroft, K. Brooks, E. B. Culbard and I. Thornton. Metal contamination of dusts and soils in urban and rural households in the United Kingdom. 1. Sampling and analytical techniques for household and external dusts. in: Hemphill, D. D. (ed.), *Trace Substances in Environmental Health* **17**, 229-235 (1983), University of Missouri, Columbia, USA.
41. World Health Organisation, Regional Office for Europe. *Air quality guidelines for Europe, European Series No. 23*. WHO Regional Publications, Copenhagen.

*Proc. 30th Int'l. Geol. Congr.*, Vol. 19, pp. 165-172
Xie Xuejin (Ed)
VSP 1997

# Sulfur Isotopic Composition of Two Asian Great Rivers and Uplift of Qinghai-Xizang Plateau

HONG YETANG, GU AILIANG, WANG HONGWEI, HONG BING

*State Key Laboratory of Environmental Geochemistry, Institute of Geochemistry, Chinese Academy of Sciences, Guiyang, Guizhou Province 550002, P. R. China*

**Abstract**

In this work, the sulfur isotopic composition of Huanghe (the Yellow River) and Changjiang (the Yangtze River) were measured. Our study reveal that the sulfur isotopic compositions of dissolved sulfate ($\delta^{34}SO_4^{2-}$) of the two large rivers show remarkably differential patterns and a scatter diagram of $\delta^{34}SO_4^{2-}$ plotted against the $SO_4^{2-}$ concentration can be used to elucidate them. Huanghe riverwater is characterized by enriching $^{34}S$ with a high $SO_4^{2-}$ concentration. In addition, the $\delta^{34}SO_4^{2-}$ values of Changjiang water is characterized by depleting of $^{34}S$ with a lower $SO_4^{2-}$ concentration. In addition, the $\delta^{34}SO_4^{2-}$ values of the Changjiang water show a fluctuating variation along the main channel. In contrast to Changjiang, the $\delta^{34}SO_4^{2-}$ of Huanghe water show progressively increase from the upper to lower reaches which reflects a comprehensive effects of different environmental conditions and sulfur sources on the river. The above-mentioned remarkable difference between Huanghe and Changjiang may be mainly attributed to their unique occurring environment, and the uplift of Qinghai-Xizang (Tibetan) Plateau has significant impact on the sulfate isotopic composition of these two large Chinese River.

*Keywords: Huanghe, Changjiang,, sulfur isotopic composition, uplift of Qinghai-Xizang Plateau*

## INTRODUCTION

The sulfur isotopic composition of dissolved sulfate in any rivers has been recognized a mixture of the sulfur derived from the soil, bedrock and the atmosphere. So the measurement of the sulfur isotopic composition in a river water can provide background information on the biogeochemical cycling of sulfur in the river basin. So far, however, relatively little work has been done on sulfur isotope systematics in the watersheds of the major rivers of the world[1]. In particular, no information of the sulfur isotope of the major rivers in Asian continent has been reported in the western literature[2,3]. Huanghe and Changjiang are two great rivers in Asian continent that by means of the chemical gravimetric method. The chemical pretreatment of $BaSO_4$ prior to the production of sulfur dioxide for mass spectrometric analysis of $^{34}S/^{32}S$ ratio was conducted by means of the reducing mixture method[4]. $\delta^{34}S$ was determined relative to meteorite troilite by a MAT-252 mass-spectrometer. The results of the determination of the sulfur isotope and the sulfate concentration are listed in Table 1.

**Table 1** Analitical Data

| Sample No.* | River | Location | Date | $SO_4$ (mg/l) | $\delta^{34}SO_4$ (‰) | Discharge ($m^3/s$) |
|---|---|---|---|---|---|---|
| 1 | Huanghe | Maqu | Sept.1993 | 7.57 | +3.59 | 475 |
| 2 | Huanghe | Xunhua | Sept.1993 | 18.0 | +3.57 | 729 |
| 3 | Huanghe | Jingyuan | Sept.1993 | 60.5 | +7.50 | 997 |
| 4 | Huanghe | Zhongwei | Sept.1993 | 54.0 | +6.94 | |
| 5 | Huanghe | Qingtongxia | Sept.1993 | 67.5 | +6.89 | |
| 6 | Huanghe | Dengkou | Sept.1993 | 92.1 | +8.45 | |
| 7 | Huanghe | Hekouzhen | Sept.1993 | 94.4 | +8.50 | 776 |
| 8 | Huanghe | .Wubao | Sept.1993 | 101.4 | +9.18 | |
| 9 | Huanghe | Hukou | Sept.1993 | 99.2 | +9.09 | |
| 10 | Huanghe | Longmen | Sept.1993 | 96.0 | +11.10 | 1060 |
| 11 | Huanghe | Sanmenxia | Sept.1993 | 104.0 | +11.12 | 1280 |
| 12 | Huanghe | Huayuankou | Sept.1993 | 104.6 | +10.71 | 1470 |
| 13 | Huanghe | Pingyin | Sept.1993 | 103.7 | +10.41 | |
| 14 | Huanghe | Luokou | Sept.1993 | 103.7 | +9.96 | 1430 |
| 15 | Huanghe | Lijin | Sept.1993 | 108.6 | +10.23 | 1370 |
| 16 | Changjiang | Shigu | Oct. 1993 | 28.4 | +2.17 | 1320 |
| 17 | Changjiang | Panzhihua | Oct. 1993 | 23.5 | +1.50 | |
| 18 | Changjiang | Yibin | Oct. 1993 | 16.7 | +4.56 | |
| 19 | Changjiang | Wanxian | Oct. 1993 | 20.0 | +3.11 | |
| 20 | Changjiang | Yichang | Oct. 1993 | 21.7 | +1.23 | 14300 |
| 21 | Changjiang | Shashi | Oct. 1993 | 23.0 | +5.94 | |
| 22 | Changjiang | Honghu | Oct. 1993 | 21.6 | +4.48 | |
| 23 | Changjiang | Ezhou | Oct. 1993 | 21.9 | -3.47 | 23400 |
| 24 | Changjiang | Jiujiang | Oct. 1993 | 20.9 | -2.71 | |
| 25 | Changjiang | Anqing | Oct. 1993 | 17.8 | +0.84 | |
| 26 | Changjiang | Datong | Oct. 1993 | 16.9 | +5.14 | 28900 |
| 27 | Changjiang | · Wuhu | Oct. 1993 | 16.9 | +4.26 | |
| 28 | Changjiang | Nanjing | Oct. 1993 | 16.9 | +2.29 | |

* Numbers coorespond to locations in Fig. 1

take the first place in terms of sediment transport and flow in Asian rivers respectively. Our studies reveal that the discharge weighted mean value of sulfur isotopic composition of the Huanghe and Changjiang are +9.2 ‰ and +1.3 ‰, respectively. This remarkable difference may be an example reflecting the great effects of the uplift of the Qinghai-Xizang plateau on the sulfur biogeochemical cycle in Asian continent.

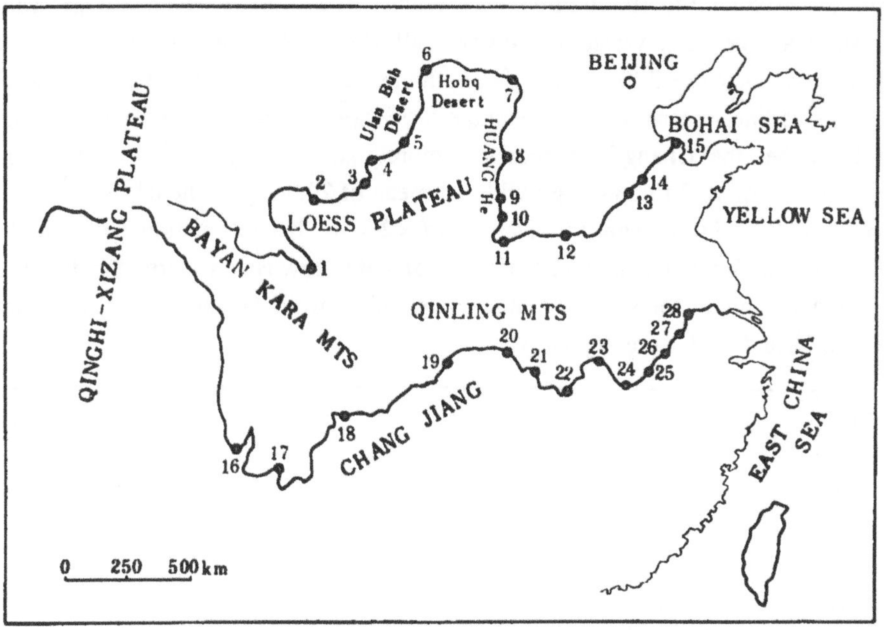


**Figure 1.** Location map of the Samples

## SAMPLING AND ANALYZING METHODS

From the last ten-day period of September to the first ten-day period of October of 1993, we carry out an environmental geology-geochemical investigation along the main channel of the Huanghe and Changjiang while the rivers have a moderate flow condition. Water samples were collected in 15 and 13 river section from the upper to the lower reaches of the Huanghe and the Changjiang respectively ( Fig. 1). Near these sampling sites there generally are national hydrometric stations so that some of hydrological data can been collected. In each sampling site 10L of bulk river water were taken from the surface waters in mid-channel of the rivers by measuring ship of the hydrometric station or by small boat. The sampling bottle made of polythene acid-cleaned in advance was washed by the river water and was filled with the river water until overflow. After sealing up with wax the samples were sent back to laboratory in two separate groups. The water samples were filtered through a 0.45μm Millipore fiberglass filter and dissolved sulfate was precipitated as $BaSO_4$ which was measured

**RESULT AND DISCUSSION**

The sulfur isotopic composition of dissolved sulfate ($\delta^{34}SO_4^{2-}$) of Huanghe and Changjiang river water show remarkable differential patterns and a scatter diagram of $\delta^{34}SO_4^{2-}$ plotted against the $SO_4^{2-}$concentration may be used to elucidate them (Fig. 2). From the Fig. 2 we can see that the $SO_4^{2-}$ concentration of the Huanghe water varies from 7.5 mg/l to 108 mg/l with the average of 81.0 mg/l. The $\delta^{34}SO_4^{2-}$ value also show a wide range variation. However, all $\delta^{34}SO_4^{2-}$ values of the Huanghe are in a positive range of +3.57 to +11.1 ‰, with the mean value of +8.4 ‰. There nearly is a positive linear correlation between the $\delta^{34}SO_4^{2-}$ and $SO_4^{2-}$. As contrasted with the Huanghe, the Changjiang shows a more narrow range variation of $SO_4^{2-}$ concentration of 16.7 to 28.4 mg/l with the mean of 20.5 mg/l. The $\delta^{34}SO_4^{2-}$ range from -3.5 ‰ to 5.9 ‰ with the average of +2.3 ‰. No correlation between the $\delta^{34}SO_4^{2-}$ and $SO_4^{2-}$ can be found. If a rate of water flow is considered the discharge weighted average of the sulfur isotopic composition of the Huanghe and Changjiang are +9.2 ‰ and +1.3 ‰ respectively.

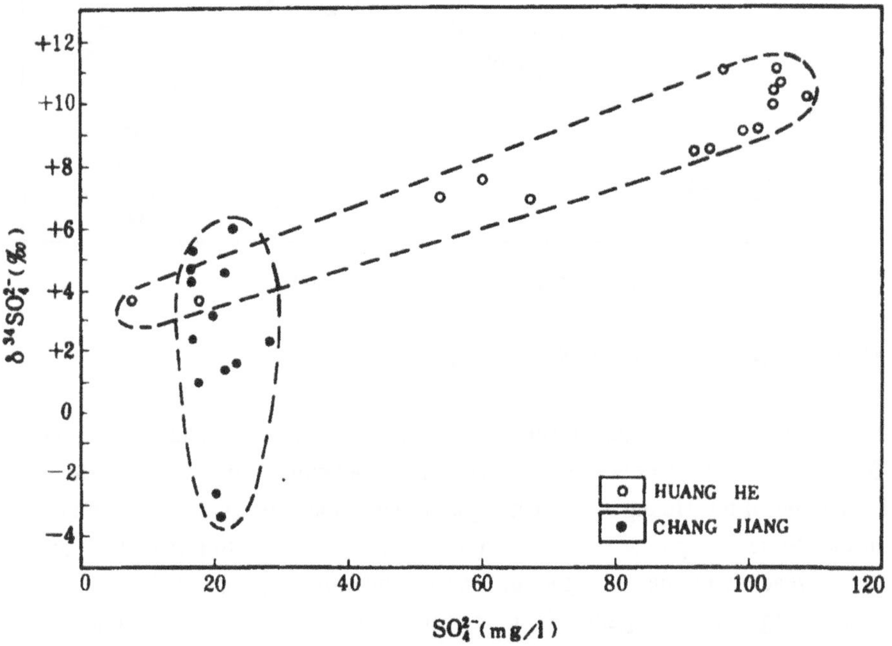


**Figure 2.** The $\delta^{34}SO_4^{2-}$ values of Huanghe and Changjiang

Comparing the $\delta^{34}SO_4^{2-}$ values of the Huanghe and the Changjiang to those of the major rivers in the world it can be seen that the Huanghe water seems to have the heaviest $\delta^{34}SO_4^{2-}$ and the Changjiang seems to have the lightest $\delta^{34}SO_4^{2-}$ in the major rivers of the world[Table 2]. The result reveals that these two great rivers, located in the north and the south of China respectively and flowing through the hinterland of the eastasian continent from west to east, are obviously different in the two indexes of the $\delta^{34}SO_4^{2-}$ and $SO_4^{2-}$ concentration. The Huanghe water is characterized by enriching $^{34}S$ with a high $SO_4^{2-}$ concentration, while the Changjiang water is Characterized by depleting of $^{34}S$ with a low $SO_4^{2-}$ concentration.

Table 2 Sulfur isotopic composition of some rivers of the world

| River | Range of $\delta^{34}SO_4^{2-}$ (‰) | Mean $\delta^{34}SO_4^{2-}$ (‰) | Reference |
|---|---|---|---|
| Huanghe | +3.57 ~+11.1 | +8.4 | This work |
| Changjiang | -3.47 ~+5.94 | +2.3 | This work |
| Amazon River | +6.0 ~+8.3 | +7.3 | [1] |
| Mackenzie River | -24 ~+20 | <+5 | [1] |
| Rivers of the Former Soviet Union | -5 ~+32 | ~+6 | [1] |
| North America rivers | | ~+4 | [1] |
| South America rivers | | ~+7 | [1] |
| Europe rivers | | ~+6 | [1] |
| Pacific Islands rivers | | ~+5 | [1] |
| African rivers | | ~+2 | [1] |

In addition, the $\delta^{34}SO_4^{2-}$ values of the Changjiang water show a fluctuant variation along the main channel. In contrast to the Changjiang, the $\delta^{34}SO_4^{2-}$ of the Hunaghe water show progressively increase from the upper to the lower reaches which reflects a comprehensive effects of different environmental conditions and sulfur sources on the rivers (Fig. 3). In the upper reaches of the river the Huanghe flow through alpine grassy marshland and grassland in the northeastern fringe of the Qinghai-Xizang plateau with an elevation of 4000 meters. Evaporate and salt lakes are widespreadle distributed. The mean $SO_4^{2-}$ concentration and the $\delta^{34}SO_4^{2-}$ value of Maqu and Xunhua river sections are only 12.8mg/l and +3.6 ‰ respectively, which is a lightest $\delta^{34}SO_4^{2-}$ in the river sections investigated of the Huanghe though it is heavier than that of the Changjiang and some major rivers of the world. After flowing out Xunhua the Huanghe goes into the loess distribution area. Both $\delta^{34}SO_4^{2-}$ and $SO_4^{2-}$ show firstly obvious increase. The mean $SO_4^{2-}$ concentration and the $\delta^{34}SO_4^{2-}$ value of the water of Jinyuan, Zhongwei and Qingtongxia river sections are 60.7mg/l and +7.1 ‰ respectively. From Qingtongxia to Hekouzhen the Huanghe flows cross the margin of the Ulan Buh desert and the Hobq desert in turn. Dry climate, large evaporating capacity and much sand from the desert of the region lead to the increase again of

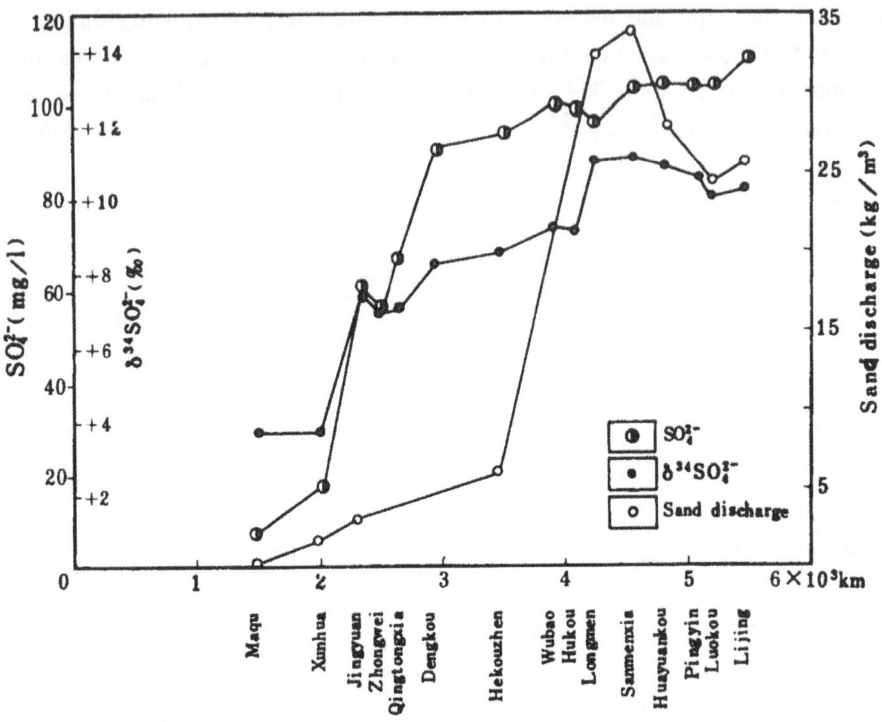


**Figure 3.** $\delta^{34}SO_4^{2-}$ values as a function of location along the main channel of the Huanghe

both $SO_4^{2-}$ and the $\delta^{34}SO_4^{2-}$ of the river water. The average $SO_4^{2-}$ concentration and the $\delta^{34}SO_4^{2-}$ value of the water of Dengkou and Hekouzhen river sections have reached 93.2mg/l and +8.5 ‰ respectively. From Hekouzhen down the Huanghe turns into the area of valley between the Shannxi and Shanxi Provinces and goes into the most serious denudation area of loess plateau. Following the rapid increase of the sand amount of the river and the release of the dissolved sulfate into the river water from mud and sand $SO_4^{2-}$ concentration and the $\delta^{34}SO_4^{2-}$ value of the Huanghe increase again on the basis of the considerably higher and heavier level. In the river section from Longmen to Sanmenxia the mean $SO_4^{2-}$concentration and the $\delta^{34}SO_4^{2-}$ values have reached 100mg/l and 11.1 ‰ respectively which is a heaviest $\delta^{34}SO_4^{2-}$ in the Huanghe. After that the Huanghe flows into its lower reaches. It is a wide alluvial plain with low elevation of 50 meters and there is no joining tributary excepting the river section from Sanmenxia to Huanyuankou. The Huanghe is kept within bounds of a great embankment and becomes a special depositional environment of mud and sand. Therefore the Huanghe water can maintain the characteristics of both high sulfate concentration and heavier sulfur isotopic until flowing into Bohai Sea.

the $\delta^{34}SO_4^{2-}$ shows a regular and slow decrease excepting Lijin site close to the river mouth. Whether this is an isotope fractionation process caused by the precipitation of the great amount of mud and sand it still remains to be studied.

The above-mentioned remarkable differences between the Huanghe and Changjiang may be mainly attributed to their unique occurring environment. At present day there has been enough evidence to reveal that since the last stage of Cretaceous period the continuous extrusion of the India landmass toward the Eurasian continent has caused the great uplift of the Qinghai-Xizang plateau, located in the west part of China continent, and the accompanying rise of the Qiling mountains, located in the central part of the China continent and keeping the Huanghe separate from the Changjiang. On the one hand, the uplift and the rise have changed the general circulation of the atmosphere of the east Asian continent and have stopped the warm and moist air current coming from both Indian and the Pacific Ocean from moving to the north of China which leads the climate of the north west of China to gradually drying[5,6]. A series of important geological processes, such as desertification, loess deposition, evaporate and formation of the Huanghe with a tremendous amount of mud and sand, occurred in succession[7,8,9]. The geochemical cycling of not only dissolved sulfate, as reported by a previous work[2], but also heavy sulfur isotopic composition composition in surface water environment. On the other hand, following the close of the ancient ocean of the Tethys caused by collision between above-mentioned two plates marine carbonate formed from Permian to Triassic Period were outcropped and widespreadly distributed in the south continent of China. It has been pointed out that marine carbonate gives a significant influence on the ion chemistry of the Changjiang water and the Changjiang becomes a typical carbonate river[2]. Unfortunately there is no available data of the sulfur isotope of these marine carbonate rocks so far. We have reason to suggest, however, that the sulfur isotopic composition of the Changjiang closely related to these marine carbonate rocks because it has been reported that the dissolved sulfate of marine carbonate formed in the warm and moist climate region generally has lighter sulfur isotopic composition[10]. Both coal and meteoric water of the south China are also characterized by depleting of $^{34}S$[11,12]. More detailed measurement of the sulfur isotopic composition of rocks and soils in the river basin still remains to be solved further.

## ACKNOWLEDGEMENTS

We thank Drs. Zhu Yongxuan and Zeng Yiqiang for much help during the measurement of sulfur isotopes. Financial support for this study was provided by NSFC NO. 49273190.

**REFENCENCES**

[1] Krouse, H. R., Grinenko, V. A., Stable Isotopes, Natural and Anthropogenic Sulfur in the Environment (SCOPE 43), John Wiley & Sons LTD, Chichestes • New York • Brisbane • Toronto • Singapore. (1991).

[2] Hu Minghui, Stallard, R. F. and Edmond, J. M., Nature 298, 550-553, (1982).

[3] Sarin, M. M. and Krishnaswami, S., Nature 312, 538-541, (1984).

[4] Thode, H. G., Monster, J. and Dunfordm, H. B., Geochim. Cosmochim. Acta, 25, 159-174, (1961).

[5] Syukuro Manaba and Tespstra, T. B. , J. Atmos. Sci., 31, 1, 3-42 (1975).

[6] Douglas, G. H. and Syukuro Manaba, J. Atmos. Sci., 32, 8 (1975).

[7] Liu Tungsheng et al., Loess and the Environment, China Ocean Press, Beijing(1985).

[8] Zhu Zengda, Wu Zheng, Liu Shu and Di Xingming, Introduction to Deserts of China, Science Press, Beijing(1980). (In Chinese).

[9] Hong Yetang, Piao Hechun and Jiang Hongbo, Science in China(Series B), 34, 7, 859-870. (1991).

[10]Grinenko, L. N., Kononova, V. A. and Grienko, V. A., Geokhimiya, 1, 66-75(1970).

[11]Hong Yetang, Zhang Hongbin and Zhu Yongxuan, Chinese Journal of Geochemistry, 12, 1, 51-59(1993).

[12]Hong Yetang, Zhang Hongbin and Zhu Yongxuan et al., Progress in Natural Science, 5, 3, 344-349 (1995).

*Proc 30th Int'l. Geol. Congr.*, Vol. 19, pp. 173-188
Xie Xuejin (Ed)
© VSP 1997

# Kinetics and thermodynamics of dissolved rare earth uptake by alluvial materials from the Nevada Test Site, Southern Nevada, U.S.A.

F.CHRISTOPHER BENEDICT Jr
*Graduate Program of Hydrologic Sciences, University of Nevada, Reno, Nevada 89512 USA*

ERIC H. DeCARLO
*Department of Oceanography, University of Hawaii, Honolulu, Hawaii 96822 USA*

MANFRED ROTH
*Zentrales Isotopenlabor, Ruhr Universität Bochum, 4630 Bochum-Querenberg, Germany*

ABSTRACT

Results are presented from thermodynamic and kinetic studies of dissolved rare earth element (REE) uptake by basin-fill alluvial materials from the Yucca Flat area of the Nevada Test Site (NTS). Aqueous/particle interactions between REE and alluvial materials were investigated through stable isotope and radiotracer batch and time-series experiments in synthetic "groundwater" solutions. Synthetic groundwater was prepared so that its composition duplicated the major ion composition resulting from the equilibration of distilled deionized water with the alluvial materials for a period of 120 hours. Synthetic groundwaters were prepared from analytical reagent-grade chemicals and 18 Mohm-cm distilled deionized water. Studies were conducted as "free-drift" experiments in which the pH was adjusted to the desired value at the beginning of the experiment and subsequently recorded at the end of the experiment. Experiments were conducted with the complete suite of REE in a concentration range of 1 nanomolar to 1 micromolar and with a suspended solid load of 30-35 g/L. Separate experiments were conducted using four different alluvial material types distinguished on the basis of predominant clast mineralogy and, with in each material type, on the basis of size fraction: fines (< 53μm), sand (53μm - 2mm), and gravel (>2mm). The results from stable isotope experiments using micromolar REE concentrations reveal an atomic number dependency on REE uptake from solution over a period of 48 hours. In three of the four basin-fill material types light REE (LREE) were preferentially sorbed relative to their heavy (HREE) counterparts. The fractionation between LREE and HREE is a result of competition between REE-complexes in solution and the formation of REE-surface complexes on the sediment particles comprising the NTS alluvium. The results from nanomolar concentration stable isotope experiments showed a much lower degree of REE fractionation, owing either to less competition for available surface sites or poor analytical resolution at such low liquid-phase concentrations. The radiotracer experiments conducted at nanomolar concentration levels, indicate that there is no systematic relationship between particle size and uptake rate, and that the uniformity of uptake rate is proportional to particle size. The radiotracer experiments indicate that REE fractionation is inversely proportional to particle size. The observations presented here are generally consistent with previous experimental studies conducted using pure, synthetic mineral phases and with solution complexation studies, indicating that the HREE have a greater tendency to form more stable solution complexes than the LREE.

## INTRODUCTION

This research investigates the solid-liquid partitioning and uptake kinetics of dissolved rare earth elements (REE) as a function of substrate mineralogy and particle size, and major ion water composition. The Yucca Flat area of the Nevada Test Site (NTS) has functioned as one of the principal underground nuclear testing areas since the early 1960's. Several of the surface subsidence features, or craters, formed as a result of underground nuclear testing have also been used as waste repository sites. Yucca Flat is a spatially heterogeneous alluvial basin where the distribution of the fill materials reflect sedimentary transport and depositional mechanisms, and the denudation sequence and locations of distinct lithologic parent units in the adjacent mountain ranges [1]. The basin-fill materials can be subdivided on the basis of the relative abundances of different source rock types. The basin-fill deposits in the northern part of Yucca Flat has been subdivided [1] into three end-member types characterized by:

a) Sediments consisting predominantly (>50%) of Paleozoic and Precambrian non-carbonate clasts ("Unit 1 - Paleozoic clastic alluvium").

b) Sediments consisting predominantly (>50%) of Paleozoic carbonate clasts ("Unit 2 - Paleozoic carbonate alluvium").

c) Sediments consisting predominantly (>95%) of Tertiary pyroclastic volcanic material ("Unit 5 - Tertiary tuffaceous alluvium").

An additional basin-fill material type has been designated [1] "Unit 3 - mixed alluvium" where no single source material type predominates. Previous research [2, 3] suggests that significant surface water infiltration is taking place as a result of crater formation, and [4] that these infiltrating waters vary significantly with regard to major ion water composition depending on the mineralogical composition of the alluvium. The behavior of dissolved fission products in the presence of geological materials is influenced by both the nature of the particle surfaces and the ionic composition of the fluid. The study of the environmental behavior of heavy fission products has often focused on the REE or lanthanide series of elements ($z = 57$ through $z = 71$) since they constitute up to 40% of the nuclide yield of a typical U or Pu fission reaction [5]. REE behavior in aqueous solution is also viewed as an analogue for the behavior of trivalent actinide elements for which fewer data are available [6, 7, 8]. In the absence of organic compounds, the solution complexation of both actinides and REE is strongly influenced by the presence of dissolved carbonate and phosphate [6, 7, 8, 9, 10]. Laboratory sorption experiments investigating REE uptake on synthetic Fe and Mn oxides [11, 12] have shown that REE partitioning onto the solid phase and the rate of uptake in the marine environment are inhibited in the presence of dissolved carbonate. The mobility of americium, discharged as an unidentified anionic complex ($\leq$ 2 nm filtrate) in treated waste water from the Los Alamos National Laboratory into a shallow aquifer, allowed it to be detected with fairly uniform activities in monitoring wells as far as 3.4 km away from the point of release [13]. The partitioning and uptake kinetics of REE from solution by geologic media is a competitive process that depends on both the nature and distribution of surface sites in addition to the composition and concentrations of complex forming ligands present in solution [e.g. 12]. Any evaluation of a potentially sorbing system should realistically be viewed from within the framework of the interrelationships between geologic media and solution composition.

## OBJECTIVES

The objectives of the present investigation are to provide an internally consistent baseline for the study of REE sorption in a heterogeneous natural environment by evaluating: (I) the solid-liquid partitioning of REE; and (ii) the uptake kinetics of REE, as a function of differences in basin-fill mineralogy and particle size, and the major ion chemistry of infiltrating waters. The incrementally trending solution behavior of REE in the presence of key complexing ligands such as carbonate [e.g. 14] should provide an incisive geochemical tool for studying their environmental mobility. An improved understanding of REE behavior under "natural" hydrogeologic conditions will enhance the ability to accurately assess potential environmental impacts related to underground nuclear testing or the storage of nuclear waste.

## APPROACH

A series of pre-nuclear test bore hole alluvium samples were collected and categorized in accordance with previous workers [1]. These samples were characterized and evaluated with regard to mineralogy, particle size distribution, cation exchange capacity (CEC), specific surface area, and associated major ion water chemistry as determined through a saturation paste extract procedure [15]. The basin-fill materials are divided into four units (Unit 1 - Paleozoic clastic alluvium, Unit 2 - Paleozoic carbonate alluvium, Unit 3 - mixed alluvium, and Unit 5 - Tertiary tuffaceous alluvium), and further subdivided based on particle size (fines - <53μm, sand - 53μm to 2mm, and gravel - >2mm). The saturation paste extract major ion composition determined for each unit (Table 1, and Figure 1) was used as the representative "groundwater" for subsequent sorption experiments. For each given basin-fill unit particle size

| | Basin-fill material | | | |
|---|---|---|---|---|
| **Constituent** | **Unit 1** | **Unit 2** | **Unit 3** | **Unit 5** |
| pH | 8.4 | 8.6 | 8.9 | 8.6 |
| $Ca^{2+}$ | $9.7\times10^{-4}$ | $8.5\times10^{-4}$ | $6.2\times10^{-4}$ | $6.8\times10^{-4}$ |
| $Mg^{2+}$ | $6.3\times10^{-4}$ | $6.4\times10^{-4}$ | $3.7\times10^{-4}$ | $4.2\times10^{-4}$ |
| $Na^{+}$ | $1.4\times10^{-3}$ | $2.1\times10^{-3}$ | $1.9\times10^{-3}$ | $4.0\times10^{-3}$ |
| $K^{+}$ | $2.3\times10^{-4}$ | $3.7\times10^{-4}$ | $2.9\times10^{-4}$ | $3.6\times10^{-4}$ |
| $Cl^{-}$ | $3.6\times10^{-3}$ | $2.5\times10^{-3}$ | $1.0\times10^{-3}$ | $2.3\times10^{-3}$ |
| $SO_4^{2-}$ | $9.8\times10^{-5}$ | $9.5\times10^{-5}$ | $1.0\times10^{-4}$ | $1.7\times10^{-4}$ |
| $HCO_3^{-}$ | $1.1\times10^{-3}$ | $2.4\times10^{-3}$ | $2.6\times10^{-3}$ | $3.4\times10^{-3}$ |
| $CO_3^{2-}$ | $1.2\times10^{-5}$ | $4.0\times10^{-5}$ | $8.7\times10^{-5}$ | $5.7\times10^{-5}$ |
| $F^{-}$ | $5.8\times10^{-5}$ | $4.1\times10^{-5}$ | $3.9\times10^{-5}$ | $1.1\times10^{-4}$ |
| $NO_2^{-}$ | $1.5\times10^{-8}$ | $1.1\times10^{-7}$ | $3.9\times10^{-7}$ | $3.3\times10^{-8}$ |
| $NO_3^{-}$ | $1.3\times10^{-8}$ | $4.4\times10^{-7}$ | $5.0\times10^{-7}$ | $3.4\times10^{-7}$ |
| $NH_4^{+}$ | $3.2\times10^{-7}$ | $6.7\times10^{-7}$ | $1.0\times10^{-5}$ | $2.8\times10^{-7}$ |
| alkalinity | 1.15 | 2.48 | 2.71 | 3.50 |

**Table 1. Major ion analyses of saturation paste extracts from Yucca Flat basin-fill materials. All values reported in mol $L^{-1}$ except alkalinity which is reported in meq $L^{-1}$.**

fraction/groundwater composition combination two groups of sorption experiments were conducted. REE stable isotope batch experiments were conducted to evaluate the "equilibrium" partitioning after 48 hours of elapsed time. Previous experimental work [11, 12] indicates that REE sorption reactions approach equilibrium after 4 to 6 hours. REE radiotracer time-series experiments were conducted subsequently to evaluate uptake kinetics.

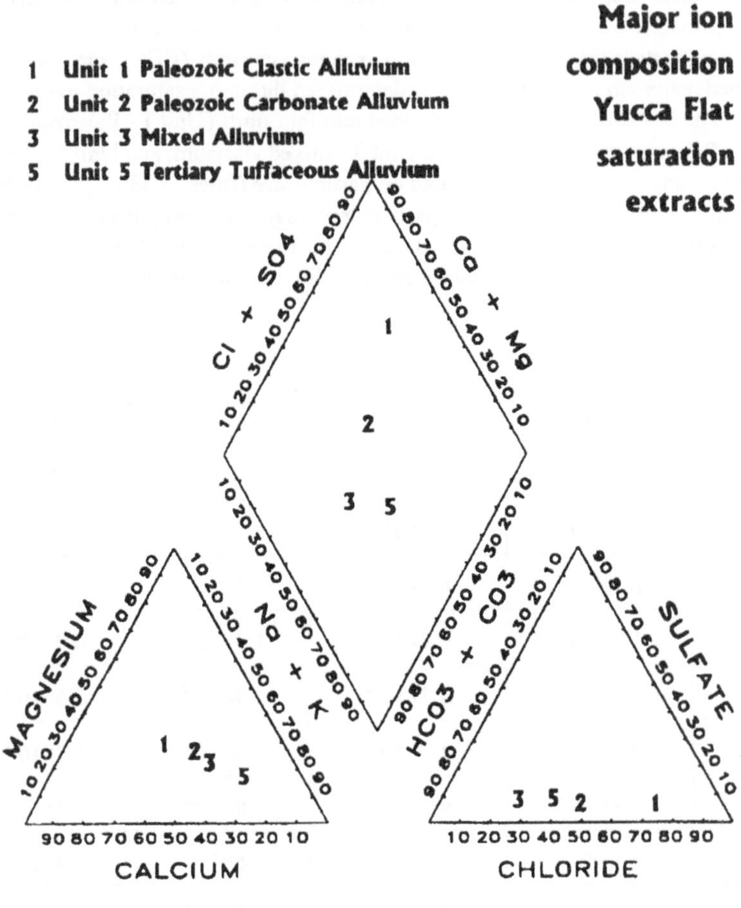


**Figure 1. Piper diagram showing major ion compositions of saturation paste extracts generated from Yucca Flat basin-fill materials. See Table 1 for data used to construct diagram.**

*Stable isotope experiments*

For each basin-fill unit, the alluvium was subdivided into the 3 particle size fractions and a 48 hour batch-type experiment was conducted using the appropriate "groundwater". "Groundwaters" for all sorption experiments were synthesized using 18 Mohm cm distilled deionized water and analytical grade reagent chemicals to match the major ion compositions of the saturation paste extracts after an equilibration time of 120 hours. The stable isotope batch tests were conducted as "free-drift" experiments in which the pH was adjusted (where necessary) to the saturation paste value at the beginning of the experiment and subsequently recorded at the completion of the experiment. Two groups of experiments were conducted with the complete suite of REE - one at a concentration of 50 nanomolar, and one at a concentration of 1 micromolar. Suspended solids loadings ranged from 30 to 35 g $L^{-1}$. For a given experiment, the appropriate volume of pH adjusted groundwater solution was divided into two approximately equal parts. The sediment was added to one half-volume and the REE spike added to the other half-volume and both allowed to equilibrate for approximately 24 hours before mixing the two half-volumes together (in the REE spiked vessel) and commencing the experiment. Reaction vessels were sealed and placed in a low speed tumbler for 48 hours of continuous agitation. Upon completion of the experiment, the liquid fraction was vacuum filtered through a 0.22μm Millipore GS filter membrane and acidified to pH < 2.0. The solid fraction (including the filter membrane from the filtration of the liquid fraction) was acidified with 50 ml of 2.8% $HNO_3$, agitated for 10 minutes then centrifuged and filtered through a 0.22μm Millipore GS filter membrane and spiked with concentrated $HNO_3$ to maintain pH < 2.0. Both fractions were subsequently analyzed by inductively coupled plasma emission spectroscopy. The liquid phases of the 50 nanomolar experiments were analyzed using a 8-hydroxyquinoline (8-HOQ) in-line column [16] flow injection apparatus (FIA) [17] on the VG PQ II-S ICP-MS to minimize groundwater matrix cation interferences on the REE measurements. Stable isotope batch experimental procedures are summarized in Figure 2.

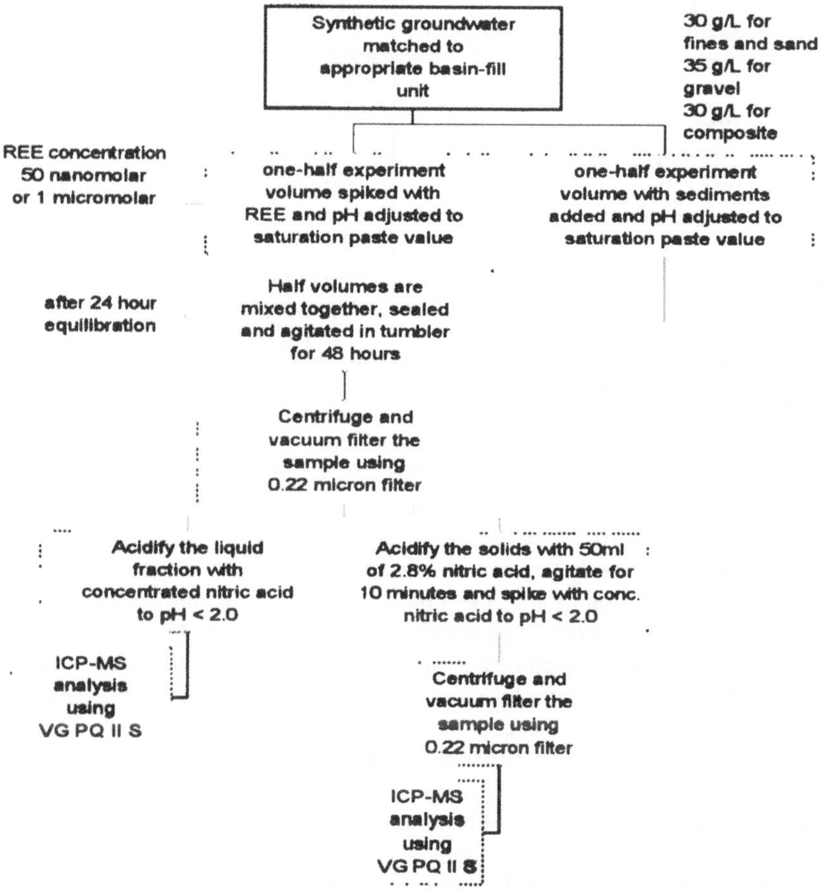


**Figure 2. Overview of REE stable isotope sorption experimental procedure.**

As a function of the REE concentrations in each phase, the partition coefficient ($K_d$) between the alluvial materials and groundwater were calculated as follows:

$$K_d = \frac{[REE]_{solid\ phase}\ (\mu g/g)}{[REE]_{liquid\ phase}\ (\mu g/g)} \qquad (1)$$

*Radiotracer experiments*

As for the stable isotope experiments, the radiotracer experiments were conducted using the 3 different particle size fractions for each of the basin-fill units. Similarly, synthesized groundwaters were used that correspond to the saturation paste extract compositions. The REE radiotracers were prepared in two groups based on: 1) the half-life of the activated nuclide; 2) the natural isotopic abundance; 3) the thermal neutron cross section; and 4) the gamma spectrum of the activated isotopes [18]. The isotopic data for the rare earth elements used in the radiotracer experiments are summarized in Table 2.

**Table 2. Isotopic data for REE used in radiotracer experiments.**

| REE | Isotopes (natural abundance) | $\sigma$ [1] (in barn) | activated isotope ($\gamma$ energy in kev) | $T_{½}$ (half-life) |
|---|---|---|---|---|
| **Group I** | | | | |
| Ce | $^{140}$Ce (88.48%) | 0.58 | $^{141}$Ce (145.4) | 32.5 days |
| Yb | $^{174}$Yb (31.8%) | 65 | $^{175}$Yb (113.8, 396.3) | 4.19 days |
| Nd | $^{146}$Nd (17.19%) | 1.4 | $^{147}$Nd (91.1, 531.0) | 10.9 days |
| **Group III** | | | | |
| Eu | $^{153}$Eu (52.2%) | 350 | $^{154}$Eu (123.1, 1274.8) | 8.5 years |
| Tb | $^{159}$Tb (100%) | 23 | $^{160}$Tb (298.57, 879.4) | 72 days |
| Tm | $^{169}$Tm (100%) | 105 | $^{170}$Tm (84.26) | 129 days |

[1] Thermal neutron cross section in barn (1 barn = $10^{-24}$ cm$^2$)

Group I consists of intermediate life radioisotopes and includes $^{141}$Ce, $^{175}$Yb, and $^{147}$Nd. Group III consists of long life radioisotopes and includes $^{154}$Eu, $^{160}$Tb, and $^{170}$Tm. Duplicate samples (2 for each group) for irradiation were prepared by placing a fragment of Amersil 20mm quartz frit into a piece of type 224 quartz tubing (both from GM Associates of Oakland, California) with one end sealed and adding exactly 1 micromole of each REE (using >99.99% pure, 10,000 $\mu g\ ml^{-1}$ spectroscopic standards from Leeman Lab Corporation) for the appropriate group. The samples were evaporated to dryness (at 60°C) and the quartz tubing was sealed and sent to the Kernforschungsanlage (Nuclear Research Center) Geesthacht (Germany) for activation. After 24 hours of "cooling" time, the quartz tube was cracked in a polyethylene bottle and the activated REE were transferred to a volumetric flask using 2.8% $HNO_3$ and diluted to 100 ml. All radiotracer experiments were conducted at an activated REE concentration of 1 nanomolar. For each group of radioisotopes a solution of complementary (the complete REE suite minus those activated) non-activated REE was added to ensure that any effects of inter-REE sorption competition and fractionation across the series could be accurately monitored. The total volume of a given experiment was divided in half with one half spiked with the radiotracers and complimentary REE solution and the sediments added to the other half. These solutions were allowed to equilibrate for 24 hours before pH adjustment (to saturation paste pH values, if necessary) and commencement of the experiment. pH monitoring was done with a glass electrode (Ingold 405-60-TT-S7) in conjunction with a digital pH meter (Knick 643). All experiments were conducted at 21 ± 1°C.

The batch-type time-series experimental procedure varied for the different particle size fractions due to handling and solid/liquid phase separation constraints. The fine size fraction (< 53µm) experiments were conducted in 1 liter Teflon bottles with suspended solids loadings of 30 g $L^{-1}$ on magnetic stirring plates using Teflon coated stir bars. Aliquots (50 ml) of the suspension were sampled with a syringe at predetermined time intervals ranging from 1 minute to 48 hours and filtered through 0.22 µm Millipore GS membranes in Swinnex holders. The gravel size fraction (> 2 mm) experiments were conducted in 2 liter polypropylene bottles with suspended solids loadings of 35 g $L^{-1}$ on magnetic stirring plates using Teflon coated stir bars. Aliquots (50 ml) of the suspension were sampled with a syringe at predetermined time intervals ranging from 1 minute to 48 hours and filtered through 0.22 µm Millipore GS membranes in Swinnex holders. The sand size fraction (53 µm to 2 mm) experiments were conducted in 50 ml polypropylene centrifuge tubes with suspended solids loadings of 30 g $L^{-1}$ agitated on a Kottermann Model 4020 reciprocating table. Samples were decanted into a 50 ml syringe at predetermined time intervals ranging from 1 minute to 48 hours and filtered through 0.22 µm Millipore GS membranes in Swinnex holders. For all experiments, the activities of the radiotracers in the filtrate and on the solids (including the filter membranes) for each time interval were counted for gamma radiation using a high resolution coaxial germanium detector coupled to an Ortec 7150 multi-channel analyzer. The measured counting rates were corrected for radioactive decay during the elapsed time between irradiation and measurement of each individual REE radioisotope. For the fine and gravel experiments, which had a filtrate mass of approximately 50 g, counting geometry corrections for the solid phase were determined by leaching the adsorbed REE radiotracers from a selected group of the extracted solids with 5-10 ml of concentrated $HNO_3$ for 15 minutes and bringing the total mass up to 50 g with 2.8% $HNO_3$. For the sand experiments, which had a filtrate mass of approximately 35 g, counting geometry corrections for the solid phases were determined by leaching the adsorbed REE radiotracers from each of the extracted solids with 1.5 ml of concentrated $HNO_3$ for 15 minutes and bringing the total mass to 36 g (includes 1 gram of solids) with 2.8% $HNO_3$. As a function of the REE activities in each phase, the partition coefficient ($K_d$) between the alluvial materials and groundwater were calculated at each time step as follows:

$$K_d = \frac{Z_{solids} \times f}{Z_{liquid}} \tag{2}$$

Where:

$Z$ = the counting rate of a given radiotracer after decay correction.

$f$ = factor to correct from solid to liquid counting geometry.

The number of counts was usually sufficient to obtain statistical counting errors less than 5%. Some problems were encountered for certain isotopes in some of the liquid phases from long duration (18 and 48 hour) experiments, where low activity levels could be compounded by short half-lives and inefficient γ emission. Experimental conditions for the radiotracer experiments are summarized in Figure 3.

## RESULTS

### *Stable isotope experiments*

The results for REE stable isotope sorption are plotted as the logarithm of $K_d$ versus atomic

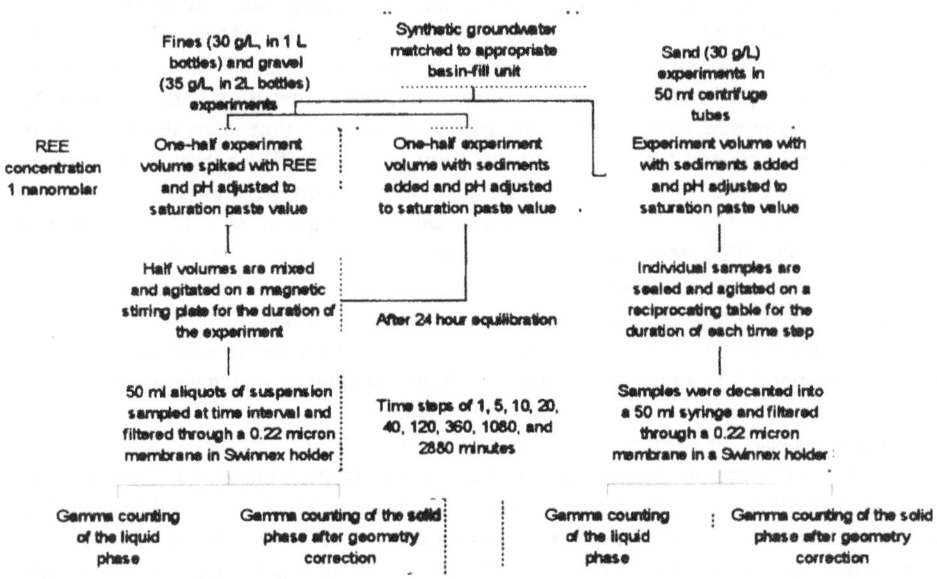


**Figure 3. Overview of REE radiotracer procedures.**

number. The data are presented as a series of plots distinguished on the basis of basin-fill unit designation and particle size fraction. Figure 4A shows the solid/liquid phase partitioning of REE from the 1 micromolar experiments relative to the four basin-fill material units with their "as collected" particle size distribution. The highest overall $K_d$ values (greatest affinity for REE sorption) were determined for Unit 2, the Paleozoic carbonate alluvium. The lowest overall $K_d$ values (lowest affinity for REE sorption) were determined for Unit 5, the Tertiary tuffaceous alluvium. REE partitioning decreases relatively smoothly for each basin-fill unit with increasing atomic number, however positive Ce and Gd anomalies are indicated for Unit 2 indicating preferential uptake of these elements relative to the adjacent elements in the series. The fractionation of REE occurs as a result of the preferential uptake of LREE from solution [19, 20]. The degree of fractionation can be approximated by the following relationship:

$$\textit{extent of fractionation} = \frac{K_d\ (La)}{K_d\ (Lu)} \qquad (3)$$

Where $K_d$ (La) is the partition coefficient for lanthanum (the lightest REE) and $K_d$ (Lu) is the partition coefficient for lutetium (the heaviest REE). The REE fractionation (as calculated using equation 3) for the 1 micromolar stable isotope experiments is summarized in Table 3.

The basin-fill unit with the lowest affinity for REE sorption (mean $K_d = 10^{2.6}$) and greatest tendency for REE fractionation ($K_d$ (La)/$K_d$ (Lu) = 18.4) is Unit 5, the Tertiary tuffaceous alluvium, which contains the highest dissolved carbonate species concentration ($3.4 \times 10^{-3}$ *m*, Table 1) in the saturation paste extract. Unit 2, the Paleozoic carbonate alluvium, exhibits the greatest affinity for REE uptake (mean $K_d = 10^{4.9}$) in addition to a significant fractionation of the REE ($K_d$(La)/$K_d$(Lu) = 16.4). The mixed alluvium of Unit 3 has the second highest dissolved carbonate species concentration ($2.6 \times 10^{-3}$ *m*) and, after Unit 5, has the second lowest order of partitioning coefficients overall (mean $K_d = 10^{3.0}$). Unit 1, the Paleozoic clastic alluvium, has a lower dissolved carbonate species concentration ($1.1 \times 10^{-3}$ *m*) with the second highest affinity for REE adsorption (mean $K_d = 10^{3.8}$) and the lowest degree of REE fractionation ($K_d$ (La)/$K_d$ (Lu) = 12.0).

| Size fraction/basin-fill unit | Unit 1 | Unit 2 | Unit 3 | Unit 5 |
|---|---|---|---|---|
| Size composite | 12.0 | 16.4 | 14.7 | 18.4 |
| Fines (< 53μm) | 0.84 | 13.2 | 20.8 | 13.8 |
| Sand (53μm - 2mm) | 5.64 | 23.5 | 10.5 | 11.7 |
| Gravel (> 2mm) | 8.46 | 18.1 | 11.8 | 20.5 |

**Table 3. REE stable isotope fractionation for the 1 micromolar experiments. The extent of fractionation is calculated using $K_d$(La)/$K_d$(Lu) and is shown for the various particle size fractions of the individual basin-fill units.**

The REE partitioning for the 1 micromolar experiments using the fine (< 53 μm) particle size fraction is plotted in Figure 4B. The order of basin-fill unit affinity for REE sorption is somewhat different from that seen in the size composites. Units 2 and 3 have essentially the same mean $K_d$ values ($10^{4.3}$) which are significantly higher than Unit 1 (mean $K_d = 10^{3.5}$) and Unit 5 (mean $K_d = 10^{3.2}$). The trends in REE fractionation are different for the fine particle size fractions than for the particle size composites. The highest degree of fractionation is seen in Unit 3 (which has the highest relative dissolved carbonate species concentration as seen in Figure 1) where $K_d$ (La)/$K_d$(Lu) = 20.8. Similar degrees of REE fractionation are present for Units 2 and 5 where $K_d$(La)/$K_d$(Lu) = 13.2 and 13.8 respectively. A value for $K_d$ (La)/$K_d$(Lu) of 0.84 for Unit 1 indicates no preferential uptake of LREE and, in fact suggests that HREE are slightly preferentially sorbed.

Figure 4C shows a plot of REE partitioning versus atomic number for the 1 micromolar batch experiments with the sand (53μm - 2mm) size fraction. The partitioning for the sands is more similar for the 4 basin-fill units than either the size composites or the fine size fraction. Mean $K_d$ values are listed in decreasing order from: $10^{3.6}$ for Unit 2; $10^{3.3}$ for Unit 1; $10^{2.8}$ for Unit 3; and $10^{2.6}$ for Unit 5. The same relative order of uptake affinity is observed for the particle size composites; this is not surprising given that the bulk (between 52% and 65% by mass) of each of the basin-fill units is comprised of sand-size material [15]. As indicated in Table 3, the

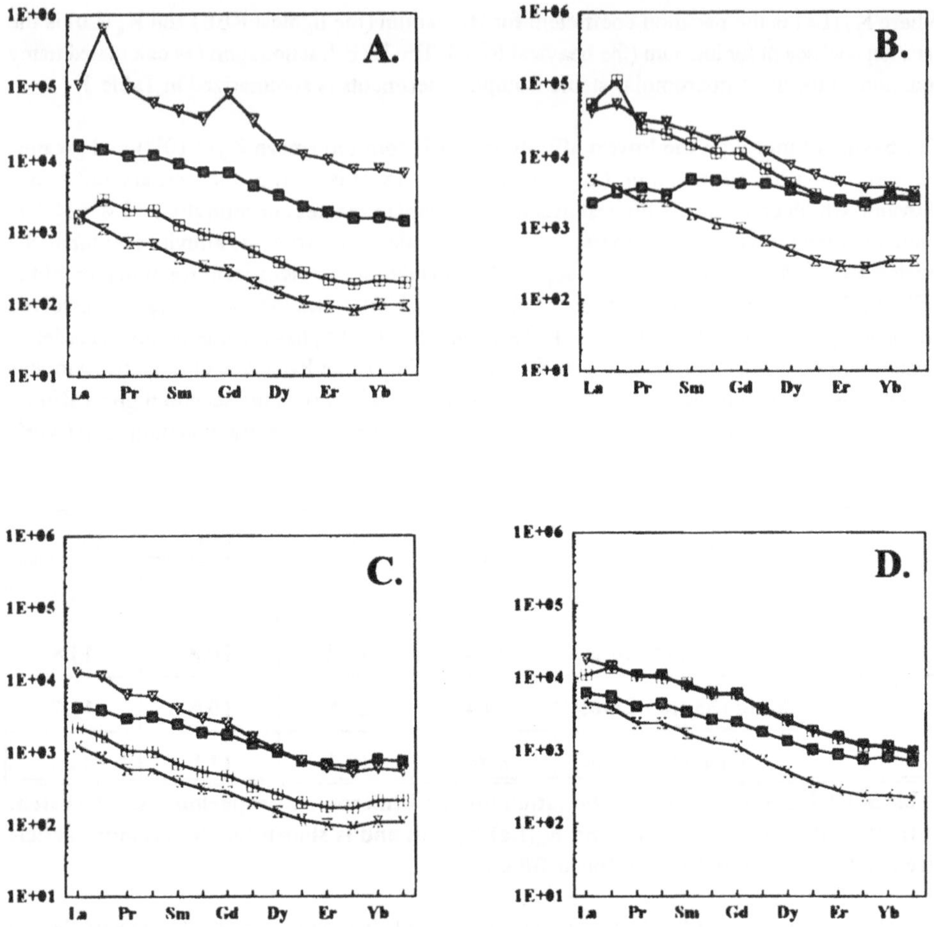


**Figure 4. REE stable isotope sorption on A) particle size composite; B) fine size fraction (< 53μm); C) sand size fraction (53 μm - 2 mm); and D) gravel size fraction (> 2mm). Dissolved REE concentration is 1 micromolar. Unit 1 is represented by the solid squares, Unit 2 by the triangles, Unit 3 by the open squares, and Unit 5 by the hourglass shapes.**

greatest REE fractionation for sand-sized materials is seen in Unit 2 ($K_d$ (La)/$K_d$(Lu) = 23.5) with similar intermediate values for Units 3 and 5 ($K_d$ (La)/$K_d$(Lu) = 10.5 and 11.7 respectively) and the lowest fractionation witnessed for Unit 1 ($K_d$ (La)/$K_d$(Lu) = 5.64).

As indicated in Figure 4D, the partitioning for the gravel size fractions shows the smallest degree of variation between the 4 basin-fill units of all the size fractions. Mean $K_d$ values are listed in decreasing order from: $10^{3.8}$ for Units 2 and 3; $10^{3.4}$ for Unit 1; and $10^{3.2}$ for Unit 5. Table 3 shows that the greatest REE fractionation for gravel-sized materials is seen in Unit 5 ($K_d$ (La)/$K_d$(Lu) = 20.5) with Unit 2 ($K_d$ (La)/$K_d$ (Lu) = 18.1) followed by Unit 3 ($K_d$ (La)/$K_d$ (Lu) = 11.8) and the lowest degree of REE fractionation is observed for Unit 1 ($K_d$ (La)/$K_d$ (Lu) = 8.46).

A series of 50 nanomolar REE concentration batch experiments were conducted using the same procedures as outlined above. The results from these experiments were inconclusive due to a number of complications. These include: near detection limit REE concentration levels in liquid phases; apparent interferences from background REE present in the minerals comprising the alluvium and dissolved during the acidification of solid phases designed to release sorbed REE; spectral interferences from "groundwater" ions which apparently were only partially eliminated by using 8-HOQ separation columns and FIA on the ICP-MS.

*Radiotracer experiments*

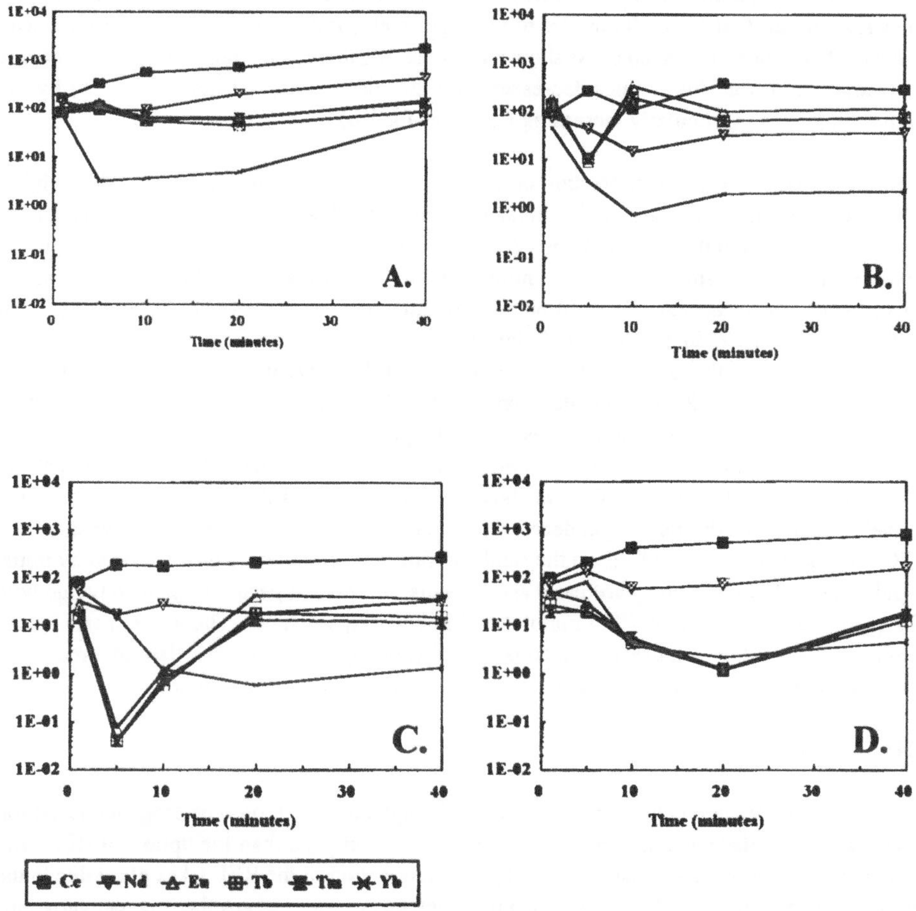


**Figure 5. Early time REE radiotracer uptake on fine grained particle size fraction (< 53 μm) from Yucca Flat basin-fill materials (A = Unit 1, B = Unit 2, C = Unit 3, and D = Unit 5). Dissolved REE concentration is 1 nanomolar and suspended solids loading is 30 g $L^{-1}$.**

The results for the 1 nanomolar REE radiotracer experiments are plotted (for a given radioisotope) as the logarithm of $K_d$ versus time. The rates of REE uptake are evaluated as a function of atomic number, basin-fill material type, and solid phase particle size. Figure 5 (A through D) show the solid-liquid partitioning of the REE radiotracers relative to the fine particle size fractions for Units 1, 2, 3, and 5 respectively. The data tend to smooth out by the 40 minute sampling interval but the initial 10 to 20 minute time intervals are somewhat erratic, particularly for the Group III radioisotopes in the Unit 2 and 3 materials. The reasons for this erratic behavior may include some non-systematic procedural variance which remains unclear at this time. Aside from the sometimes erratic behavior of the Group III radioisotopes (Eu, Tb, Tm) during the earliest times of the experiments, the rates of REE uptake show an atomic number dependency for all units. The degree of REE fractionation for this group of experiments appears to increase for the first 10 to 20 minutes of the experiment, and reaches a more or less constant value by 20 minutes elapsed time. As shown in Figure 5A, there is an indication that the degree of fractionation for Unit 1 fines decreases after 20 minutes elapsed time as evident by the converging lines (particularly increasing $K_d$ values for Yb).

The influence of particle size fraction on rate of uptake is shown in Figure 6 (A through D). These figures show the rate of Tm uptake on the individual particle size fractions for the different basin-fill units. Unit 1 (Figure 6A) shows rapid uptake by the gravel size fraction, approaching a constant value by 120 minutes with a slight gradual decline to the end of the experiment. The uptake rates for sorption onto the fines and sand size fractions are less regular with an initial rapidly increasing rate followed by a sharp decrease between 5 and 20 minutes of elapsed time, with a gradual increase to the end of the experiment. Similar behavior was observed in the Unit 2 experiments (Figure 6B). The partitioning for the gravel fraction approaches equilibrium after 360 minutes, while the partitioning for the fine and sand fractions continues to increase slightly to the end of the experiments. The particle size-dependent Tm uptake rate for Unit 3 (Figure 6C) also shows similar features, although in this case even the gravel size fraction exhibits a slight decrease in uptake rate at 360 minutes, after leveling off at 120 minutes. Partitioning of REE on the sand and fines size fractions is characterized by initial rapid uptake followed by sharp decreases and a gradual increase in partitioning to converge with the gravel size fraction at 2880 minutes (48 hours) elapsed time. The gravel particle size fraction for Unit 5 (Figure 6D) attains equilibrium by 120 minutes elapsed time, with partitioning for the fine and sand size fractions leveling off after 360 minutes.

## DISCUSSION

The results as presented above indicate that: I) the uptake of REE onto heterogeneous natural materials takes place at a slower rate and results in lower $K_d$'s than the uptake of REE onto synthesized monomineralic substrates [11]; ii) fractionation of the REE takes place due to the preferential sorption of the LREE; and, iii) REE uptake occurs in stages that suggest variations in rate controlling steps resulting from differences in substrate properties and/or dissolved complexing ligand concentrations.

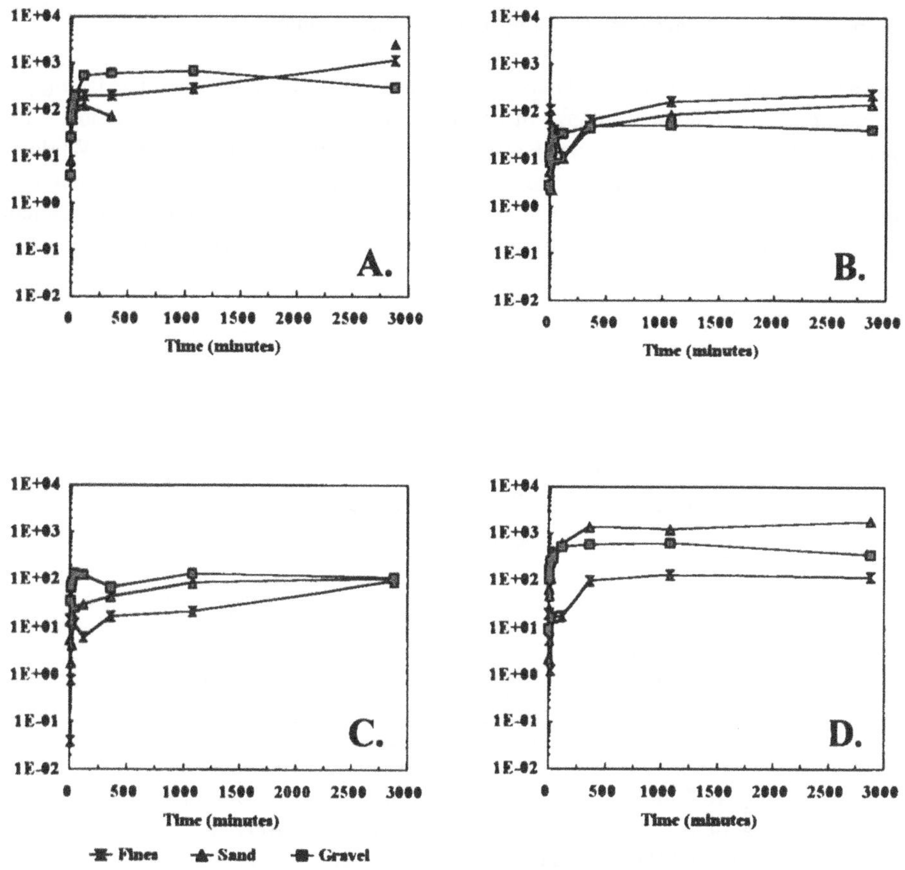


**Figure 6. Tm radiotracer uptake on: A - Unit 1; B - Unit 2; C - Unit 3; and, D - Unit 5 Yucca Flat basin-fill materials as a function of particle size fraction. Dissolved REE concentration is 1 nanomolar, suspended solids loadings is 30 g $L^{-1}$ for fine and sand size fractions, and 35 g $L^{-1}$**

The rates of REE uptake and degrees of solid-liquid partitioning determined in this study are lower than those determined [11, 12, 17] for REE uptake onto synthesized monomineralic substrates in seawater systems. $K_d$ values for REE uptake onto heterogeneous basin-fill materials are 2-3 orders of magnitude lower than those for sorption onto amorphous FeOOH in seawater. Uptake onto heterogeneous natural basin-fill was also slower, reaching a plateau in 2 to 6 hours. The interpretation of early time uptake data is complicated by sharp reversals in uptake curves, particularly for the finer grained size fractions, suggesting that experimentally induced mineral phase transformations might be taking place. Uptake curves for the gravel size fractions and most of the sand fractions are typically smoother, indicating a more stable substrate with a lower susceptibility to breakdown or transformation during aqueous suspension and abrasion. Regardless of particle size fraction, the kinetic uptake curves all tend to smooth out after 20 to 40 minutes elapsed time suggesting that the sediment particles had attained physical equilibrium with the experimental mixing and agitation conditions.

As seen in both the stable isotope and radiotracer experiments, REE fractionation is taking place as a result of the preferential uptake of LREE by solid phases [11]. Fractionation is most apparent in the 1 nanomolar fine size fraction radiotracer experiments (Figure 7). While not directly comparable, since La and Lu were not included as radiotracers in the time-series experiments, the trends in Figure 4C indicate that the degree of fractionation for the 1 nanomolar fine size fraction experiments is higher than that for the 1 micromolar stable isotope

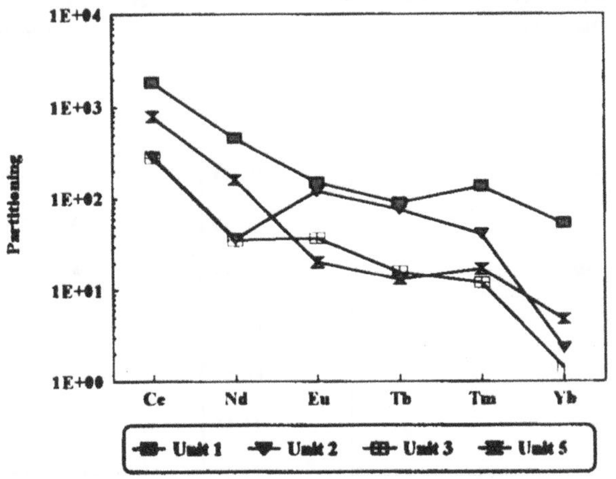


**Figure 7. REE radiotracer fractionation for the fine size fractions (< 53 μm) of Yucca Flat basin-fill materials. Dissolved REE concentrations are 1 nanomolar and suspended solids loadings are 30 g $L^{-1}$.**

experiments. While the sand size fraction in the 1 micromolar experiments (Figure 4C) indicate preferential sorption of LREE, the results from the 1 nanomolar experiments (after 40 minutes) indicated in Figure 8 indicate preferential sorption of the middle REE (MREE). Fractionation is also very apparent in the 1 micromolar stable isotope experiments using the gravel particle size fraction (Figure 4D). Procedural complications with the gravel size fraction Group I radiotracer experiments resulted in poor data quality. This precludes a reasonable assessment of fractionation.

The kinetic data indicate that REE uptake rates are not only time dependent, but also depend on the properties of the basin-fill materials. As shown in Figure 6(A-D), the uptake rates fluctuate markedly for the fine size fractions, and to a lesser extent for the sand size fractions. This suggests that uptake steps may be closely related in these instances to mineral particle transformations that occur under experimental conditions. After roughly 20 to 40 minutes elapsed time, the uptake curves smooth out suggesting that the solid particles have equilibrated to experimental conditions. The uptake curves for the gravel size fractions are much smoother. This indicates that the gravel sized particles are more stable under our experimental conditions. The steps and/or plateaus in the uptake curves indicate multi-step kinetics with different rate determining steps. Additional study is planned to further evaluate the specific uptake processes observed in this work.

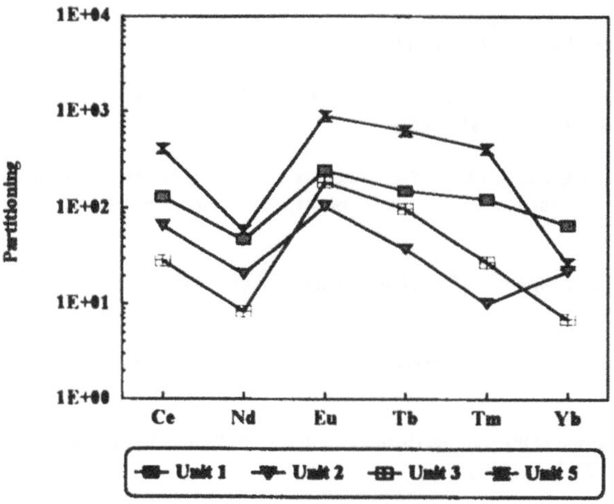


**Figure 8. REE radiotracer fractionation for sand size fractions (53 μm - 2 mm) of Yucca Flat basin-fill materials. Dissolved REE concentrations are 1 nanomolar and suspended solids loadings are 30 g $L^{-1}$.**

The results presented from this study are consistent with theoretical predictions [9, 14, 20] of REE complex behavior and experimental studies [11, 12] of REE sorption on pure synthesized mineral phases. The heterogeneous natural materials used in this study have introduced complexities not seen in prior studies. The behavior of the different particle size fractions for the different basin-fill materials suggest differences in specific uptake mechanisms. The influence of carbonate complexation on REE solution behavior is very apparent with the decreased partitioning, enhanced fractionation, and slower uptake kinetics in the presence of high dissolved carbonate levels. High partitioning values in the presence of abundant solid-phase carbonate minerals suggests that sorption of REE on to carbonate substrates can be a significant process.

## ACKNOWLEDGMENTS

We would like to gratefully acknowledge the cooperation of the staff at the Kernforschungsanlage Geesthacht for irradiation of the REE samples. We would like to thank the staff at the Zentrales Isotopenlabor, especially Bodo Schalwat, Peter Bodzian, and Michael Siebert, for technical support during the course of the radiotracer experiments. We would also like to thank Wen Xi Yuan for his assistance with the stable isotope experiments. This research was supported through a grant to UNR from the United States Department of Energy (sponsored project number DE-FC08-93NV11359).

## REFERENCES CITED

1 J.L. Wagoner. Making Sense of the Mixed Alluvium in the Yucca Flat Basin. *Fourth Symposium on the Containment of Underground Nuclear Explosions, Vol.2, Proceedings*. CONF- 870961.Lawrence Livermore National Laboratories. Livermore, CA (1987).

2. G.M. Pohll and J.J. Warwick. Coupled surface/subsurface hydrologic model of a nuclear subsidence crater at the Nevada Test Site. *Jour. Hydrol.* **186**:43-62 (1996).

3. S.W. Tyler, W.A. McKay, and T.M. Mihevc. Assessment of soil moisture movement in nuclear subsidence features. *Jour. Hydrol.* **139**:159-181 (1992).

4. F.C. Benedict Jr, E.H. DeCarlo, and X.Y. Wen. Studies of the interaction between dissolved rare earth elements and alluvial materials from the Nevada Test Site (abstract). *Eos, Trans. Am. Geophys. Union.* 76, no **46**:F275 (1995).

5. E.K. Hyde. *The Nuclear Properties of the Heavy Elements, Volume III, Fission Phenomena.* Prentice Hall, New Jersey. 519 p (1964).

6. G.R. Choppin. Speciation of trivalent f elements in natural waters. *Jour. of the Less-Common Metals.* **126**:307-313 (1986).

7. G.R. Choppin. Comparison of the solution chemistry of the actinides and the lanthanides. *Jour. of the Less-Common Metals.* **93**:323-330 (1983).

8. F. David. Thermodynamic properties of lanthanide and actinide ions in aqueous solution. *Jour. of the Less-Common Metals.* **121**:27-42 (1986).

9. R.H. Byrne, J.H. Lee, and L.S. Bingler. Rare earth complexation by $PO_4^{3-}$ ions in aqueous solution. *Geochim. Cosmochim. Acta.* **55**:2729-2735 (1991).

10. K.H. Johannesson, W.B. Lyons, and D. Bird. Rare earth element concentrations and speciation in alkaline lakes from the western U.S.A. Geophys. Res. Lett. **9**:773-776 (1994).

11. D. Koeppenkastrop and E.H. DeCarlo. Uptake of rare earth elements from solution by metal oxides. *Environ. Sci. Technol.* **27**:1796-1802 (1993).

12. D. Koeppenkastrop and E.H. DeCarlo. Sorption of rare-earth elements from seawater onto synthetic mineral particles:An experimental approach. *Chem. Geol.* **95**:251-263 (1992).

13. W.R. Penrose, W.L. Polzer, E.H. Essington, D.M. Nelson, and K.A. Orlandini. Mobility of plutonium and americium through a shallow aquifer in a semiarid region. *Environ. Sci. Technol.* **24**:228-234 (1990).

14. S.A. Wood. The aqueous geochemistry of the rare-earth elements and yttrium. *Chem. Geology.* **82**:159-186(1990)

15. F.C. Benedict Jr, E.H. DeCarlo, and W.B. Lyons. Basin-fill alluvium from Yucca Flat, Nevada: The influence of sediment variability on contaminant attenuation potential. *Soil Science Society of America Journal.* In review.

16. K. Orians and E. Boyle. Determination of picomolar concentrations of titanium, gallium, and indium in seawater by inductively coupled plasma spectrometry following an 8-hydroxyquinoline chelating resin preconcentration. *Anal. Chim. Acta.* **282**:63-74 (1993).

17. L. Ebdon, A. Fisher, H. Handley, and P. Jones. Determination of trace metals in concentrated brines using inductively coupled plasma mass spectrometry on-line pre-concentration and matrix elimination with flow injection. *Jour. Anal. At. Spectrom.* **8**:979-981 (1993).

18. D. Koeppenkastrop, E.H. DeCarlo, and M. Roth. A method to investigate the interaction of rare earth elements with metal oxides in aqueous solution. *Jour. Radioanal. Nuclear Chem.* **152**: 337-346 (1991).

19. R.H. Byrne and K. Kim. Rare earth element scavenging in seawater. *Geochim. Cosmochim. Acta.* **54**:2645-2656 (1990).

20. K.J. Cantrell and R.H. Byrne. Rare earth element complexation by carbonate and oxalate ions. *Geochim. Cosmochim. Acta.* **51**:597-605 (1987).

*Proc. 30th Int'l. Geol. Congr.*, Vol. 19, pp. 189-206
Xie Xuejin (Ed)


# Evaluation of Metal Retention in a Wetland Receiving Acid Mine Drainage

GYŐZŐ JORDÁN[1,2], ANDREA SZŰCS[1,2], ULF QVARFORT[1] and BALÁZS SZÉKELY[3]
[1] *Uppsala University, Inst. for Earth Sciences, Norbyvägen 18B, Uppsala S-75236, Sweden*
[2] *present address: Hungarian Geological Survey, Stefánia út 14, Budapest 1143, Hungary*
[3] *Geophysical Research Group, Hungarian Academy of Sciences, Ludovika tér 2, H-1083 Budapest, Hungary*

**Abstract**

The behaviour of sedimentary Pb, Ni, Cu, Zn, Fe and Mn in a natural wetland (central Sweden) impacted by acid mine drainage was studied to determine the potential for long-term retention of these metals in the sediments. Partial abundances of metals in peat sediments were determined by a sequential extraction procedure giving five fractions potentially sensitive to changes in environmental conditions. Sampling design was carried out using landscape geochemical and geographic information system techniques. Geochemical abundances, geochemical gradients and geochemical flow patterns were analysed and modelled and the wetland-acid mine drainage interface was interpreted as a landscape geochemical barrier. Geochemical data interpretation and modelling was carried out using the robust techniques of Exploratory Data Analysis to treat multimodal data populations and accommodate information inherent in outlying values. The interpretation of results was enhanced by Multivariate Statistical Analysis. Results show that Pb, Cu and Zn are efficiently retained in stable metal-organo-complexes, but Mn and Ni remain mobile or bound as labile outer-sphere complexes in the reducing environment. Fe migrates with groundwater and diffuses upwards along concentration gradients and precipitates in top layers of peat sediments as amorphous iron oxides. Micro-environments are important in sulphate reduction and sulphide formation in the peat sediments. Under the acidic conditions encountered in the studied mire only copper sulphide may precipitate.

*Keywords: acid mine drainage, statistical data analysis, geographical information system, landscape geochemistry, metal speciation, wetland*

## INTRODUCTION

Acid mine drainage (AMD) flowing from abandoned ore mines is a major environmental pollution problem in mining areas in central Sweden. If left untreated, AMD can have adverse effects on natural ecosystems and human health. Natural wetlands offer inexpensive treatment alternatives to chemical neutralisation and precipitation processes [1]. However, the effectiveness of wetland treatment depends on many chemical and biological factors, and the long-term capacity of wetlands to retain metals has to be assessed by the modelling of metal-sediment interactions and by evaluating the stability of the wetland system as a whole.

The purpose of this research was to describe the behaviour of six metals, Pb, Ni, Cu, Zn, Fe and Mn, in a natural wetland affected by AMD in an effort to determine if wetlands can act as permanent metal sinks. A sequential extraction procedure of Tessier et. al. [2] was adopted to determine metal speciation in the sediments. Despite that sequential extraction is only an operational definition of solid materials and does not identify specific mineral phases [3, 4], it is still a useful means of studying metal associations provided the data are interpreted with care.

The specific objectives of the investigation were: 1. to determine the relative partitioning of metals into five sedimentary fractions likely to be affected by various environmental conditions; 2. to evaluate the stability of various metal forms within the sediments by correlating the spatial distribution of chemical parameters and to determine the relation between these parameters and samples using statistical techniques; and 3. to evaluate the stability of the wetland system. Since

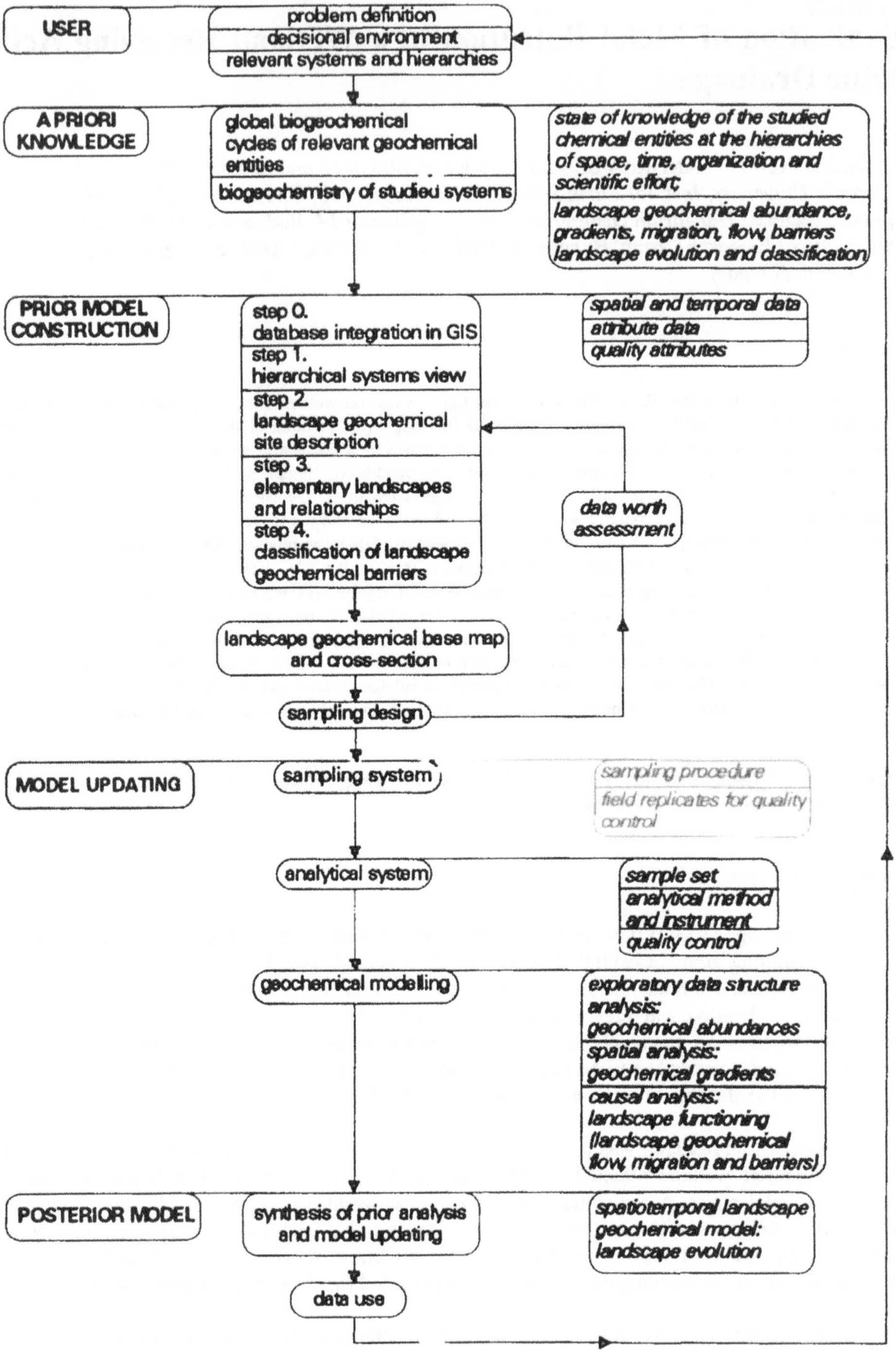


**Figure 1.** The modelling scheme of the investigation. Arrows show the flow of information. Shaded parts are discussed in detail in the text.

the chemistry of the contaminated groundwater discharging into the studied mire is well documented [5] the area offers a good opportunity to analyse and model metal behaviour in this environment and estimate the potential for long-term metal retention.

## METHODS

The investigation followed the geochemical modelling procedure shown in Fig. 1. According to this scheme landscape geochemical methodologies [6] and GIS techniques are used for the design of sampling strategy. The local geochemical flow patterns are analysed by the classification of elementary landscapes (areas which are geochemically homogeneous with respect to matter flow patterns) according to landscape geochemical principles [7, 8]. Relationships between adjacent landscapes are analysed by the construction of matter transport models, such as run-off models and watershed models, derived from the digital elevation model (DEM). Then landscape geochemical barriers along elementary landscape boundaries are defined and classified following Perel'man's scheme [9]. In this way a landscape geochemical base map and cross-section of the area is produced showing elementary landscapes and their relationships by matter flow. Next, data obtained in the sampling and analytical systems is modelled. In geochemical data modelling the studied super-aqual mire landscape is considered as a system where the input is the positive extra landscape flow (+ELF) of metals in the AMD groundwater from the neighbouring eluvial geochemical landscapes. The output is the spatiotemporal metal distribution in the mire.

In this application biogeochemical processes are considered stochastic processes and numerical modelling uses statistical methods. In accordance with landscape geochemical methods, first patterns in *geochemical abundances* are analysed. The univariate robust techniques of Exploratory Data Analysis [10-12] and Multivariate Statistical Analysis, such as Cluster Analysis (CA) and Principal Component Analysis (PCA) are applied for the analysis in the present study. Next, in the modelling procedure, *landscape geochemical gradients* are studied and modelled using geostatistical methods. *Geochemical migration* and *flow patterns* are then analysed and correlation analysis and multivariate Canonical Correlations and Multiple Regression Analysis are implemented to reveal significant relationships among the measured variables and groups among samples. All steps of the statistical analyses and results are controlled interactively.

Thermodynamic modelling was not considered in this study because *1.* natural water systems are open, dynamic, non-equilibrium systems, *2.* many redox reactions are very slow and are often biologically mediated, hence they depend on biological activity, *3.* redox reactions often proceed independently of one another, thus, there may be several different 'apparent' oxidation-reduction levels in the same location and *4.* redox levels probably vary from one microenvironment to the next [13].

## LOCATION OF STUDY AREA AND SITE DESCRIPTION

The location of the studied Kambamyran mire in the Slättberg area in central Sweden (60° 48'N, 15° 5'E) is shown in Fig. 2. The inset of the figure is a composite of regional geological and physical geographical maps and climatic data. Regional data show that the area has a background geochemical composition characteristic of island-arc igneous rocks and associated ore deposits which received upper-amphibolite facies metamorphosis [14]. The low annual average temperature and long winter indicate that the role of vegetation and microbial activity is limited in element mobilisation and cycling in the area. Wind directions suggest that the major sources of atmospheric heavy metal contamination in top layers of soils throughout Scandinavia are the southern Scandinavian and continental industrial regions. This was confirmed by the observation of regional N-S metal pollution gradients [15, 16].

The local geology is characterised by early- to syn-orogenic Proterozoic granitoid and, to a lesser extent, by basic intrusive rocks (Fig. 2). The main geological feature of the area is an amphibolite dyke which can be followed for about 1600 m. The 1.5 m wide mineralisation is concentrated in the middle of the dyke and the ore consists of pyrrhotite ($Fe_{1-x}S$), pyrite ($FeS_2$), pentlandite ($(Fe,Ni)_9S_8$) and bravoite ($(Fe,Ni)S_2$), millerite (NiS), linnaeite ($(Co,Ni)_3S_4$) and chalcopyrite ($CuFeS_2$) [17]. The bedrock is covered with bouldery-sandy glacial till deposits and glaciofluvial sediments.

The ombrotrophic bog is slightly raised with swampy parts along wetland streams and a 3-5 m wide shallow marginal open water zone. In the marginal waters iron-oxihydroxide precipitates are observable (bog iron). According to field observations the groundwater table is about 50 cm below surface in the dry season and 20 cm when highest in the autumn, giving a 30 cm wide zone of fluctuating reducing-oxidising conditions. The thickness of the sediments is 2-2.5 m on average, which is characteristic of bogs in central Sweden [18]. The bog is fed and intersected by a braiding river and a smaller stream, both crossing the mineralization and accompanying mining area

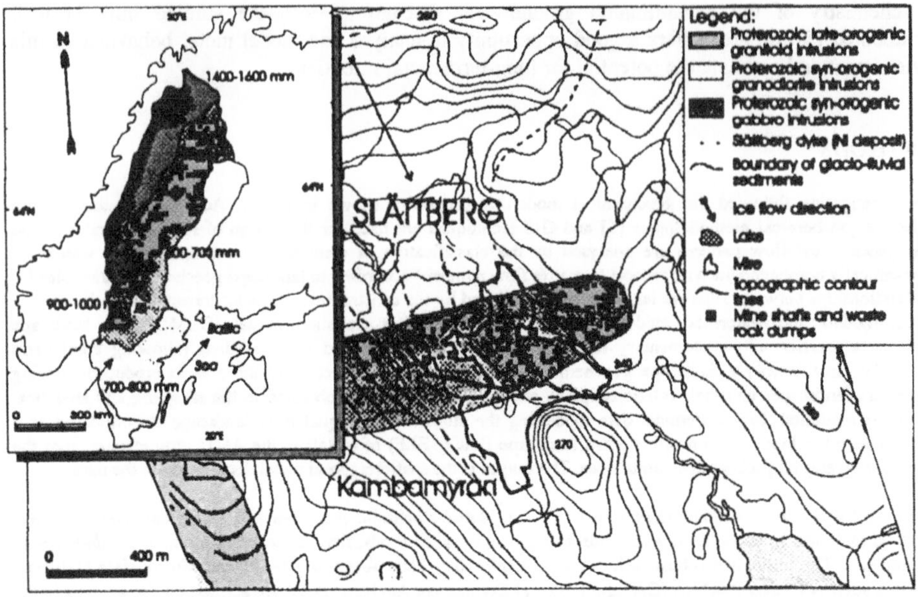


**Figure 2.** Location of study area. *Large map:* a result of GIS overlay operation of topographical-, hydrological-, geological and quaternary deposits layers, also showing the location of mine shafts and waste rock dumps; arrow indicates ice movement during the latest glaciation. *Inset:* the study area (solid box) is situated in the 'southern boreal zone' of Scandinavia and falls into the region of the Svecofennian island arc lithologies of Early Proterozoic age (dotted line). Arrows indicate wind directions in winter; figures are average annual precipitations in millimetres.

upgradient to the mire. The vegetation is typically hummocky bog vegetation characterised by Sphagnum species, primarily *Sphagnum fuscum*, with sparse pines.

The most important imprint of human activity in the area is a series of small, abandoned mine shafts and waste rock dumps along the dyke upgradient to the mire (Fig. 2). The mines were worked primarily for Ni in the first half of the century. The waste material has been exposed to atmospheric oxidation and heavy metals, such as Ni, Pb, Zn and Cu have been leached from the dump into the groundwater and discharged as acid mine drainage into the mire for about 70 years. The groundwater is characterised by low pH (pH <3), elevated $SO_4$ (3,750 mg/L) and high trace metal concentrations (Cu: 40 mg/L, Zn: 18.5 mg/L, Ni: 32.1 mg/L, Fe: 1,110 mg/L, Pb:0.29 mg/L, Mn: 4.4 mg/L). These concentrations are considerably attenuated by the point where AMD discharges into the mire (pH: 3.96, $SO_4$: 460 mg/L, Cu: 4.4 mg/L, Zn: 4.6 mg/L, Ni: 2.5 mg/L, Fe: 0.03 mg/L, Pb: <0.05 mg/L, Mn 1.98 mg/L) (groundwater samples collected in May 1993) [5].

## SAMPLING DESIGN

The goal of sampling was to reconstruct the spatial extent of AMD contamination plume in the mire and analyse the controlling processes of metal transportation and retention. For the design of sample locations the local geochemical flow patterns were analysed by the classification of elementary landscapes. The permanently water-saturated wetland was classified as a reducing gley super-aqual landscape, and the neighbouring areas as eluvial-accumulative landscapes [7, 8] (Fig. 3). A DEM using triangular irregular network (TIN) interpolation was constructed and geochemical relationships between adjacent landscapes were analysed by the overlay of the classified elementary landscape map, watershed model, slope- and slope-break model and run-off model derived from the DEM. Landscape geochemical barriers were classified as complex reducing gley-adsorption barriers where slightly acidic groundwater is discharging into the water-logged mire sediments [9]. By the superimposition of potential metal contamination sources, such as ore mines and waste rock dumps, the path of AMD from sources to end at complex reducing hydrogen sulphide-adsorption barriers was traced. Although the water flow model is based on surface transport directions, relative homogeneity of the hydrological properties of the aquifer over the study area was assumed and the surface run-off directions could be regarded as reasonable approximations of groundwater

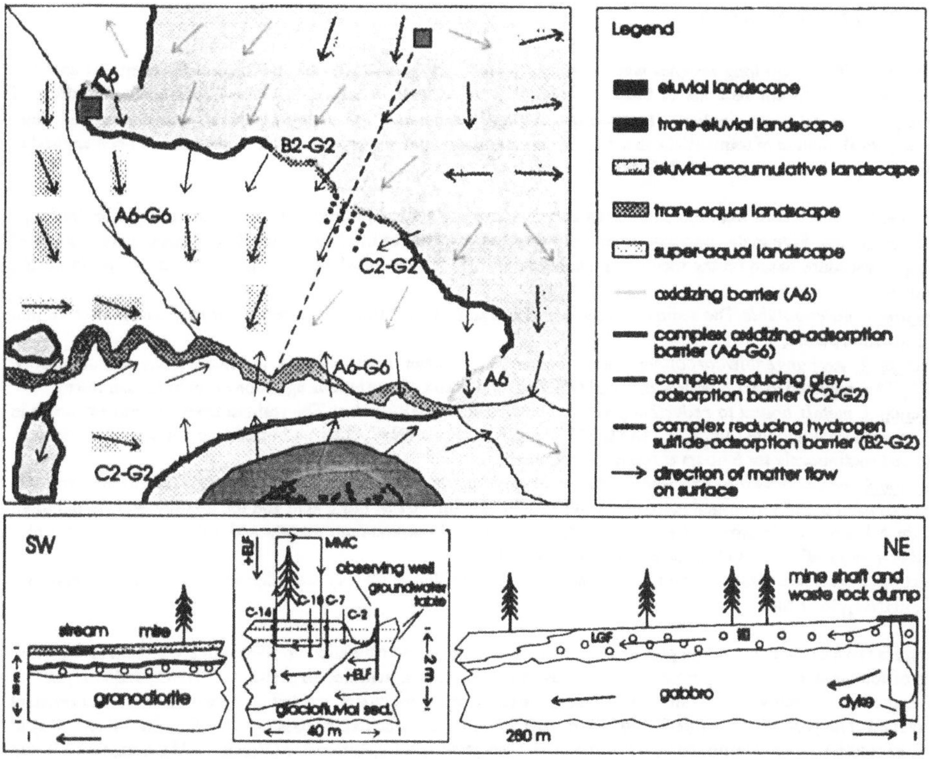


**Figure 3.** Sampling design based on landscape geochemistry and GIS technique. The landscape geochemical base map shows elementary landscapes, the classified landscape geochemical barriers and matter flow directions on the topographic surface. Dots are locations of peat sediment cores in the mire. The cross-section indicates matter transport directions and patterns within and between landscapes (MMC: main migration cycle; LGF: landscape geochemical flow; ELF: extra landscape flow). Sampling locations are also shown.

flow as well. Based on the model systematic sampling was designed along an oblique grid where AMD discharges into the mire (Fig. 3).

Horizontal sampling density was higher (5 m separation in the groundwater flow direction) close to the edge of the mire in order to detect abrupt changes in concentration gradients due to spatially separated processes. Samples further along the flow direction were placed 10 m apart. All samples were located 15 m apart from each other parallel to the edge of the mire (Fig. 3). Vertical sampling density was designed to characterise geochemical conditions in the peat profiles above, at and below the groundwater fluctuation zone representing oxidising, fluctuating oxidising-reducing and permanent reducing conditions (Fig. 3).

## SAMPLING SYSTEM

Fifteen peat cores 5 cm in diameter were taken using a hand-held steel tube corer in the last week of May in 1994. Cores were sliced into 10 cm sections and were extruded into plastic Ziploc bags in the field and transported to the laboratory on ice. The samples were kept at 4 °C and were extracted within five days. Plastic tools were used throughout the whole sampling procedure to minimise contamination and only the inner parts of cores were used later in the analysis.

## ANALYTICAL SYSTEM

Altogether fifty 10 cm long samples were selected 30 cm apart from each other in the cores for chemical analysis. The uppermost 10 cm sections of the cores were not selected for analysis in order to eliminate the effect of atmospheric metal contamination. The solid speciation of trace metals in the peat sediments was determined using the empirical method of sequential extraction of operationally defined fractions. Such methods have been applied to solids from various environments [19-22].

Samples from each core were forced through a 2 mm sieve. A 10 g test portion of each was separated after thorough homogenisation. Five metal fractions potentially sensitive to changes in environmental conditions were elucidated using a procedure based on the method of Tessier et al. [2]. The chemical extractions were made in the following sequence:

Fraction 1. *water soluble* The sediment sample was extracted in 1:2 solid:deionized water suspension after 20 min equilibration period;

Fraction 2. *exchangeable/carbonate bound phase* The residue from (1) was leached in ammonium acetate ($NH_4OAc$) adjusted to pH 5 with acetic acid (HOAc) for 5 hours of continuous agitation at room temperature;

Fraction 3. *metals bound to reducible phases (amorphous Fe-Mn oxides)* The residue from (2) was extracted in 0.3 M natrium dithionite ($Na_2S_2O_6$) and 0.175 M natrium citrate ($NaC_6H_5O_7$) adjusted to pH 4.8 with acetic acid, agitated continuously for 6 hours at room temperature;

Fraction 4. *metals bound to oxidisable phases (organic matter and amorphous sulphides)* The residue from (3) was extracted in 30% hydrogen peroxide buffered to pH 2 with 0.02 M nitric acid and the mixture was heated to 85 °C for 5 hours with intermittent agitation. After cooling 3.2 M ammonium acetate and 20% (v/v) nitric acid plus double volume of water was added and agitated for 30 min;

Fraction 5. *residual (detrital and lattice metals)* The residue from (4) was digested in hot aqua regia 3:1 ($HCl:HNO_3$) for 1 hour.

Samples were extracted in 50 ml polyethylene tubes followed by centrifugation at 3000 rpm for 15 min to minimise loss of solid material. The supernatant was removed with a pipette, acidified to pH<2 with $HNO_3$ and analysed for trace metals, whereupon the residue was resuspended in deionized water, centrifuged again and the supernatant discarded. The procedure was performed after each successive extraction. Metals (Pb, Ni, Cu, Zn, Fe and Mn) in the test solution were determined by a flame atomic absorption spectrophotometer (Varian 1275). Sediment pH was determined using 1:2 weight suspension of homogenised fresh sediment and deionised water after 10 min equilibration period while loss-on-ignition was determined on dried samples (105 °C for 48 hours) at 550 °C for 6 hours.

It is noted that the distribution of a given metal between various fractions obtained by the above sequential extraction procedure does not necessarily reflect the scavenging action of discrete sediment phases, but rather is operationally defined [2]. No attempt was made to distinguish between ion-exchangeable and carbonate-bound metals since carbonates precipitate only from neutral or alkaline waters and under anaerobic conditions, carbonates are not expected to be present in significant quantities in the studied sediments. In fraction 3, dithionate-citrate mixture was used although it is less effective in extracting Fe-Mn oxide-bound metals than, for example, hydroxylamine-hydrochloride-acetic acid mixture. No attempt was made to separate metals adsorbed by different forms of organic matter [23]. Uncertainty in the metal content of the oxidisable phase (fraction 4) arises from the fact that some humic and fulvic acids are extremely stable and the oxidation of all forms of organic matter may not be complete. More efficient method is the use of cc. $HNO_3$ and hydrochloric or perchloric acid but these partially attack silicate lattices. It should also be noted that extraction with aqua regia is capable of extracting only about the 70-90% of the residual material, hence the results underestimate the true concentrations. Uncertainty in the pH measurements may arise from the increase in pH due to the loss of volatile acids (for example $CO_2$ degassing) upon sample processing and from photodecomposition of organic matter which lowers pH. In sulphide-rich peats the oxidation of sulphides is also a source of error [24].

## RESULTS OF GEOCHEMICAL MODELLING

### *Exploratory Data Structure Analysis: Geochemical Abundances*

According to data in Table 1, most of the Cu (around 50% of the total concentrations) is found in the oxidisable phase demonstrating its high affinity for organic and/or sulphide complexation. Manganese is probably in its reduced Mn(II) state which explains its higher proportion in the exchangeable and water soluble phases (70-80% of total concentrations). Ni is also abundant in the exchangeable-phase (around 30% of total) indicative of its mobility in the peat environment. Iron is mostly found in the reducible phase (70-80% of total) which is somewhat surprising in an acidic reducing gley environment. This can be explained, however, by the improper selectivity of the extraction procedure 3 (see below) and the existence of resistant crystalline iron oxides which is supported by the high concentrations of iron in the residual phase. One would expect lead to be preferentially bound to organic matter or sulphides, but in this case it is found mostly in the reducible phase (40-50% of total). Since its concentrations are also relatively high in the residual fraction Pb seems to be associated with iron which can only be explained, regarding their otherwise very different geochemistry, by assuming that Pb is found in resistant sedimentary iron oxides and/or clays. Correlation analysis discussed below reinforces this assumption. All the metals are greatly enriched in the exchangeable or water soluble phase in samples close to the AMD discharge showing the location and extension of contamination.

The relative variability of Pb, as measured by the IQR/M (inter quartile range/median) ratio, was the lowest among the metals in all phases, particularly in the water soluble fraction showing that Pb is the least mobile element in the peat. In accordance with the above, Mn and Ni had the highest variability in the water soluble and exchangeable phases. Copper was the most variable metal in the exchangeable, oxidisable and reducible phases followed by Zn. The relatively high IQR/M ratio for Pb in fraction 2 (exchangeable) may suggest that this is the pool in which local geochemical conditions cause the most significant changes in Pb adsorption processes. In general, most of the metals were least variable in the residual phase, with the exception of Fe and Mn and in the water soluble phase apart from Ni and Mn. It seems plausible that the high variability of Fe and Mn is because they are in different mineral phases, while Mn and Ni are the most loosely adsorbed metals controlled by even small alterations of local adsorption conditions.

If the AMD input were the only major source of heavy metals in the studied mire then the ratios of median total concentrations to the groundwater concentrations would be the enrichment (or accumulation) factors due to retardation which are in this case in the order: Fe(530)>Pb(320)>>Cu(32)>Zn(18)>Ni(4.8)>Mn(1.4). The enrichment of iron is hardly feasible since it is mostly mobile in reducing peat sediments [25]. Although Pb has high affinity to organic matter, it is difficult to explain how it can have a much higher enrichment factor than the even more strongly-bound Cu [26] which is more than 100 times more abundant in the discharging AMD. Clearly, Fe and Pb have an other, probably common, source, too, which is also suggested by their relatively high concentrations in the residual phase. Manganese seems to migrate with groundwater almost intact in the peat sediments.

In order to study the univariate structure of variables and obtain information about their populations, stem-and-leaf displays were constructed for each variable and outliers and far-outliers were identified using the robust inner-fence outer-fence criteria [10]. Separation of populations in the geochemical data was enhanced by the use of Sinclair's graphical method based on probability graphs [27]. For copper, as shown in figure 5, two widely separated populations are apparent: the one with smaller values contains sample locations further away from the input ('background population') and the other all consists of higher or outlier values and cover samples close to the AMD input ('anomalous population'). This suggest that there are basically two spatially separated regions in the course of Cu migration in which different processes are responsible for Cu distribution. (Obviously, if one process or group of processes in a laterally homogeneous matrix, such as peat could be assumed, one population would result.) Water soluble and exchangeable phase variables were most skewed probably representing the effect of contaminant plume while residual phase distributions were most symmetric for all metals. Symmetric distributions in the residual and oxidisable phases for Pb, Mn and Fe indicate that these metals are uniformly distributed in the sampled mire region because they represent the geochemical background dominating in the time of mire formation. It is also interesting that a sub-population within the anomalous population of copper is only apparent in the water soluble and oxidisable phases (and to a lesser extent in the reducible phase) in the samples from 10-20 cm depth. Beyond implying that these two phases are related to each other, this may suggest that the oxidisable phase for the

**Table 1.** Letter-value displays of measured variables (sample size: 50). All concentrations are in ppm, Loss-on-ignition (LOI) is in percent of dry (105 °C) sample.

| Letter-value display | Pb | Ni | Cu | Zn | Fe | Mn |
|---|---|---|---|---|---|---|
| **Fraction 1** | | | | | | |
| Upper eight | 0.04 | 0.9 | 1.82 | 3.14 | 0.86 | 0.54 |
| Upper quartile | 0.02 | 0.58 | 0.98 | 1.38 | 0.34 | 0.42 |
| Median | 0 | 0.17 | 0.38 | 0.66 | 0.25 | 0.1 |
| Lower quartile | 0 | 0.04 | 0.16 | 0.28 | 0.16 | 0.02 |
| Lower eights | 0 | 0.04 | 0.06 | 0.18 | 0.12 | 0.02 |
| **Fraction 2** | | | | | | |
| Upper eight | 2.59 | 3.61 | 36.4 | 62.65 | 3.4 | 2.77 |
| Upper quartile | 1.16 | 3.08 | 9.8 | 36.75 | 2.14 | 2.14 |
| Median | 0.335 | 2.295 | 0.37 | 11.375 | 1.35 | 1.33 |
| Lower quartile | 0.11 | 0.18 | 0.18 | 5.6 | 0.77 | 0.95 |
| Lower eights | 0.04 | 0.07 | 0.14 | 3.15 | 0.6 | 0.6 |
| **Fraction 3** | | | | | | |
| Upper eight | 4.55 | 8.4 | 26.5 | 37.1 | 130.9 | 0.77 |
| Upper quartile | 4.55 | 6.65 | 10.85 | 28 | 71.05 | 0.39 |
| Median | 3.85 | 4.2 | 1.385 | 17.675 | 45.5 | 0.32 |
| Lower quartile | 3.5 | 2.45 | 1.09 | 7.7 | 21.7 | 0.25 |
| Lower eights | 3.5 | 1.82 | 0.98 | 5.95 | 16.1 | 0.18 |
| **Fraction 4** | | | | | | |
| Upper eight | 3.05 | 1.26 | 110.95 | 40.95 | 13.58 | 0.21 |
| Upper quartile | 2.38 | 0.95 | 21.7 | 31.15 | 7.74 | 0.21 |
| Median | 1.89 | 0.65 | 1.805 | 20.475 | 5.235 | 0.18 |
| Lower quartile | 1.51 | 0.46 | 0.91 | 5.25 | 3.82 | 0.11 |
| Lower eights | 1.23 | 0.39 | 0.7 | 1.58 | 3.01 | 0.11 |
| **Fraction 5** | | | | | | |
| Upper eight | 1.52 | 0.58 | 7 | 14 | 15.6 | 0.32 |
| Upper quartile | 1.36 | 0.44 | 1.9 | 10.8 | 6.6 | 0.2 |
| Median | 1.12 | 0.34 | 0.58 | 6.7 | 3.99 | 0.09 |
| Lower quartile | 1.04 | 0.28 | 0.44 | 4 | 2.38 | 0.02 |
| Lower eights | 0.92 | 0.26 | 0.36 | 3.6 | 1.8 | 0.02 |
| | **pH** | **LOI** | | | | |
| Upper eight | 4.03 | 98.6 | | | | |
| Upper quartile | 3.87 | 98 | | | | |
| Median | 3.74 | 97 | | | | |
| Lower quartile | 3.64 | 95 | | | | |
| Lower eights | 3.54 | 89.64 | | | | |

anomalous population actually represents two processes in Cu retention below and above the groundwater table. Assuming again the horizontal homogeneity of the peat sediments with respect to organic matter and sedimentary mineral distribution, and noting that the operationally defined oxidisable phase may include both organic-bound and sulphide-bound metals, it can be interpreted that Cu retention might be controlled by sulphide formation below groundwater and by organic matter in the more oxidising peat layers above. Similar separation is absent in the exchangeable and residual phases indicating that these processes are homogeneous in the peat with respect to depth. This hypothesis will be further tested in the causal analysis. pH has no major role in the separation of the samples in the phases of copper because it has only one well defined population. Ash content was also roughly divided into two populations coinciding with those of Cu.

Separation of populations in the multidimensional variable space was studied by CA and PCA using log-transformed data [28]. Cluster analysis for Cu using Euclidean distance as a measure of similarity and average linkage method also confirmed the separation of a background cluster (n=32) and an anomalous one (n=18) with a sub-cluster for samples at 20 cm depths in the latter which readily merged into the background cluster at higher similarity levels. This implies that these samples are geochemically closer to the background cluster, which supports indirectly the hypothesis that in the background only organic matter controls Cu retention similarly to the relatively oxidising anomalous samples at shallow depth. In the anomalous population strong sub-clusters were apparent showing the higher inhomogenity of this region. In the plane of the first two principal components (93% of the total variance) the same groups and sub-groups became apparent (Fig. 4). If water soluble phases represent the propagating plume, then it is obvious from the figure that separation of samples is due to differences in the adsorption-retardation processes represented by the first principal component.

Although the presence of sub-populations was observed for the other metals too, distinct separation such as for Cu was not so obvious. Slightly overlapping sub-populations were most apparent for Ni and Zn, coinciding with that of Cu, and virtually absent for Mn, which had the most symmetric distributions. These patterns may imply that Ni and Zn behave similarly to Cu, while Mn seems to be the most inert metal in this environment. Iron showed relative uniformity too, with a remarkable difference in the reducible phase where high outliers were all at depth 10-20 cm, indicating that Fe is enriched in oxidised form in the peat sediments above groundwater table. These findings were also confirmed by CA and PCA.

*Spatial Analysis and Modelling: Geochemical Gradients*

Horizontal metal distributions were studied by the construction of isoconcentration maps for each fraction of each metal in three horizons representing peat sediments above, at and below groundwater table fluctuation (10-20, 40-50 and 70-80 cm below mire surface, respectively). As shown in Fig. 4, it was possible to reconstruct the shape of the plume by the collected samples, verifying the appropriateness of the sampling design. It can also be seen that the highest gradient for copper is within about 10 m from the input. The typical 'plume-shape' pattern similar to that shown in Fig. 4 was found in all phases and depths for Cu and to a lesser extent for Ni and Zn, but was virtually absent for the other metals with exceptions of the exchangeable phase for Mn, Fe and Pb and oxidisable phase for Pb. Uniformly for all metals at all depths, the water soluble phase and to a lesser extent the exchangeable phase provided the best means to study the spatial distribution of the contaminant plume by yielding distribution patterns similar to that shown in Fig. 4. For copper, the gradients were far higher in the anomalous population at all depths and phases compared to the background population showing that copper retention is more efficient in the first 10 m of contaminant propagation. Also, gradients were higher in the anomalous population for oxidisable phase-bound Cu below groundwater table indicating that copper is retained more efficiently in this phase.

The high concentration gradients are also a result of the relatively sharp hydraulic gradient present at the discharging point between till sediments, having hydraulic conductivity of around $10^{-6}$ m/s [29], and peat sediments, characterised by hydraulic conductivity in the range of $10^{-8}$-$10^{-10}$ m/s [30]. Thus, the till-peatland interface acts as a physical barrier for groundwater and transported dissolved salts. Because of the shallow depth of the mire (2-2.5 m on average) it is likely that the main path of contaminant propagation is actually beneath the bog.

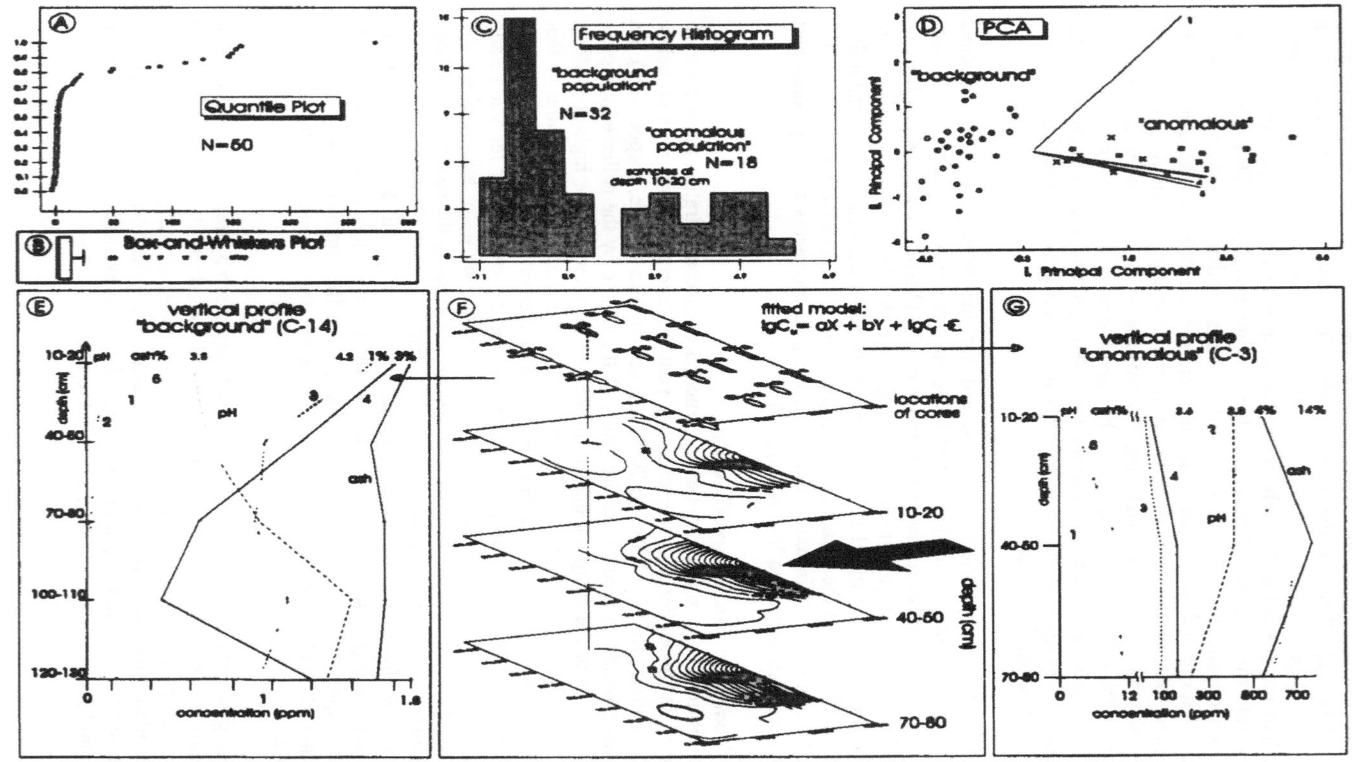


**Figure 4.** Geochemical modelling of metal behaviour in the studied mire demonstrated on Cu. 'Background' and 'anomalous' samples are shown as empty circles and solid boxes, respectively, and numbers from 1 to 5 in the figure indicate the five metal fractions (see text). *A-C.* Univariate exploratory data structure analysis. *A.* Quantile plot for Cu in the oxidisable phase. *B.* Box-and-whiskers plot for Cu in the oxidisable phase. *C.* Frequency histogram for Cu in the oxidisable phase after log-transformation. *D.* Combined Q- and R-mode PCA display for Cu. Samples in the 'anomalous' population at depth 10-20 cm are marked with 'x'. *E-G.* Spatial analysis. *F.* Horizontal distributions of total Cu (fraction 1+..+5) at different depth as shown by contour lines using kriging interpolation. *E and G.* Typical vertical profiles of partial abundances of Cu in the peat sediments.

Vertical profiles for copper, nickel and ash content in the anomalous population show a characteristic peak at depth 40-50 cm probably as a result of remobilisation and deposition of reduced forms at and below groundwater fluctuation zone [31] (Fig. 4). The vertical profile of Cu in peatlands following ash content was observed in other wetlands, too [32]. This peak is absent in the background population which may imply that similar remobilisation processes are insignificant further away from the AMD input location. Iron has its highest concentration at 10-20 cm depth and drops with depth because iron is in its mobile reduced Fe(II) form below the groundwater table and it is likely to move along concentration gradients or be uplifted by plants to the oxidising top sediments.

Spatial numerical modelling of metal distributions in the contaminant plume was carried out separately for the different populations by log-transforming the data first to obtain stationarity in the variance and reach linearity, then regressing concentrations on the spatial coordinates with the least-squares method (Fig. 4). The assumption of exponential trend was supported by the fact that concentrations decrease exponentially in groundwater plumes due to hydrodynamic dispersion. In this model it was assumed that the additive parameter cannot be separated from the error term, that is metal concentrations already present in the sediments before the AMD input are smaller than the random variance of the contamination process. The residuals were checked for autocorrelation by fitting semi-variograms. All the residuals followed nugget-effect-type variograms, that is no significant autocorrelation remained in the variables. This is most probably due to the high spatial variability of the geochemical processes responsible for metal distributions and that samples were located out of the range [33]. Although normality for the residuals was not confirmed by chi-square tests, they had zero mean and the distributions were approximately symmetric as measured by letter-value displays [34].

*Modelling of Causal Relationships: Element Migration and Geochemical Flow*

Potential associations were first identified by the analysis of Spearman correlation matrices and construction of Draftsman's plots [12]. All the correlations (Pearson linear correlation coefficient) were graphically examined to avoid spurious correlations. Variables were log-transformed to stabilise their variance and robustness was achieved by the use of interactive outlier rejection. No more than 10% of the samples were excluded in any case but in the majority of cases it was below 5%. Pairwise regressions discussed were all significant at the 0.05 level.

High correlations found between reducible and oxidisable phases (Pb: 0.59; Ni: 0.88; Cu: 0.89; Zn: 0.52; Fe: 0.87; Mn: 0.43) may indicate the poor selectivity of the procedure applied for the reducible phase [35]. Most of these metals form stable thiosulphate complexes where thiosulphate forms as a result of rapid disproportionation of dithionite into sulphite and thiosulphate ions in acidic solutions. Hence, it is possible that under the experimental conditions thiosulphate desorbed and complexed the relatively loosely bound metals from organic ligands.

The high negative correlations between pH and Ni (-0.76), Mn (-0.76) and Zn (-0.63) in fraction 1 suggest that the mobility of these metals is most dependent on pH in peat sediments which is in accordance with their higher relative concentrations in this (and exchangeable) phase compared to the other phases. The strong association of Ni, Cu, Zn and Mn as expressed by their mutually high correlations (between 0.72 and 0.90) in this phase shows that their main common source is the AMD discharge.

Direct relationship between exchangeable metals and pH was not evidenced by correlation analysis. The correlation of 0.73 between Ni and Mn in the background samples shows their similar geochemistry in this phase in the organic matter-rich reducing environment. Also, the correlation coefficient of 0.87 found between Pb and Cu in this fraction for the anomalous samples (and in the oxidisable fraction for the background population, see below) derives from their very similar adsorption properties even under different geochemical conditions. As it could be expected, no significant correlations were found in fraction 3 since hydrous Fe-Mn oxides are unlikely to control metal behaviour in the acidic reducing peat sediments.

Negative correlations between pH and oxidisable Pb (-0.75), Cu (-0.56) and Ni (-0.54) were found only for samples identical to the background population of copper. The correlation can be explained with the pH dependent 'hydrolysis + adsorption' model [36] represented in a general form:

$$(H_2O)_5M(OH_2)^{z+} + SOH \Leftrightarrow SOM(H_2O)_5^{(z-1)} + H_3O^+. \quad (1)$$

More interesting is the absence of similar correlations in the anomalous region. This shows that processes regulating metal concentrations in the oxidisable phase here do not influence or depend on pH directly. Partial correlation on water soluble and oxidisable fractions and pH caused the elimination of the related correlations which means that these variables are mutually interrelated in the studied peat system. Correlation between ash content and metals in the anomalous samples in fraction 4 was found only for Cu (0.87) which means that ash content represents primarily oxidisable phase Cu in this area. However, in the background population ash content showed significant correlation with metals in the oxidisable phase (Pb:0.62; Ni:0.51; Cu:0.54; Zn:0.45; Fe:0.78; Mn:0.68) which suggest, together with the relationships between pH and this phase, that ash content in this region represents mainly the organic bound metals together. Correlations between metals in this phase were found only between Pb and Cu (0.6) and Ni and Cu (0.6) in the background region only. The close association represented by the high correlation (0.94) of Ni and Cu was observed in the area of anomalous Cu samples.

The pH dependence of the water soluble phase and the oxidisable phase metals and the absence of correlation with exchangeable metals suggest that metals are either in the pore solution or form strong inner-sphere complexes with organic ligands. That is the exchangeable outer-sphere positions are relatively unimportant in the adsorption process at the given pH. This may imply that metal adsorption to organic substances is kinetically fast.

In the operationally defined residual phase (fraction 5) significant correlation were found between Ni and Cu (0.63), Ni and Fe (0.64) and Pb and Fe (0.54). Correlation coefficients with ash content were generally smaller in this fraction (Pb:0.5; Ni:0.52; Cu:0.7; Fe:0.55) than between ash and fraction 4. Correlation between pH and residual phase was found only for Cu (-0.5) and Ni (-0.67) in the background samples but not for Fe and Pb. Since insoluble minerals, such as clays and oxides present in the residual phase, and ash content, do not influence pH appreciably, Fe and Pb are likely to be present in this fraction in the mineral phase. The pH dependence of residual Cu and Ni in the background population can be explained by the presence of extremely stable Cu and Ni organo-complexes forming according to equation (1).

The strong negative correlation (-0.71) between pH and ash content in the background shows that metals in pH dependent phase are most responsible for the ash content in this region. Considering the results discussed above, this implies that the ash content for these samples represents organically bound metals mainly. The absence of similar relationships in the anomalous sample group shows, in accordance with the finding in fraction 4, that high ash content and metal retention is mainly caused by a process related to the oxidisable phase in association with Cu.

In order to study and model metals and control processes together in their interactions in the peat sediments, MSA was implemented on the geochemical database. Since the AMD-peat system is an open system, that is metals are continuously input into any given volume of sediment and assuming that fraction 1 (water soluble phase) represents metals in the pore water, the relative effectiveness of metal retardation by various pools or sinks (cation exchange, organic complexation, sulphide precipitation, oxide scavenging) can be assessed by the correlation of fraction 1 (input) and the other fractions (sinks). Steady state conditions are also assumed. However, because metals in the solution and adsorption sites are "competing" they are modelled together by canonical correlations, where the first canonical variable represents the "solution" (fraction 1) and the second combined variable represents the metals adsorbed by the given phase, similarly to the concept of ionic strength. High canonical correlation was found only for the oxidisable phase ($r=0.91$) confirming our hypothesis of previous modelling that retention of metals from solution is primarily due to organic adsorption. The canonical correlation coefficient is far higher than any individual correlation justifying the use of canonical variables. It is interesting that samples from both the background and anomalous region are scattered along the same line indicating that complexation by organic matter is the main process even in the latter population.

Based on the above concept we examined whether the studied metals together can account for changes in pH. Multiple regression of oxidisable phase metals on pH yielded zero correlation coefficient for the anomalous population and $r=-0.67$ for the background. This means that the measured metals account only for 44% of total variation in pH in the background; in other words, pH is influenced by other processes as well, such as adsorption-desorption of other chemical elements and compounds, for example.

## DISCUSSION

Results show that precipitation of insoluble Cu sulphides might be a major process close to the location of contaminant discharge while adsorption by organic matter alone may control metal distribution further away along the migration path. There are basically three limiting conditions for sulphide formation in peat sediments: *1.* reduction of $SO_4$, for which, in turn, sulphate in right quantities and appropriate biogeochemical conditions are needed; *2.* the quantity and availability of reactive metal forms and *3.* the pH dependent solubility of metal sulphides [37-40]. In our case it seems that the quantitative requirements are satisfied because there is abundant metal and sulphate in the discharging AMD. Reduction of sulphate is conditioned on the presence and activity of sulphate-reducing bacteria, availability of decomposable organic matter and low redox potential. Low pH may seem an inhibiting factor for dissimilatory reduction of $SO_4$ by anaerobic heterotrophic sulphate reducing bacteria, such as the genera of *Desulfovibrio* and *Desulfotomaculum*, since the overall pH (3.74) in the studied peat sediments is well below their tolerance level (pH>6) [41]. However, their presence can not be excluded because of the existence of micro environments. Microsites have more hospitable conditions where $SO_4$ reduction itself tends to cause alkalinization and respiration produces carbonic acid, such micro-environments would tend to be "self-buffering" with respect to pH [42]. The annual average temperature of the area (-7 °C in winter and 14 °C in summer on average) is also well out of the optimal temperature range for sulphate-reducers (30-40 °C) and long winters in the boreal zone may also limit microbial activity in the peat bog, but the low heat conductivity and high thermal capacity of water-saturated organic matter-rich peat sediments can provide more hospitable conditions below surface. Because $O_2$ is quickly consumed below groundwater in the sediments and $NO_3$ is not a major constituent of AMD they are unlikely to act as a terminal $e^-$ acceptors in the anoxic zone. Fe and Mn are already in reduced forms in the AMD, hence they cannot serve as $e^-$ acceptors for organic matter decomposition [43]. Beyond the narrow zone of $O_2$ consumption, redox potential progressively drops and sulphate reduction commences provided that redox potential is sufficiently low (<-250 mV). In this zone, depending on the stability of sulphides under acidic bog conditions, metals not withheld from precipitation by organic complexation are preferentially complexed by sulphide ions generated by bacterial sulphate reduction.

In summary, while sulphate-reducing bacteria are necessary catalysts for sulphate reduction to occur in peat sediments and require specific environmental conditions for optimal growth, they can be active in these acidic peat soils as evidenced by the strong $H_2S$ odour of samples below the groundwater table.

Microbially mediated sulphide formation occurs in two steps:

$$SO^{2-}_4 + 2(CH_2O) + 2H^+ \rightarrow H_2S^o + 2CO_{2(g)} + H_2O \quad (2)$$
$$M^{2+} + H_2S \rightarrow MS_{(s)} + 2H^+. \quad (3)$$

Detailed column studies have shown that metal sulphides can form as a result of the sequential microbially mediated inorganic redox reactions in aquatic systems, but the precipitation of sulphides is strongly pH dependent [44]. In the pH range encountered in the studied mire (pH 3.5-4) the reactive $H_2S$ is the stable form of sulphur. As it can be seen, sulphide precipitation can be inhibited by low pH. However, the high content of Cu in the oxidisable phase, the presence of two distinct populations in variables of Cu also influencing relationships of the other metals, the distinction of samples at shallow depth in the anomalous region from samples below the groundwater table for Cu, the pH independence of oxidisable Cu in the anomalous population as opposed to its pH dependence in the background population, all provide indirect evidence that the most insoluble copper sulphides, probably $Cu_2S$ may precipitate closer to the discharge location. The high ash content of samples in the anomalous population is in accordance with Lett and Fletcher's [45] findings that where microbial $SO_4$ reduction takes place peats are rich in ash. Further supporting evidence is that the $H_2S$ gas formation was observable only for cores in the background region and was absent close to the AMD input. This is quite unexpected because one would expect more intensive sulphate reduction where the AMD reaches the mire. A plausible explanation is provided by assuming that sulphate reduction and therefore copper sulphide precipitation occurs in microclaves within the sediments. A minor part of the dissolved $SO_4$ is turned to $H_2S$ at the microsites where most of the $H_2S$ readily reacts with the Cu to precipitate insoluble sulphides, hence preventing the escape of $H_2S$ gas. However, the majority of sulphate is not available for reduction and remains in solution migrating towards the mire interior with the flowing groundwater. This is consistent with other's findings that while $SO_4$ reduction and the presence of $H_2S$ is obvious in peat sediments the majority of S is actually in sulphate form [46]. This is also supported by the observation that $SO_4$ reduction takes place in micro-environments within the sediments [47]. X-ray diffraction analysis of the samples did not confirm the presence of crystalline sulphides in the sediments. Although sulphides have considerable adsorption capacity for metals they are generally adsorbed only at higher pH (for example Pb at pH 6-7; Zn at pH 6-8; Cd at pH 8-10) following their hydrolysis constants [48]. Since copper sulphide is overwhelmingly abundant in this sample population according to this model, it is not unexpected that only copper showed correlation with ash content in the oxidisable phase. Also, because sulphide precipitation and co-precipitation is not simply related to pH according to eq. (2) and (3), in contrast to eq. (1), the absence of correlation between pH and metals in fraction 4 is not surprising.

As Cu is efficiently retained by organic adsorption and sulphide precipitation, the migrating groundwater becomes depleted with respect to copper within 10-15 meters in its course in the mire (Fig. 4). Beyond this distance there is little copper sulphide formation, and organic complexation controls primarily metal retardation. In this region $SO_4$ is also reduced at microsites but the generated HS remains in gas form in the absence of Cu, as evidenced by the field observation of $H_2S$ only in samples in this area. Here metals extracted in the oxidisable fraction are bound to organic matter and give the ash contents which explains the significant correlations between pH and these variables for samples from this region.

It seems from the analytical results that Fe and Pb were already present in higher quantities in the peat sediments prior to the beginning of AMD discharge. As shown by the analysis of the reducible and residual phases, they might originate from the background and were transported and deposited in the sediment phase by background surface drainage, probably as Fe/Mn or other

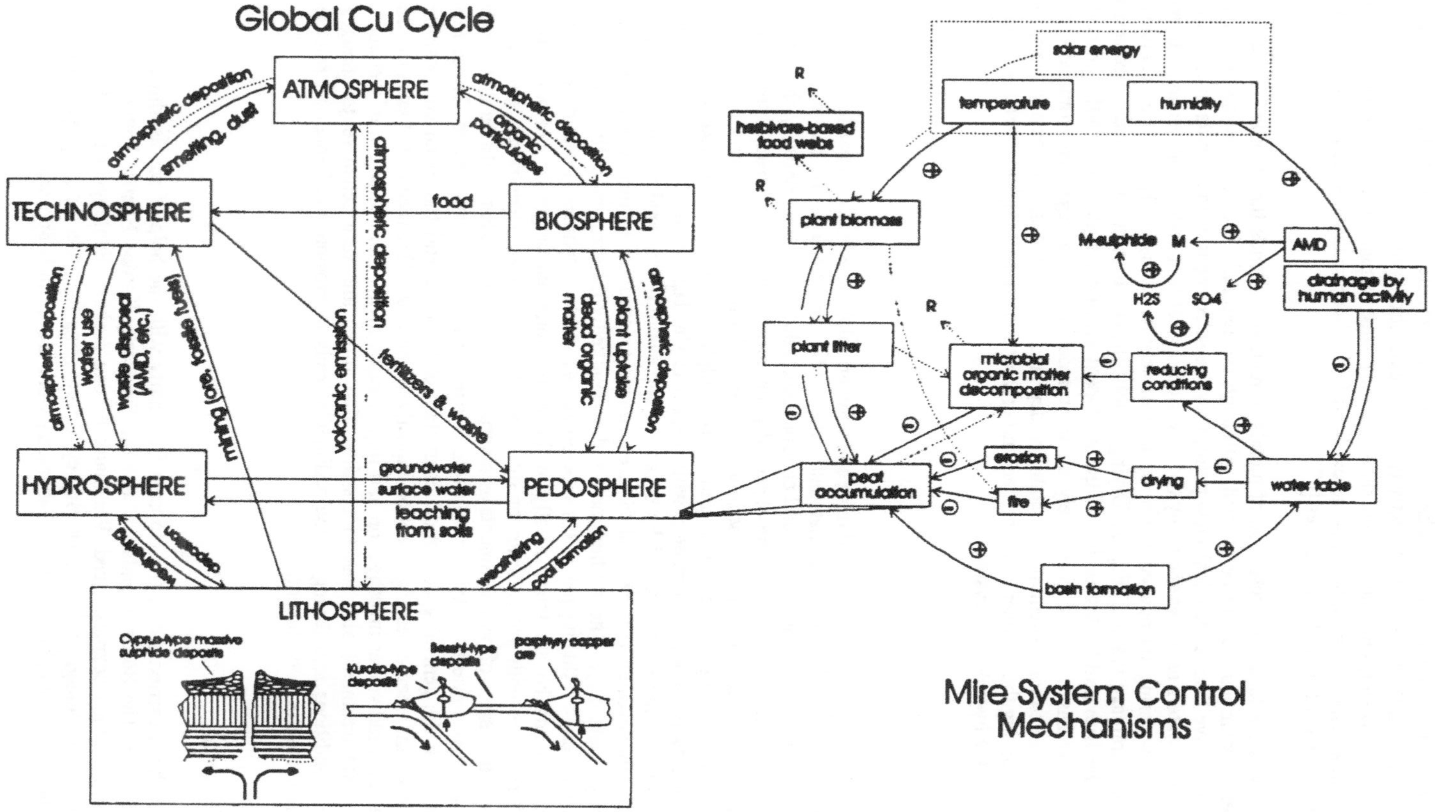


**Figure 5.** System control mechanisms in a mire ecosystem and its relation to the global Cu cycle. Dashed lines in this part indicate energy flow (R: respiration). The two diagrams are related by their similar organisation as well. Compiled from [49-53].

partly crystallised hydrous oxides and in clay minerals. The relatively high concentrations of Pb in these phases compared to Cu and Ni is consistent with the fact that Pb is several times more abundant in felsic rocks than in basic rocks. The high concentrations of these metals in fraction 3. may well be the result of the poor selectivity of the extraction procedure but, as we saw the importance of micro-environments in the reducing gley super-aqual mire landscape, it is not unreasonable to assume the presence of some oxidised Fe in these sediments, as it was reported to coexist with $SO_4$-reduction in tidal marshes [47].

In order to assess the long-term retention of metals in natural wetlands the stability of the whole peat sediment system also has to be studied. Although the naturally high adsorption capacity of wetlands is attractive from a technical view point, they are selective towards metals usually found together in waste waters as it was shown by the case study. Most prominently, mires are extremely vulnerable to dewatering introduced either by human activity or naturally, such as changes in climate or local drainage systems (Fig. 5). Drying is not balanced by natural feedback mechanisms and leads to the irreversible destruction of the mire-peat sediment system [49, 50]. Since their metal retention is based almost exclusively on their reducing waterlogged environments, their long-term metal retention potential and stability depends only on a single, easily changeable variable. Upon drying the sediments are eroded and organic matter and sulphides are fast oxidised and released to the hydrological system.

## CONCLUSIONS

Where AMD discharges into the studied mire an hydrogen sulphide-adsorption barrier forms in the peat sediments. It was also revealed that under the acidic conditions in the bog only the most stable copper sulphides may precipitate and control metal adsorption in a relatively short distance of plume propagation (about 10 m). The observed phenomena can be most readily interpreted by the assumption of microsites in peat sediments where, despite the bulk geochemistry of the sediments, microbial sulphate reduction and copper sulphide formation can take place. Where migrating groundwater becomes depleted with respect to Cu, $H_2S$ remains in gaseous form and metal retardation is mostly controlled by organic complexation. The analysis and modelling of geochemical data revealed that Fe, and Pb are primary constituents in the mire most probably bound in oxides and clays. The studied metals show markedly different mobility and adsorption stability in the acidic reducing peats: Cu precipitates as stable sulphides, Pb is mostly organic bound while Mn and Ni are found mostly in the more labile exchangeable and water soluble phase, $Fe^{2+}$ migrates with groundwater or diffuses upwards along concentration gradients and precipitates in top layers of peat sediments as amorphous iron oxides. This confirms that wetlands are highly selective towards metals usually found together in mine waste waters. These results indicate that both the general geochemical characters of the mire and the metal composition of AMD are important when assessing retention capacity of wetlands. This means that each treatment proposal requires individual investigation and impact assessment. The evaluation of wetlands for long-term AMD treatment also has to consider that mires are very sensitive to changes in hydrological conditions and drying of the sediments leads to erosion and hence the release of adsorbed metals to the environment.

### *Acknowledgements*

Research has been carried out at the Uppsala University with the help of scholarships from the Swedish Institute, The Soros Foundation and the Szádeczky-Kardos Elemér Foundation. Their assistance is gratefully acknowledged. We thank Emőke Jocháné Edelényi, István Horváth and György Tóth at the Hunagrian Geological Survey for providing facilities for the preparation of this paper. The help of Inger Pahlson with the laboratory work is gratefully appreciated.

## REFERENCES


1. W. J. Tauritis, R. T. Unz, and R. P. Brooks. Behaviour of sedimentary Fe and Mn in a natural wetland receiving acidic mine drainage, Pennsylvania, U.S.A., *Applied Geochemistry* 7, 77-85 (1992).
2. A. Tessier, P. G. C. Campbell, and M. Bisson. Sequential extraction procedure for the speciation of particulate trace metals, *Analytical Chemistry* 51, 844-851(1979).
3. C. Kheboian and C. F. Bauer Accuracy of selective extraction procedures for metal speciation in model aquatic sediments, *Anal. Chem.* 59, 1417-1423 (1987).
4. P. M. V. Nirel, and F. M. M. Morel. Pitfalls of sequential extractions, *Water Resources* 24, 1055-1056 (1990).
5. R. B. Herbert JR., Metal transportation in groundwater contaminated by acid mine drainage, *Nordic Hydrology* 25, 193-212 (1994).
6. J. A. C. Fortescue. *Environmental geochemistry: a holistic approach* Springer-Verlag, Inc., New York (1980).
7. M. A. Glazovskaya. On geochemical principles of the classification of natural landscapes, *International Geological Review* 5, 1403-1430 (1963).
8. A. I. Perel'man. *Landscape Geochemistry* (in Russian). Vysshaya skhola, Moscow (1972).
9. A. I. Perel'man. Geochemical barriers: theory and practical application, *Applied Geochemistry* 1, 669-680 (1986).
10. J. W. Tukey. *Exploratory Data Analysis.* Addison-Wesley Publishing Company, Reading (1977).
11. D. C. Hoaglin, F. Mosteller, and J. W. Tukey. *Undertanding Robust and Exploratory Data Analysis.* John Wiley and Sons, Inc., New York (1983).
12. J. M. Chambers, W. S. Cleveland, B. Kleiner, and P. A. Tukey *Graphical Methods for Data Analysis.* Wadsworth International Group, Belmont, Calif., Duxbury Press, Boston, MA. (1983).
13. W. Stumm and J. J. Morgan. *Aquatic Chemistry.* John Wiley and Sons, New York, N.Y., 2nd ed. (1981).
14. G. Gaál. Tectonic styles of early proterozoic ore deposition in the Fennoscandian Shield, *Precambrian Research* 46, 83-114 (1990).
15. A. Ruhling and G. Tyler. Regional differences in the deposition of heavy metals over Scandinavia, *Journal of Applied Ecology* 8, 497-507 (1971).
16. A. Ruhling and G. Tyler. Heavy metal deposition in Scandinavia, *Water Air Soil Pollution,* 2, 445-455 (1973).
17. N. H. Magnusson. *Malm geology.* Jernkontoret (1953).
18. L. G. Franzen. *Peat in Sweden - an important energy resource or just a paranthesis?.* In: Proceedings of the Peat and Environment '85 International Peat Society Symposium, The Swedish National Peat Commitee, Stockholm (1985).
19. A. Tessier, P. G. C. Campbell and M. Bisson. Particulate trace metal speciation in stream sediments and relationships with grain size: implications for geochemical exploration, *Journal of Geochemical Exploration* 16, 77-104 (1982).
20. A. Tessier, F. Rapin and R. Carignan. Trace metals in oxic lake sediments: possible adsorption onto iron oxyhydroxides. *Geochim. Cosmochim. Acta* 49, 183-194 (1985).
21. W. A. Mahler. Evaluation of a sequential extraction scheme to study associations of trace metals in estuarine and oceanic sediments, *Bull. of Environmental Contamination and Toxicology* 32, 339-344 (1984).
22. K. C. Jones. The distribution and partitioning of silver and other heavy metals in sediments associated with an acid mine drainage stream, *Environmental Pollution, Ser. B* 12, 249-263 (1986).
23. S. Karlsson, B. Allard and K. Hakansson. Chemical characterization of stream-bed sediments receiving high loadings of acid mine effluents, *Chemical Geology* 67, 1-15 (1988).
24. D. Postma. Pyrite and siderite oxidation in swamp sediments, *Journal of Soil Science 34,* 163-182 (1983).
25. J. Henrot and R. K. Wieder. Processes of iron and mangeneses retention in laboratory peat microcosms subjected to acid mine drainage, *J. Environ. Qual.* 19, 312-320 (1990).
26. Alloway, B., J., (Ed), *Heavy Metals in Soils.* Blackie, London (1990).
27. A. J. Sinclair. Selection of threshold values in geochemical data using probability graphs, *Journal of Geochemical Exploration* 3, 129-149 (1974).
28. R. J. Howarth and R. Sinding-Larsen. *Multivariate analysis.* In: R. J. Howarth (Ed), Statistics and Data Analysis in Geochemical Prospecting. Handbook of Exploration Geochemistry, Vol. 2. Elsevier, Amsterdam (1983).
29. L. Lundin. *Mark- och grundwatten: moränmark och mark typens betydelse för avrinningen* (in Swedish). UNGI Report 56, University of Uppsala (1982).
30. W. D. Reynolds, D. A. Brown, S. P. Mathur and R. P. Overend. Effects of in-situ gas accumulation on the hydraulic conductivity of peat, *Soil Science* 153, 397-408 (1992).
31. P. Pakarinen, K. Tolonen and J. Soveri. Distribution of trace metals and sulfur in the surface peat of Finnish raised bogs. *Proc. 6th Int. Peat Congr.*, Duluth, U.S.A. (1980).
32. M. Sillanpää. Distribution of trace elements in peat profiles, *Proc. 4th Int. Peat Congr.*, Otaniemi 5, 1972.
33. R. G. Garrett. *Sampling methodology,* In: R. J. Howarth (Ed), Statistics and Data Analysis in Geochemical Prospecting. Handbook of Exploration Geochemistry 2, Elsevier, Amsterdam (1983).

34. P. F. Velleman and D. C. Hoaglin. *Applications, Basics and Computing of Exploratory Data Analysis.* Duxbury Press, Boston (1981).
35. R. Chester and M. J. Hughes. A chemical technique for the separation of ferromanganese minerals, carbonate minerals and adsorbed trace elements from pelagic sediments, *Chemical Geolology* **2**, 249-262 (1967).
36. W. Stumm (Ed). *Aquatic Surface Chemistry.* John Wiley and Sons, Inc., New York (1987).
37. R. W. Boyle. Cupriferous bogs in the Sackville area, New Brunswick, Canada, *Journal of Geochemical Exploration* **8**, 495-527 (1977).
38. K. A. Brown. Sulphur distribution and metabolism in waterlogged peat, *Soil Biology and Biochemistry* **17**, 39-45 (1985).
39. J. W. M. Rudd, C. A. Kelly and A. Furutani. The role of sulfate reduction in long term accumulation of organic and inorganic sulfur in lake sediments, *Limnology and Oceanography* **31**, 1281-1291 (1986).
40. I. Bodek. W. J. Lyman. W. F. Reeh, and D. H. Rosenblatt (Eds). *Environmental Inorganic Chemistry.* Pergamon Press Inc., New York (1988).
41. R. A. Berner. Sedimentary pyrite formation. An update, *Geochimica et Cosmochimica Acta* **48**, 605-615 (1984).
42. M. C. Rabenhorst and B. R. James. *Iron sulfidization in tidal marsh soils.* In: H. C. W. Skinner and R. W. Fitzpatrick (Eds), Biomineralization Processes of Iron and Manganese, Modern and Ancient Environments. Catena Supplement, **21**, 203-217 (1992).
43. D. R. Lovley and E. J. P. Phillips. Manganese inhibition of microbial iron reduction in anaerobic sediments. *Geomicrobiology Journal* **6**, 145-155 (1988).
44. U. von Gunten, and J. Zobrist. Biogeochemical changes in groundwater-infiltration systems: Column studies, *Geochimica et Cosmochimica Acta* **57**, 3895-3906 (1993).
45. R. E. W. Lett and W. K. Fletcher. The secondary dispersion of transitional metals through a copper-rich hillslope bog in the Cascade Mountains, British Columbia, *Proc. 7th Int. Geochem. Explor. Symp.*, Denver, Colo. (1979).
46. W. Shotyk. Review of the inorganic geochemistry of peats and peatland waters, *Earth-Sciences Reviews* **25**, 95-176 (1988).
47. M. C. Rabenhorst and K. C. Haering. Soil micromorphology of a Chesapeake Bay tidal marsh: implications for sulfur accumulation, *Soil Science* **147**, 339-347 (1989).
48. G. E. Jean and G. M., Bancroft. Heavy metal adsorption by sulphide mineral surfaces, *Geochemica et Cosmochimica Acta* **50**, 1455-1463 (1986).
49. P. D. Moor. The ecology of peat-forming processes: a review, *International Journal of Coal Geology*, **12**, 89-103 (1989).
50. C. C. Cameron, J. S. Esterle and C. A. Palmer. The geology, botany and chemistry of selected peat-forming environments from temperate and tropical latitudes, *International Journal of Coal Geology* **12**, 105-156 (1989).
51. S. S. Butcher, R. J. Charlson, G. H. Orians, and G. V. Wolfe (Eds). *Global Biogeochemical Cycles.* San Diego, Academic Press Ltd. (1992).
52. W. H. Schlesinger. *Biogeochemistry, An Analysis of Global Change.* Academic Press, San Diego (1991).
53. I. M. Szabó. *Microbiology of the Geosphere I-III* (in Hungarian). Akadémia Kiadó, Budapest (1991).

*Proc. 30th Int'l. Geol. Congr.*, Vol. 19, pp. 207-222
Xie Xuejin (Ed)
VSP 1997

# Interaction of $Fe^{3+}$, $Ca^{2+}$, and $SO_4^{2-}$ at the jarosite/alkaline coal fly ash interface

M.DING[1], H.A.van der SLOOT[2] and M. GEUSEBROEK[2].

[1] *Department of Geochemistry, Faculty of Earth Sciences, Utrecht University, P.O.Box 80021, 3508 TA Utrecht, the Netherlands.*
[2] *Netherlands Energy Research Foundation (ECN), P.O.Box 1, 1755 ZG Petten, the Netherlands.*

**Abstract:** Radio tracer diffusion experiments have been carried out to study self-sealing/healing isolation and immobilization of two chemically contrasting wastes: acidic jarosite and alkaline coal fly ash. Precipitation of iron oxyhydroxide FeO(OH) and gypsum $CaSO_4 \cdot 2H_2O$ occurs at the interface between jarosite and fly ash due to diffusion and interaction among the principal reacting solutes $Fe^{3+}$, $SO_4^{2-}$, $Ca^{2+}$, and $OH^-$. A layer with low permeability is formed at this interface because of pore filling by such precipitates causing a change in the transport properties of the bulk system. Precipitation at the interface between jarosite and fly ash affects also the mobility of chemical constituents including contaminants from the wastes. Co-disposal of chemically contrasting wastes may therefore lead to a self-forming barrier layer between them, which, next to isolating the wastes, also immobilizes contaminants from their leachates. Such systems will have a wide application in waste management.

***Key words:*** *interface precipitation, self-sealing/healing isolation, jarosite, alkaline fly ash, effective diffusion coefficient, permeability, hydraulic conductivity.*

## INTRODUCTION

Studies dealing with multicomponent-multiphase reactive transport in porous media have been receiving increasingly more attention due to their inherently closer similarity to realistic systems[1]. Some research emphasis has been placed on the study of interfaces between media in which large physical or chemical gradients occur[2]. It can be anticipated that chemical reactions at such interfaces must influence the transport properties of a system and the mobility of its constituents, especially in the presence of solutions.

A new concept was introduced by Coté and van der Sloot[3] who developed "self-forming and self-repairing seals", as a waste isolation technique utilizing interface induced chemical reactions. The principle of this technique involves forming a layer with low permeability to groundwater and leachate at the interface of waste-soil or waste-waste based on diffusion and precipitation of reacting solutes. Damage to such seal will self-repair by continued diffusion and precipitation. Additional studies employing mathematical formulations for different types of chemical reactions, which dominate waste-soil interfaces have been carried out by Hockley *et al*[4]. Their results indicated that migration of contaminants from both waste and soil not only depends on the property of the waste but is also strongly influenced by the type of interface reaction between the waste and its surroundings.

The results of these pioneering researches provided the impetus to address the solution of a serious environmental problem: the disposal of acidic jarosite and alkaline coal fly ash. We surmised that there could be a potential benefit by co-disposal of these two chemically contrasting wastes. The process we envisioned was that by putting them adjacent to one another, precipitation would happen at the interface. As result a decrease in the permeability of the bulk system occurs together with a decrease in mobility of its constituents including contaminants from both wastes. Support for our working hypothesis came from the discovery of hematite and gypsum between a layer of acidic waste and limestone in a waste sulfuric acid lake of a $TiO_2$-plant at Armyansk, Crimea, Ukraine[5]. Combining this observation with our preliminary batch and flow through experimental results[6] on these two wastes reinforced the idea that the interaction of $Fe^{3+}$, $SO_4^{2-}$, $Ca^{2+}$, and $OH^-$ at the jarosite/fly ash interface would lead to precipitation. These newly formed minerals might eventually completely fill the pores, leading to the formation of a seal and hence to waste isolation.

In light of the above, this study has three goals. The first one is to determine the nature of the precipitates at the interface between acidic jarosite and alkaline fly ash. The second one is to investigate the effects of precipitation reactions at this interface on the transport properties of the bulk system and on the mobility of its constituents. The third one is to explore the quantitative relation between the change in transport properties of the bulk system and the microstructural change caused by pore filling precipitates at this interface.

Our principal result is that reactions involving $Fe^{3+}$, $Ca^{2+}$, $SO_4^{2-}$, and $OH^-$, tend to form a layer with low permeability to groundwater and leachate at the interface between acidic jarosite and alkaline coal fly ash. Constituents including contaminants from both wastes are thus immobilized at the interface. Such combined systems may provide a superior solution to the disposal of acidic jarosite and alkaline fly ash.

## THE PRINCIPLE OF SELF-SEALING/HEALING ISOLATION

Isolation is the most common environmental option to dispose hazardous waste[7]. The key issue addressed in this technique is how to create a barrier between waste and its surroundings. This barrier can consist of compacted soil with low permeability to form a so-called liner, as commonly used in Modern Sanitary Landfills[8]. However such liner can also be developed by decreasing the permeability at the interface between waste and soil or waste and waste by chemical reactions such as precipitation, which results in filling the pores at the interface causing so-called self-sealing/healing[3]. This precipitation occurs due to chemical interaction either between existing reacting solutes at the waste-waste or waste-soil interface[5] or by introducing chemical discontinuities in the pore fluid of one of the materials[3]. It is obvious as stated by van der Sloot[9] that the necessary condition for self-sealing is that the concentration of the reacting solutes in both materials ought to be high enough to initiate a precipitation reaction.

Figure 1 shows the principle of self-sealing/healing isolation. Suppose there are two materials adjacent to one another. Reacting solute A occurs in one material and solute B in the other. Due to the concentration gradient of solutes A and B between these two materials, diffusion of A and B will occur in opposite direction. Once the concentrations of A and B are sufficiently high, a precipitation reaction involving A and B will start. This precipitation will cause a local lowering of the concentration of A and B in the pore fluid, thus maintaining a concentration gradient, which will lead to further diffusion of A and B towards the interface, resulting in more precipitation. These coupled processes of diffusion and precipitation will continue until the pores are completely filled with precipitate, resulting in self-sealing layer formation at the interface. As long as sufficiently large amount of solutes A and B are present at opposite sites of the interface, this coupling process will be maintained and any mechanical damage to the interface, i.e. the liner, will cause further diffusion and precipitation thus healing the liner. In this manner the interface seals as well as heals hence the term self sealing/healing layer. This built in security of healing upon damage can without doubt be considered as the principal advantage of the self-sealing/healing isolation technique.

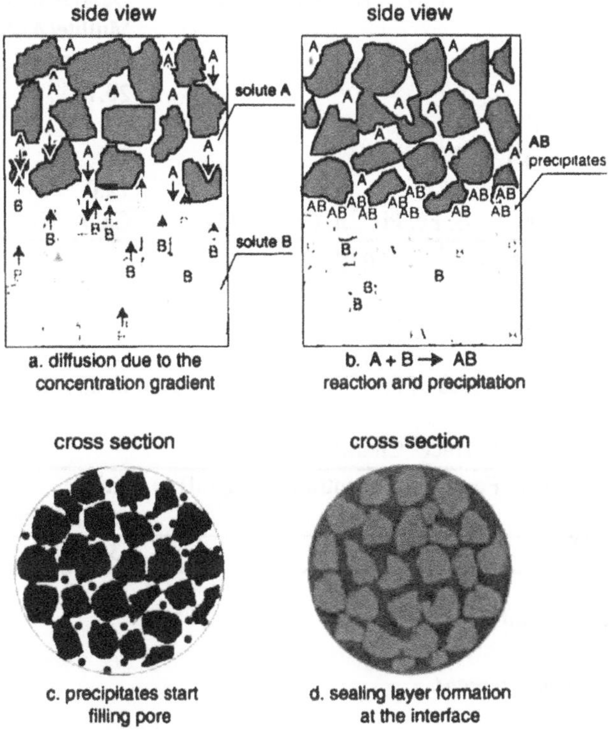


**Figure 1.** The Principle of Self-sealing/healing Isolation.

## EXPERIMENTAL METHODS AND MATERIALS

### Materials

#### *Jarosite and alkaline fly ash*

Jarosite is one of the most important members of a large group of isostructural minerals. The general formula of this group[10] is $MA_3(BO_4)_2(OH)_6$, where M is a monovalent or divalent cation such as $Na^+$, $K^+$, $NH_4^+$, or $Pb^{2+}$, A stands for iron or aluminium, and B for sulphur, phosphorous, or arsenic. The jarosite used in this study comes from a zinc manufacturing plant and is the result of removing Fe from acidic, sulphate-rich solutions [11]. The alkaline fly ash used in this study is the residue from a power plant. It consists mainly of an amorphous phase, with small amounts of quartz ($SiO_2$), mullite ($Al_6Si_2O_{13}$), and hematite ($Fe_2O_3$). The physical and chemical characteristics of jarosite and alkaline fly ash are summarized in Table I.

**Table I.** Physical and Chemical Characteristics of Jarosite and Alkaline Fly Ash

| | Jarosite | Alkaline Coal Fly Ash |
|---|---|---|
| Composition | $[M]Fe_3(SO_4)_2(OH)_6$<br>[M=$Na^+$, $K^+$, $NH_4^+$, and $Pb^{2+}$] | Amorphous Phase<br>Quartz($SiO_2$)<br>Mullite($Al_6Si_2O_{13}$)<br>Hematite($Fe_2O_3$) |
| Particle Size (μm) | 5.2 | 20.1 |
| Dry Density (kg/m³) | 2.9 | 2.0 |
| Bulk Density (kg/m³) | 2.2 | 1.2 |
| Natural pH* | 1.7 | 12.0 |
| **Principal Leachates** | **S, Fe, Pb, Zn, Ca, As, Mg** | **Ca, K, Na, S** |

** Determined at a Liquid/Solid ratio of 10*

#### *$^{59}Fe$ tracer*

In Table II the characteristics of the $^{59}Fe^{3+}$ radiotracer used in this study are presented.

**Table II.** $^{59}Fe^{3+}$ Tracer Characteristics

| Nuclide | Chemical Form | Concentration (μg/ml) | Initial Activity (counts/min•ml) | Half-life ($T_{1/2}$) |
|---|---|---|---|---|
| $^{59}Fe$ | $Fe^{3+}$ | 9.21 | $2.33*10^7$ | 45.1 days |

### Experimental methods

#### *Diffusion tube measurement*

A schematic of a diffusion tube measurement is depicted in Figure 2[12]. In these small diffusion tubes (Ø8 mm, L50 mm), either one material (jarosite or fly ash) or a combination of two materials (jarosite with fly ash) were compacted. One half-space contained the material to be labelled with a radiotracer, the other half-space contained the tracer-free material. Both ends of the tube were closed by pistons. These diffusion tubes remained stored in a constant moisture atmosphere at room temperature and zero fluid flow condition. After one day exposure, both the labelled and unlabelled segments were sliced and counted with L 1282 Compugamma Universal Gamma Counter.

***Preparation and slicing of diffusion tube contents***

Figure 3 shows the preparation and slicing of the diffusion tube contents[12]. Thirty grams of jarosite and fly ash were mixed with 12 ml and 9 ml distilled water respectively, to form a non-labelled paste. Meanwhile, another 30 grams of jarosite and fly ash were mixed with 12 ml and 9 ml diluted tracer solution respectively, to form the labelled paste. Half of the tube was filled with the non-labelled paste, the remaining half with the labelled one. The amount of solution used to mix with solid proved experimentally to be the best scheme to avoid any liquid spilling from the tubes. The degree of dilution was chosen such that $10^{-2}$ % of the originally added activity level (counts/min.g) could be detected in each sliced sample. The 50 mm long tube was sliced into about 40 slices.

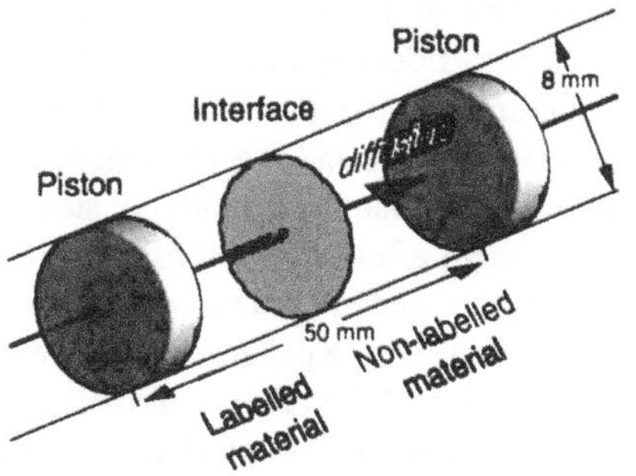


**Figure 2.** Schematic View of a Diffusion Tube Measurement.

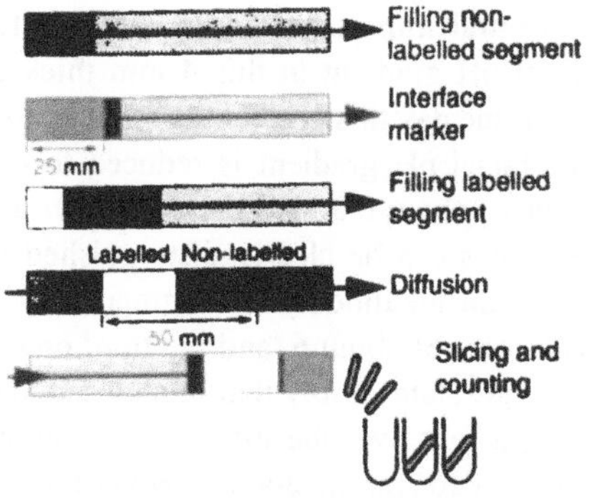


**Figure.3.** Preparation and Slicing of Diffusion Tube Contents.

***pH gradient measurement***

pH tube measurements were designed to track the variation in the pH gradient during transport and chemical reaction in the layered jarosite/fly ash system. The preparation and slicing of the pH tube content is exactly the same as that for the diffusion tube experiment except that no tracer was added in either of the segments. The procedure consists in adding a specific amount of distilled water to each sliced segment for a liquid to solid ratio of 10. The measured pH refers to the pH of such solution.

***Hydraulic conductivity measurement***

Hydraulic conductivity measurements were carried out in a triaxial cell (Ø50 mm, L140 mm) under controlled three-dimensional stress conditions (flexible wall) to prevent leakage as well as flow along the wall. One half part of the cell was filled with 70 mm jarosite with a dry weight of 293 g, the other part with 70 mm alkaline fly ash with a dry weight of 234 g. The fly ash was placed on top of the jarosite, the fluid flow direction was from the bottom upwards. The samples were compacted mechanically and saturated with pure water.

### *X-ray power diffraction*

X-ray power diffraction of jarosite, fly ash, and interface samples were determined with a Philips Power Diffractometer PW 1050/PW 1730 and ADM Software Packet.

## RESULTS

### pH development in the layered jarosite and alkaline fly ash system

The variation in pH as a function of time and distance across the interface in the layered jarosite/alkaline fly ash system is shown in Figure 4. It is essential to pay particularly attention to the pH variation in the 2 mm zone at the both sides of the interface because of its central importance in self-sealing formation. Inspection of Figure 4 shows that the initial pH gradient in this 4 mm thick zone (2 mm at both sides) is about 10 pH units. After one day this gradient is reduced to 7 units, and continues to decrease with time. This interfacial pH gradient is reduced to 4 units in about one month and the pH reached a value of about 6 across the zone after about two months. Note that three relatively steady pH states can be clearly distinguished in this layered system after two months: the first one in the alkaline fly ash segment with a pH of about 10, the second in the interface zone with a pH of about 6, and the third one in the jarosite segment with a pH around 4. These three pH states imply that in a closed system, in which the size of the tubes is limited, the pH gradient over the interface will continuously decline as long as the interface has not been sealed completely. Moreover the existence of a zone of precipitation causes the jump in gradient to appear non-linear, as illustrated in Figure 4 where the second pH region extends about 2 mm into the alkaline fly ash zone. If, on the other hand, the pores are fully filled, a discontinuous steady state across the interface will develop. We also anticipate that with the use of longer tubes or in an open system, pH gradients will be maintained longer.

### Radio tracer experiment

The radiotracers we have been using include all the principal reacting solute components such as $^{59}Fe^{3+}$, $^{45}Ca^{2+}$, and $^{35}SO_4^{2-}$ as well as $^{22}Na^+$ and $^3H^+$. The choice of the last two was predicated by their supposedly inert nature. In this paper we shall focus only on the behavior of $^{59}Fe^{3+}$, as representative of the behavior of one of the principal reacting solutes, $Fe^{3+}$, in the jarosite/fly ash system.

***$^{59}Fe^{3+}$ diffusion in jarosite***

Figure 5 presents the $^{59}Fe^{3+}$ diffusion profiles in jarosite. Two principal results can be inferred from inspection of Figure 5. Firstly, the diffusion of $^{59}Fe^{3+}$ in jarosite slows down with increasing exposure time. As indicated from our radiotracer data for $^{59}Fe^{3+}$, the measured effective diffusion coefficients of $^{59}Fe^{3+}$ in jarosite decreases in about two months from $10^{-10.1}$ $m^2/s$ to $10^{-11.5}$ $m^2/s$[13]. Secondly, the characteristic diffusion pattern of $^{59}Fe^{3+}$ in jarosite does not coincide with the anticipated standard diffusion pattern which should result in a horizontal line through the point at $C/C_0$ equal to 0.5 under the same controlled boundary condition. The difference between the measured $^{59}Fe^{3+}$ diffusion profile and the ideal diffusion profile is due to the time dependent isotopic exchange reaction which can occur between $^{59}Fe^{3+}$ introduced into the pore water and stable $Fe^{3+}$ in the jarosite matrix.

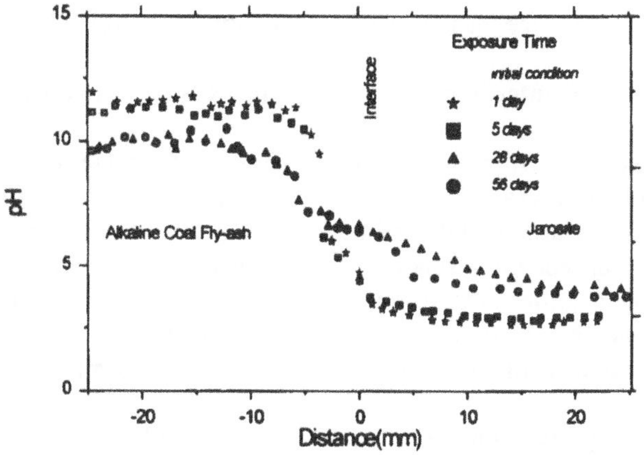


**Figure 4.** pH Development in the Layered Jarosite/Fly ash System.

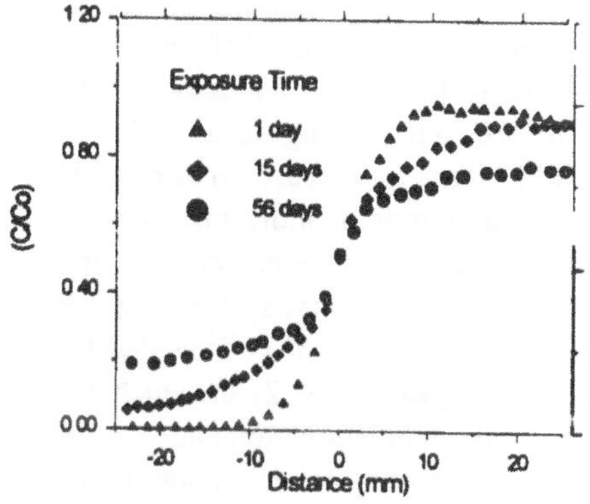


**Figure 5.** $^{59}Fe^{3+}$ Diffusion Profiles in Jarosite.

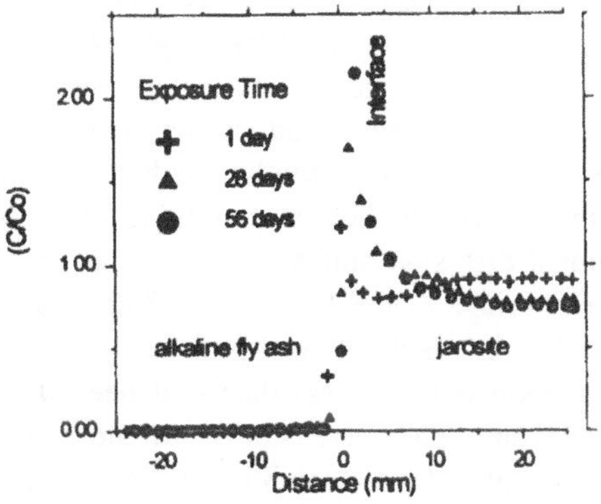


**Figure 6.** Profiles of $^{59}Fe^{3+}$ Diffusion from Jarosite to Alkaline fly ash.

***$^{59}Fe^{3+}$ diffusion from jarosite to alkaline fly ash***

Figure 6 shows $^{59}Fe^{3+}$ diffusion profiles in a combined jarosite and alkaline fly ash layered system. Our principal results, as can be seen in Figure 6, are: firstly, $^{59}Fe^{3+}$ tracer accumulation at the interface of jarosite/fly ash where a sharp peak can be observed, which indicates that iron oxyhydroxide precipitation occured; secondly, a curious first day profile showing just before and after the midpoint of the total length of labelled jarosite, a gradual decline in Fe-59 tracer concentration followed by a rapid rise near the jarosite/fly ash interface, and a sharp drop in the alkaline fly ash. Integrating the results of $^{59}Fe^{3+}$ diffusion and pH profile in the layered jarosite/fly ash system suggests that two processes are characteristic for the behavior of $Fe^{3+}$ at the interface. The diffusion of $Fe^{3+}$ is caused by its concentration gradient between jarosite and fly ash. The sharp peaks appearing at the interface are caused by the exceedingly low solubility of $Fe^{3+}$ due to the marked pH jump at the interface. The thermodynamics of ferric hydrolysis[14] indicates that due to its extremely low solubility at room temperature between pH 5 and 9 , $Fe^{3+}$ in pore water will be completely transferred to the insoluble state in the form of FeO(OH).

**Gypsum formation at the interface of the layered jarosite/fly ash system**

Figure 7 shows X-ray power diffraction results of jarosite, of the interface, and of fly ash. Inspection of this Figure reveals that the major component in fly ash is glass, together with small amounts of quartz, mullite and hematite. The jarosite used in the study, was mineralogically pure except for a small amount of quartz. In the interface gypsum as major phase could be identified next to a minor amount of bassanite. The interface sample also contains some jarosite and quartz which are contaminants from the starting materials due to the experimental difficulty in selecting a pristine interface sample.

**Hydraulic conductivity development in the layered jarosite/fly ash bulk system**

The variation in hydraulic conductivity as a function of time and hydraulic head needed to maintain a flow through the system is depicted in Figure 8. As can be seen, the hydraulic conductivity of the first system(Test 1) decreases slowly in the first 15 days, followed by a faster decrease in the next five days, from an initial value of $10^{-7}$ m/s to $5*10^{-9}$ m/s. The hydraulic head increases concomitant with this decrease in hydraulic conductivity. The increase in hydraulic head causes some damage to the self-forming layer, with as consequence that the hydraulic conductivity after 20 days starts to increase up to roughly $10^{-8}$ m/s. This situation is maintained for about one more month, indicating that no additional diffusion and precipitation occurred. We interpret this phenomenon as being due to shortage of the $Fe^{3+}$ source. Note that the starting pH in the first system was 6.6 rather than 2.0 as mentioned previously for jarosite. Thus the amount of $Fe^{3+}$ needed for precipitation is much less than that needed in the second system(Test 2) in which the starting conditions were similar to those of the diffusion tube measurement. As we mentioned already, the solubility of $Fe^{3+}$ at a pH above 4 is very low in comparison to that at pH 2. Therefore, the hydraulic conductivity in the second system continuous to decrease after 75 days.

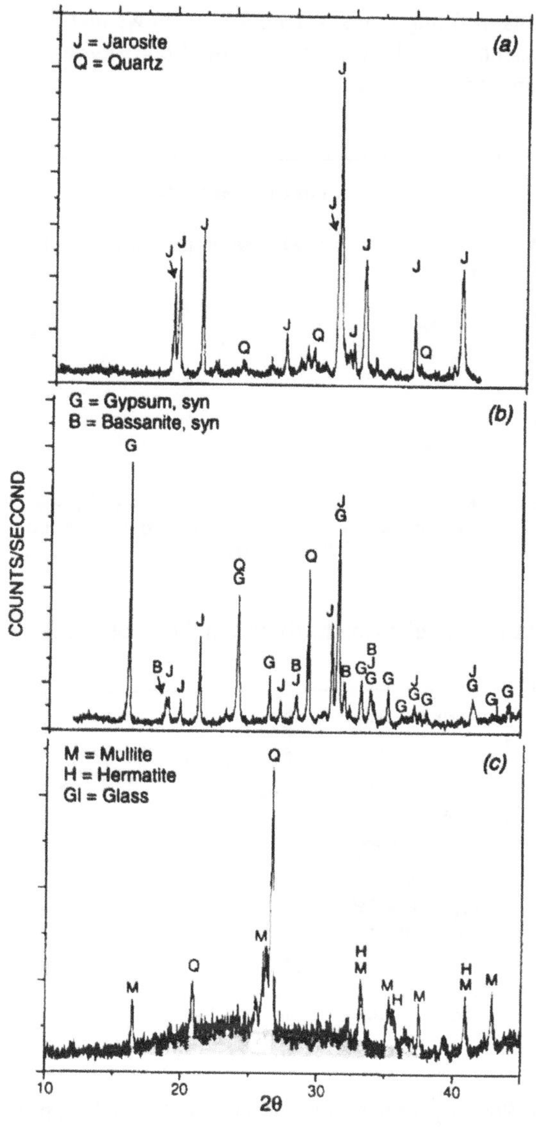


**Figure 7.** X-ray Power Diffraction of (a): Jarosite; (b): Interface; (c): Fly ash.

**Porosity variation in the jarosite/fly ash system**

One of our most interesting results, the porosity variation in the jarosite/fly ash system, is illustrated in Figure 9. This Figure shows that the decrease in porosity occurs at the interface slightly penetrating the fly ash side. This decrease in porosity (hence the microstructural change) at the interface depends on reaction time and is due to precipitation, which results in filling pores. Figure 9 also shows that the disadvantages of diffusion tubes measurements are a lack of unique initial porosity among different tubes and limited resolution imposed by slicing of the tubes.

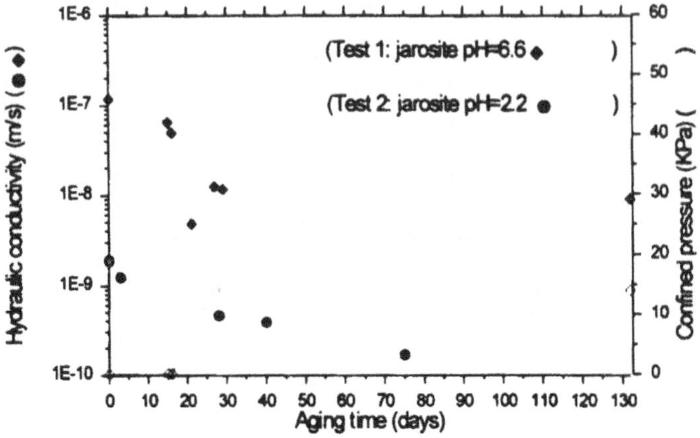


**Figure 8.** Development of Hydraulic Conductivities in the Jarosite/Fly ash Bulk System.

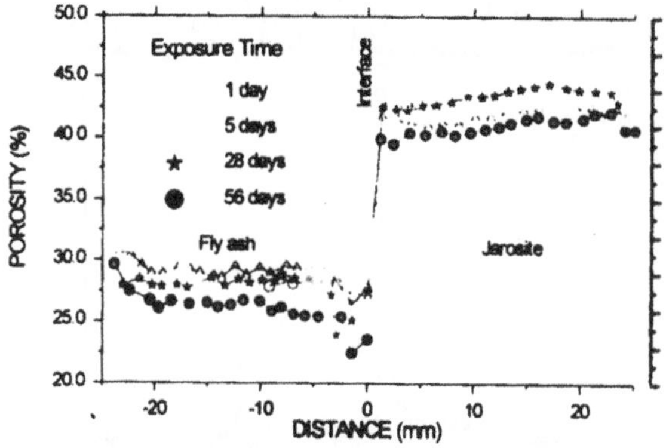


**Figure 9.** Porosity Variation in the Jarosite/Fly Ash System.

## DISCUSSION

### The effect of precipitation on the mobility of constituents in the jarosite/fly ash system

#### *Theoretical considerations*

One of the major effects of chemical reactions in porous media is to change the mobility of reactive solutes, as can be inferred from our particular results for the jarosite/fly ash system. According to our pH measurement results, a pH gradient exists in the bulk system, and in particular a pH switch occurs at the interface. This phenomenon requires special attention because of its potential effect on immobilizing constituents including contaminants from the wastes. Recalling some thermodynamics of cation hydrolysis in aqueous solution, the general equilibrium equation for oxyhydroxide or hydroxide compound formation can be expressed as:

$$M(OH)_n \Leftrightarrow M^{n+} + nOH^-, \qquad K_s=[M^{n+}]\cdot[OH^-]^n \qquad \text{(Eq. 1)}$$

Where $K_s$ is the solubility product and n is the formal charge of the cation (M). Thus, the solubility of a cation is given by:

$$[M^{n+}]=K_s/[OH^-]^n \qquad \text{(Eq. 2)}$$

This formula implies that for basic and amphoteric oxides, the higher the pH of the aqueous solution, the lower the solubility of cations. Table III lists the hydroxide solubility product of ferric iron ($Fe^{3+}$), aluminum ($Al^{3+}$), zinc ($Zn^{2+}$), and lead ($Pb^{2+}$), which are encountered as principal components in the leachates from both jarosite and alkaline fly ash. It can be inferred from Eq. 1 together with the data from Table III that $Fe^{3+}$ will become immobile ([M]<$10^{-6}$ mol/L) at a pH above 4.5 , followed by $Pb^{2+}$ at a pH above 5, $Al^{3+}$ above 7.5, and $Zn^{2+}$ above 8. Referring to our pH determinations in the layered jarosite/fly ash system, we expect that immobilization of these four constituents would occur in different zones from just before or at the interface, to the interface slanted towards the fly ash side, and finally into the fly ash segment.

**Table III.** Solubility Product of Ferric Iron, Aluminium, Zinc, and Lead Hydroxide[15].

| | $Fe(OH)_3$ | $Al(OH)_3$ | $Zn(OH)_2$ | $Pb(OH)_2$ |
|---|---|---|---|---|
| **log $K_s$** | **-38.58** | **-24.57** | **-16.16** | **-19.85** |

***Effect of interface precipitation on the $Fe^{3+}$ mobility in the jarosite/alkaline fly ash***

Considering $^{59}Fe^{3+}$ as an example, the observed $^{59}Fe^{3+}$ profile (Figure 6) reflects, as mentioned previously, the coupled processes of diffusion and precipitation. In other words, the observed $^{59}Fe^{3+}$ diffusion profile can be divided into a mobile and immobile ferric cation profile, as shown in Figure 10. The mobile $^{59}Fe^{3+}$ profile is based on knowledge about the diffusion as well as thermodynamics of ferric hydrolysis which states that at a pH of about 4.5, $^{59}Fe^{3+}$ will be transferred completely to the insoluble state. The pH is about 5 at the interface of jarosite/fly ash, indicating that the concentration of mobile $^{59}Fe^{3+}$ at this point ought be zero. Additionally, diffusion occurs only when a concentration gradient exists in the fluid state, implying that the lowest point of the observed $^{59}Fe^{3+}$ profile is still caused by mobile $^{59}Fe^{3+}$. Once the concentration starts to increase, precipitation of ferric oxyhydroxide occurs. Linking these two points, results in the mobile $^{59}Fe^{3+}$ profile. The profile of immobile $^{59}Fe^{3+}$ can be calculated by subtracting the mobile $^{59}Fe^{3+}$ profile from the measured one.

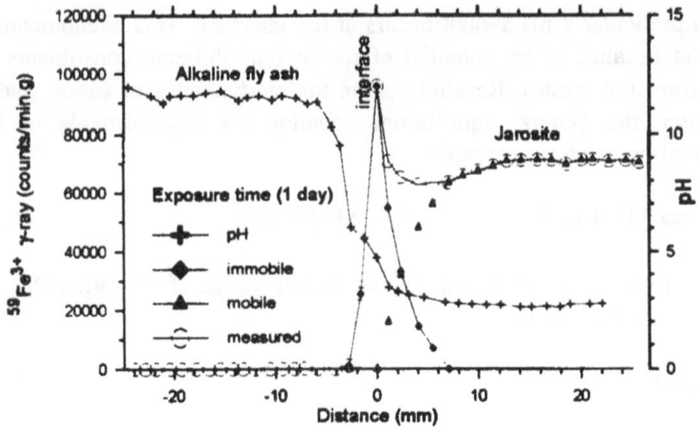


**Figure 10.** $^{59}Fe^{3+}$ **Accumulation at the Interface due to Coupling between Diffusion and Precipitation.**

Combining the mobile $^{59}Fe^{3+}$ profile in layered jarosite/fly ash system with $^{59}Fe^{3+}$ profile in jarosite only, gives our most important result, as illustrated in Figure 11. This Figure shows that the mobility of $Fe^{3+}$ in the jarosite/fly ash system decreases with respect to that observed for $Fe^{3+}$ in jarosite. This decrease is due to ferric oxyhydroxide precipitation at the jarosite/fly ash interface leading to a substantial concentration of immobile $Fe^{3+}$ even after only one day.

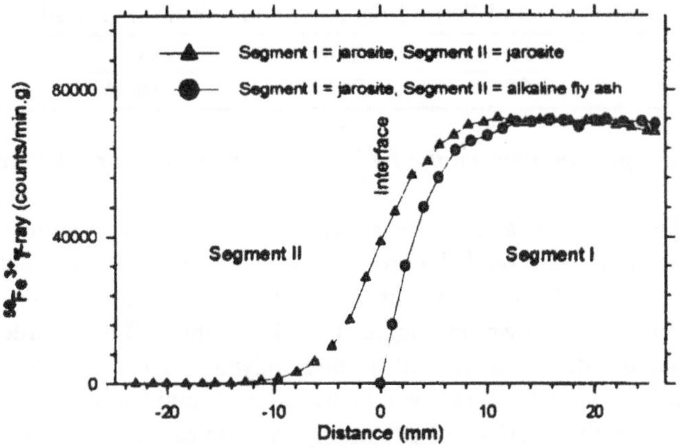


**Figure 11.** Effect of Interface Precipitation on $Fe^{3+}$ Mobility.

**The effect of precipitation on the transport properties of the jarosite/fly ash system**

Of crucial concern in this study is to assess the relation between microstructural changes caused by precipitation leading to pore filling in porous media and transport properties, such as hydraulic conductivity, and effective diffusion coefficient of inert components in the bulk system. We shall use as porosity parameter the symbol $\theta$ which is characteristic of the microstructure in a porous medium. Note that the transport properties of ions in porous media are determined not only by porosity but include many other factors, such as tortuosity, particle size, method of compaction, the manner of pore filling, *inter alia*. Consequently an explicit mathematical formulation taking these parameters into account becomes next to intractable. Therefore, most studies use empirical or semi-empirical formulas to rationalize their experimental results.

***Hydraulic conductivities and effective diffusion coefficients***

Van der Sloot *et al.*[9] conducted a study on the relation between hydraulic conductivity and effective diffusion coefficients for inert components. They tested a wide range of materials with different porosities. Their results suggest that within a specified range the relation between K, the hydraulic conductivity (m/s), and De, the effective diffusion coefficient ($m^2/s$), can be summarized as follows:

$$\log K = a \log De + b \qquad \text{Where, } a \approx 1.06,\ b \approx 2.10, \qquad \text{(Eq. 3)}$$

$$10^{-12} < K(m/s) < 10^{-8.4}, \qquad 10^{-13} < De(m^2/s) < 10^{-9.73}.$$

Our predictions concerning the variation of the effective diffusion coefficients for inert components in the jarosite/fly ash system based on this equation are collected in Table IV.

**Table IV.** Hydraulic Conductivity K (m/s) and Effective Diffusion Coefficient De ($m^2/s$) in the Jarosite/Alkaline Fly ash.

| Aging Time (days) | $K_1$(m/s, Test 1) | $De_1$($m^2/s$) | $K_2$(m/s, Test 2) | $De_2$ ($m^2/s$) |
|---|---|---|---|---|
| 0 initial condition | 1.2E-07 | 3.0E-09* | 1.9E-09 | 6.2E-11 |
| 3 | | | 1.2E-09 | 4.1E-11 |
| 15 | 6.5E-08 | 1.7E-09* | | |
| 16 | 5.0E-08 | 1.3E-09* | | |
| 21 | 4.8E-09 | 1.5E-10 | | |
| 26 | | | 4.6E-10 | 1.6E-11 |
| 40 | | | 3.9E-10 | 1.4E-11 |
| 132 | 8.9E-09 | 2.7E-10 | | |

* outside the range for which Eq. 3 is applicable.

Inspection of Table IV shows that the effective diffusion coefficient of an inert component will decrease in about one month from $6.2*10^{-11}$ $m^2/s$ to $1.4*10^{-11}$ $m^2/s$ in the jarosite/fly ash system, and will continue to decrease with decreasing bulk hydraulic conductivity.

***Permeability, hydraulic conductivity, and porosity***

Two models which relate permeability and porosity show a good agreement with the permeability derived from our experimental hydraulic conductivity measument results. The relation between permeability k ($m^2$) and hydraulic conductivity K (m/s) is given by[16]:

$$k = K \cdot \mu / \rho \cdot g \qquad \text{(Eq. 4)}$$

Where k is the permeability ($m^2$); K the hydraulic conductivity (m/s); $\mu$ the viscosity of the fluid (g/cm.s); $\rho$ the density of the fluid (g/ml); and g the gravitational acceleration (9.8 $m/s^2$). In our calculation, we have considered water as the fluid.

The relation between porosity and permability is according to Model 1[16]:

$$k = \frac{\theta^3}{(1-\theta)^2} \cdot \frac{d_m^{\ 2}}{180} \qquad \text{(Eq. 5)}$$

In which, k is the permeability ($m^2$), $\theta$ the porosity (dimensionless), and $d_m$ is the representative particle diameter.

According to Model 2[17] this relation is :

$$\frac{k_s}{k} = \frac{10\theta}{(1-\theta)^3} \qquad \text{(Eq. 6)}$$

Where $k_s = \dfrac{2r^2}{9\theta}$ is called the Stokes permeability, and r is the mean sphere radius in meters.

Note that both models contain a parameter which describes the geometry of the particle, which cannot be determined in the jarosite/fly ash system. One way to avoid this difficulty is to consider the relative value of k. If we assume that changes in the value of $d_m$ as well as r are negligible during the experiment, we obtain the following functions:

$$\frac{k_0}{k_i} = \frac{\theta_0^{\ 3}}{\theta_i^{\ 3}} \cdot \frac{(1-\theta_i)^2}{(1-\theta_0)^2} \qquad \text{(Eq. 7)} \qquad\qquad \frac{k_0}{k_i} = \frac{\theta_0^{\ 2}}{\theta_i^{\ 2}} \cdot \frac{(1-\theta_i)^3}{(1-\theta_0)^3} \qquad \text{(Eq. 8)}$$

Eq. 7 is derived from model 1, whereas Eq. 8 is derived from model 2. From both equations we are able to calculate $k_i$ as a function of $\theta_i$. Note that $k_0$ and $\theta_0$ represent the initial values of the system. Porosities used in this calculation are obtained from our previous porosity measurement results. Figure 12 shows the permeability change as a function of time in the jarosite/fly ash system. Our result indicates that in the begining stage of aging time, the measured permeability declines with time and agrees well with the models. However, after aging for 8 days, the measured permeability decreases faster with time than is indicated by the models. The reasons for this discrepancy are , as mentioned before, firstly, limited resolution imposed by slicing of the tubes, which implys that the actural porosity should be lower than our measured result; secondly, in real life the permeability of the system does not depend solely on porosity.

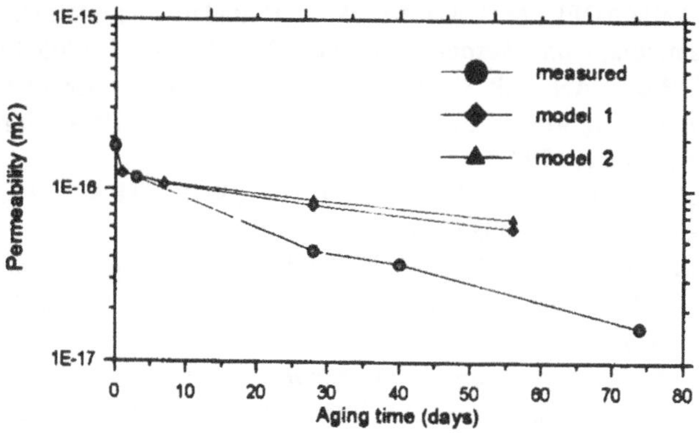


**Figure 12.** Permeability as a Function of Time.

## CONCLUSION AND SIGNIFICANCE

A layer with low permeability was formed at the interface of acidic jarosite and alkaline fly ash, which is due to a coupling between diffusion and precipitation of principal reacting chemical components in leachates from the two wastes. The diffusion of $Fe^{3+}$, $Ca^{2+}$, $SO_4^{2-}$, and $OH^-$ is caused by their concentration gradient between jarosite and fly ash. Precipitation occurred between these species at the interface, leading to pore filling(hence the porosity decrease), resulting in a hydraulic conductivity decrease of the bulk system. Immobilization of principal reacting solutes such as $Fe^{3+}$ in the wastes occurred due to newly formed minerals at this interface, and that of contaminants from the wastes presumably because of coprecipitation or surface adsorption. The results of our study indicate that interfacial chemical reactions in porous media exert a crucial effect on the transport properties of the bulk system as well as on the mobility of environmentally harmful constituents. Particularly, one can make use of the interfacial phenomena in acidic jarosite/alkaline fly ash system as a means of relatively safe and effective environmental technique to neutralize, immobilize and isolate them.

## ACKNOWLEDGEMENTS

Work was carried out at the Soil and Waste Research Laboratory, Netherlands Energy Research Foundation (ECN), Petten. Professor Dr. R.D. Schuiling is gratefully acknowledged for financial and scientific support. Special thanks go to Professor Dr. B. H.W.S. de Jong for critical reading of the manuscripts and smoothing the language. P.van

Oudenallen of the Audiovisual Service at the Institute of Earth Sciences at Utrecht is thanked for making beautiful poster and slides.

## REFERENCES


[1] LICHTNER, P.S., STEEFEL, C.I., OELKERS, E.H.eds. *Reactive transport in porous media,* Reviews in Mineralogy Volume 34. Mineralogical Society of America, Washington DC, 438p (1996).

[2] HASSANIZADEH, M.S., GRAY, W.G. Boundary and interface conditions in porous media, *Water Resources Reseach*, Vol. 25, No. 7, 1705-1715 (1989).

[3] COTÉ, P.L., VAN DER SLOOT, H.A. *Process and barrier for the contaminante of wastes*, Canadian Patent Application (1989), European Patent Approved, June, 1994.

[4] HOCKLEY, D.E., VAN DER SLOOT, H.A., WIJKSTRA, J. *Waste-soil interface*, ECN-R-92-003 (1992)

[5] SCHUILING, R.D., VAN GAANS, P.F.M. Self-sealing of the waste sulfuric acid lake of the $TiO_2$-plant at Armyansk, Crimea, Ukraine. Part I, *Applied Geochemistry*, (in press).

[6] DING, M. *Immobilization of contaminants at the interface between jarosite and alkaline fly ash,* MSc Thesis, International Institute for Hydraulic and Environmental Engineering, Delft, the Netherlands. 55p (1994).

[7] SCHUILING, R.D. Geochemical engineering: principles and case studies. *Geochemical Approaches to Environmental Engineering of Metals*, edited by Rudolf Reuther, Springer-Verlag Berlin Heideberg. 221p (1996).

[8] ARTIOLA, J.F. Waste disposal. *Pollution Science,* edited by PEPPER, I.L., GERBA, C.P., MARK, L., Academic Press, Inc. 397p (1996).

[9] VAN DER SLOOT, H.A., PEEREBOOM, D., MCGREGOR, R., *et al. Properties of self-forming and self-repairing seals*, ECN-RX-95-032.(1995).

[10] STOFFREGEN, R.E. Stability relations of jarosite and natrojarosite at 150 - 250 °C, *Geochimica et Cosmochimica Act,*. Vol. 57, 2417-2429 (1993).

[11] ELGERSMA, F. *Integrated hydrometallurgical jarosite treatment*, Ph.D. Thesis, ISBN 90-370-0073-8, Technology University Delft, the Netherlands, 288p (1992).

[12] VAN DER SLOOT, H.A., DE GROOT, G.J., HOEDE, D. *Mobility of trace elements derived from combusion residues and products containing these residues in soil and groundwater*, ECN-C-91-059, 190p (1991).

[13] DING, M., VAN DER SLOOT, H.A., GEUSEBROEK, M. *Effect of interface precipitation on the bulk diffusivity in jarosite and alkaline fly ash*, Progress Report, ECN&Utrecht University, the Netherlands, 85p (1996).

[14] BAES, C.F., MESMER, R.E. *The hydrolysis of cations*, John Wiley, NY, pp. 489 (1976).

[15] *Handbook of Chemistry and Physics*, 61st edition (1980-1981), CRC Press.

[16] FREEZE, R.A., CHERRY, J.A. *Groundwater*, Prentice-Hall, Inc. (1976).

[17] SAHIMI, M. *Flow and transport in porous media and fractured rock*: from classical methods to modern approaches, VCH, 482p (1995).

# EXPLORATION GEOCHEMISTRY

*Proc. 30th Int'l. Geol. Congr.*, Vol. 19, pp. 225-239
Xie Xuejin (Ed)


# Tin-Rare Metal Mineralization Near Mersa Alam, Eastern Desert, Egypt

MOHAMED ABDEL KADER MORSY
*Geology Department,Faculty of Science,Alexandria University,Alexandria,Egypt.*

**Abstract**

Geochemical study was carried out in three tin-rare metal mineralized localities namely : Igla, Abu Dabbab and Muelha in the Eastern Desert of Egypt. Forty-three samples were collected from the vicinity of these mineralized areas and analyzed for major, trace and rare earth elements. The results were treated with simple and sophisticated (factor and discriminant) statistical analytical methods. A graphical method was used for representing the multi-element data to get a clear picture about their enrichment and depletion. Tin shows the greatest degree in Abu dabbab and Pb, Nb and Rb are enriched in Muelha. Copper, Cs, Cr, Zn, Pb and Hf show some enrichment in the Igla and Abu Dabbab data. REE analyses with the chondrite-normalized plots show the enrichment of HREE in these mineralized areas with a negative europium anomaly in the Muelha rock data. This may suggest the use of REE geochemistry as a useful criterion for Sn-rare metal prospecting. Factor analysis was helpful in revealing four basic associations which are interpreted as representing variation in 1. mineralization, 2. parent rock type, 3. metasomatic, and 4. weathering, and this confirms most relationships identified during basic single-element interpretation. Lastly, discriminant analysis was successfully used in distinguishing between these mineralized and other barren types.

*Keywords: Geochemistry, Tin-Rare Metal, REE, Factor and discriminant analyses, Eastern Desert*

## INTRODUCTION

The present geochemical study was carried out in three well known tin mineralization localities in the Eastern Desert of Egypt namely: Igla, Abu Dabbab and Muelha. The mineralized veins in these areas are made up of massive milky quartz, cassiterite, wolframite, topaz, beryl, yellow mica, tourmaline, fluorite, sericite and chlorite,which traverse various types that include metavolcanics, metasediments, diorites and granites [1]. The location of the main Sn-W deposits in the central Eastern Desert was given by El-Ramly et al. (Fig. 1), [5].

There has been considerable controversy concerning the geochemical character of granites responsible for Sn-mineralization. Most controversy concerns the concentration of tin in the whole rock or other secondary products. Generally, Sn-bearing granites have tin contents ranging from 15 to 37 ppm [2], compared with an average of 3 ppm [25], ranging up to 8 ppm in barren granites [16]. As the difference between barren and tin-bearing granites are relatively small and may be difficult to detect using regional geochemical methods involving low-cost and high-productively analytical methods. Moreover, granite complexes may be variously enriched and depleted in Sn [6], depending on the extent and intensity of hydrothermal alteration associated with ore deposition. So, several trace-element ratios and associations other than Sn have thus proposed to identify such granites and other secondary products [2, 8, 20, 26]. In the present study, multi-element analyses of panned heavy mineral concentrates, soil and stream sediments at Igla and Abu Dabbab localities, besides the rock geochemistry in Muelha are presented and investigated by the use of simple and multivariate statistical techniques in geochemical interpretation.

## LOCATION AND GEOLOGICAL SETTING

The Igla (25° 06′ N, 34° 39′ E), Abu Dabbab (25° 20′ 27″ N, 34° 32′ 20″ E) and Gabal Muelha (24° 52′ N, 43° 00′ E) areas are located in the central part of the Eastern Desert at 32, 47 and 100 Km respectively from Mersa Alam at the Red Sea coast (Fig. 1). In the Igla area, a terrain of acid to intermediate volcanics and tuffs is intruded by apogranite with disseminated cassiterite. Mineralized quartz veins are related to this granite and occurring in association with its contact [1]. Panned alluvial samples from wadis draining the granite contain cassiterite, tantalite-columbite and monazite [14]. The Abu Dabbab deposit is related to a mass of apogranite that contains essentially Ta-Nb-Sn mineralization and is intruded into metasediments [22]. The gabal Muelha [11] is a dome-like mass of albitized biotite granite with greisen zones intruded into metasediments. Cassiterite, columbite and beryl were identified in panned samples and as disseminations in the albitites and greisens. It had been been shown [18] that the Muelha granite is a plumasitic type in which anhydrous volatiles played a major role in its genesis.

## SAMPLING AND ANALYTICAL METHODS

Four types of samples were collected from the study localities to assess the effectiveness of various sampling media in geochemical prospecting : panned heavy mineral concentrates (PC), soil (S), stream sediments (SS) and rock (R). Forty-three samples were taken in the vicinity of the Sn-mineralization at Igla (n=16),Abu Dabbab (n=13) and Muelha (n=14). Twenty-eight samples were analyzed for 30 elements by XRF (X-ray fluorescence) and the REE (La, Ce, Pr, Nd, Sm, Eu, Gd, Dy, Er, Ho, Yb and Lu) contents of eight samples were determined by ICP (Inductively Coupled Plasma) at the Technical University of Berlin. Fifteen samples were analyzed for 45 elements including REE (La, Ce, Sm, Eu, Yb and Lu) by INAA ( Instrumental Neutron Activation Analysis) at the Delft Nuclear Institute.

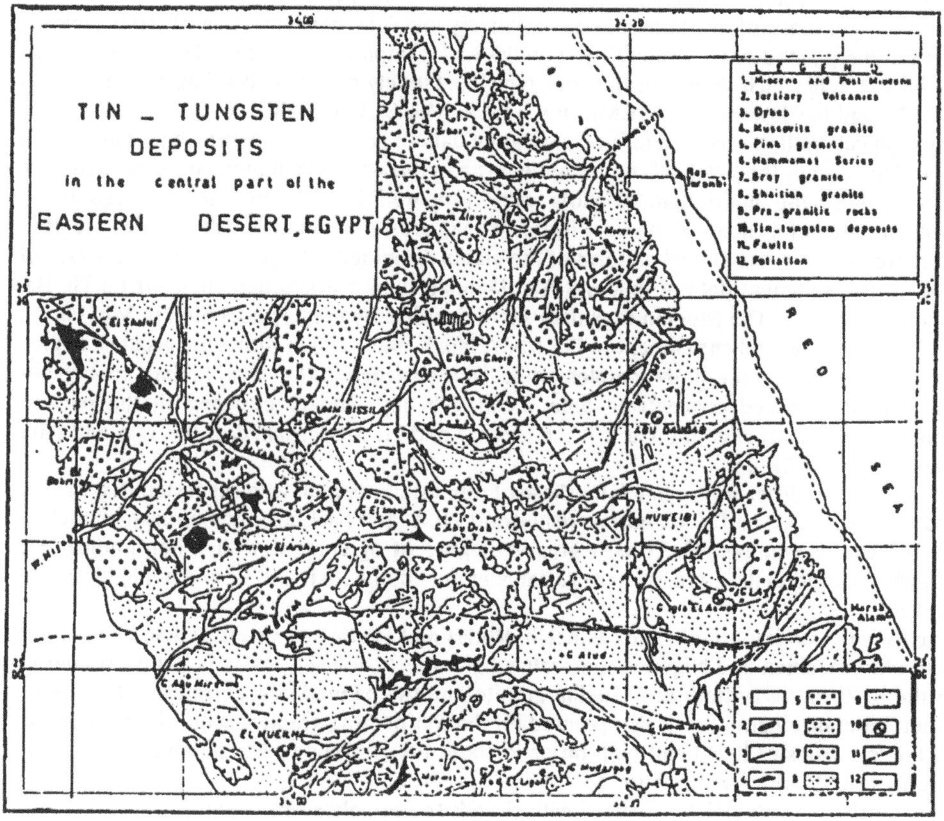


**Fig. 1. Distribution of Sn–W deposits in central Eastern Desert.**

**( after El Ramly et al. 1970 )**

1. Miocene and Post Miocene; 2. Tertiary Volcanics; 3. Dykes; 4. Muscovite granite; 5. Pink granite; 6. Hammamal Series; 7. Grey granite; 8. Shaitian granite; 9. Pre-granitic rocks; 10. Tin-tungsten deposits; 11. Faults; 12. Foliation.

## COMPARISON BETWEEN PANNED CONCENTRATES, SOIL, STREAM SEDIMENTS AND BED-ROCK GEOCHEMICAL TYPES

The statistical parameters of the chemical data in the four sampling media are studied, where the first three types (panned concentrates, soil and stream sediments), which include secondary weathering products are compared with the average soil [27]. This study indicates that these types are relatively rich in Na, Fe, Ca, Mg, Sn, Zn, Co, Sc, Cr, Ni and Cu, while the concentrations of Si, Ba and Sr are relatively low. The concentration of Rb is typically high in the panned concentrates. The muelha rock samples are compared with the average granite [15,28],which show that this rock type is relatively rich in Si, Na, Rb, Sn, Zn, Co, Cr and Ni and has low contents of Mn, Fe, Ti, Ca, Mg, Sr,V and Cu.

A graphical method was used for presenting the data to get a clear picture about enrichment and depletion in the levels of multi-elements analyzed within different sampling media. In this method, the multi-element plots were prepared by dividing the observed mean concentration of a chemical element by its average crustal abundance value, which are used to represent the geochemical patterns of the media examined. The panned concentrates, soil and stream sediment data of Igla and Abu Dabbab show marked depletion for La,Th, Ba, P and K, Table 1. The Muelha rock data show significantly greater depletion in Mg, Ba, Ti, P, Sr, Ca and Fe, and enrichment in Pb, Rb,Y, Nb, K and Zn. Tin shows the greatest degree of enrichment in all the study samples, while Ta is enriched in the Abu Dabbab area within the soil and panned concentrate data. Lead, Nb and Rb are enriched in the panned concentrates of Muelha and Cu, Cr, Zr, Zn, Pb and Hf are enriched in the Igla and Abu Dabbab data (Fig. 2).

These findings differ somewhat from those of other authors [26], who suggested that Ti, Fe, Mg, Ca, Ni, Cr, Co and V are impoverished and Th, U and REE are systematically enriched in Sn-bearing granites. There is a greater degree of agreement with other findings [7, 20],who showed that Sn granites are variably enriched in K, Rb and Li and depleted in Mg and Sr.

The use of some important ratios as K:Rb, Rb:Ba and Rb:Sr was also tested and compared with other Arabian and Nigerian mineralized granites (Table 2). There is a close similarity in the K:Rb ratios of both the Muelha granite and the panned concentrates with those of other mineralized granites. It was pointed [6] that Sn-bearing granitoids may be recognized by the low ratios of K:Rb and high ratios for both the Rb:Ba and Rb:Sr, which is the case in only the Muelha granite data.

Rare earth elements (REE) analyses with chondrite-normalized plots (Fig. 3), clearly show the enrichment of Gd, Dy, Ho, Er, Yb and Lu (HREE) in Igla and Abu Dabbab data compared with average granite [15]. It is important to mention that the REE patterns in the panned concentrates, soil and stream sediments reflect their behavior in the Sn granites [10, 20]. REE curves for the Muelha granite and the panned concentrates are characterized by heavy REE enrichment and a much negative europium anomaly. The Igla and Abu Dabbab data contain somewhat higher levels of REE (particularly HREE), but without showing the negative Eu as in the Muelha samples. The enrichment of REE and especially the HREE in such mineralized granite and associated secondary weathering products may suggest that application of REE geochemistry as a useful criterion in the prospecting for tin deposits.

TABLE 1

Panned concentrates(PC), soil(S), stream sediments(SS) and rock data normalized to average crustal abundance for Igla, Abu Dabbab and Muelha samples

| Element | IGLA | | | ABU | DAB | BAB | Mue | Lha |
|---|---|---|---|---|---|---|---|---|
| | PC | S | SS | PC | S | SS | PC | R |
| Si | 1.06 | 1.03 | 1.00 | 0.86 | 0.98 | 1.16 | 1.07 | 1.23 |
| K | 0.52 | 0.62 | 0.58 | 0.50 | 0.52 | 0.64 | 1.08 | 1.45 |
| Na | 0.91 | 0.73 | 1.25 | 0.61 | 0.50 | 0.64 | 1.28 | 1.38 |
| Al | 0.79 | 0.84 | 0.95 | 0.80 | 0.84 | 0.95 | 1.01 | 0.89 |
| P | 0.34 | 0.65 | ---- | 0.37 | 0.45 | ---- | 0.25 | 0.02 |
| Mn | 1.12 | 0.79 | 0.69 | 2.00 | 1.35 | 1.35 | 0.57 | 0.32 |
| Fe | 0.87 | 0.78 | 0.69 | 1.26 | 0.82 | 0.96 | 0.40 | 0.08 |
| Ti | 0.92 | 1.05 | 0.83 | 0.94 | 0.89 | 1.00 | 0.56 | 0.02 |
| Ca | 1.10 | 0.86 | 1.07 | 1.06 | 0.74 | 0.85 | 0.42 | 0.07 |
| Mg | 0.77 | 0.84 | 0.87 | 1.92 | 1.86 | 1.99 | 0.70 | 0.001 |
| Cs | 11.20 | 7.50 | 3.00 | 4.80 | 4.80 | 3.20 | ---- | ---- |
| Rb | 0.79 | 0.50 | 0.33 | 1.20 | 1.00 | 0.46 | 2.13 | 3.44 |
| Sn | 110.50 | 65.50 | 71.25 | 171.60 | 109.70 | 87.50 | 1567.50 | 16.90 |
| U | 0.53 | 2.80 | 0.52 | 1.33 | 1.11 | 0.86 | ---- | ---- |
| Th | 0.35 | 0.37 | 0.22 | 0.33 | 0.41 | 0.41 | ---- | ---- |
| Ta | 0.79 | 0.60 | 0.41 | 40.20 | 8.80 | 1.00 | ---- | ---- |
| Pb | 3.70 | 1.39 | ---- | 3.39 | 4.40 | ---- | 7.96 | 4.33 |
| Hf | 2.60 | 2.19 | 1.00 | 2.46 | 1.55 | 1.20 | ---- | ---- |
| Zr | 2.50 | 1.89 | 1.60 | 1.23 | 1.55 | 1.27 | 0.48 | 0.29 |
| La | 0.20 | 0.19 | 0.13 | 0.12 | 0.18 | 0.19 | 1.32 | 1.01 |
| Ba | 0.48 | 0.50 | 0.73 | 0.50 | 0.39 | 0.67 | 0.20 | 0.02 |
| Sr | 0.79 | 0.84 | 1.24 | 1.02 | 0.93 | 1.30 | 0.60 | 0.04 |
| Y | 1.20 | 1.20 | ---- | 0.82 | 0.56 | ---- | 1.25 | 2.37 |
| Zn | 2.10 | 1.30 | 0.90 | 3.90 | 1.56 | 1.34 | 1.20 | 1.41 |
| Co | 0.62 | 0.35 | 0.65 | 2.66 | 1.00 | 1.42 | 0.18 | 0.12 |
| Sc | 0.77 | 0.90 | 0.76 | 1.11 | 0.85 | 1.00 | 0.79 | 0.21 |
| V | 0.39 | 0.50 | 0.53 | 1.10 | 0.71 | 0.90 | 0.33 | 0.08 |
| Cr | 2.59 | 2.93 | 2.39 | 5.50 | 7.80 | 6.20 | 1.79 | 0.15 |
| Ni | 1.11 | 1.25 | 1.26 | 3.00 | 3.40 | 3.00 | 0.67 | 0.17 |
| Cu | 9.10 | 3.48 | 4.83 | 3.00 | 5.58 | 5.27 | 0.19 | 0.11 |
| Nb | 0.90 | 0.66 | ---- | 0.75 | 1.23 | ---- | 6.03 | 2.08 |

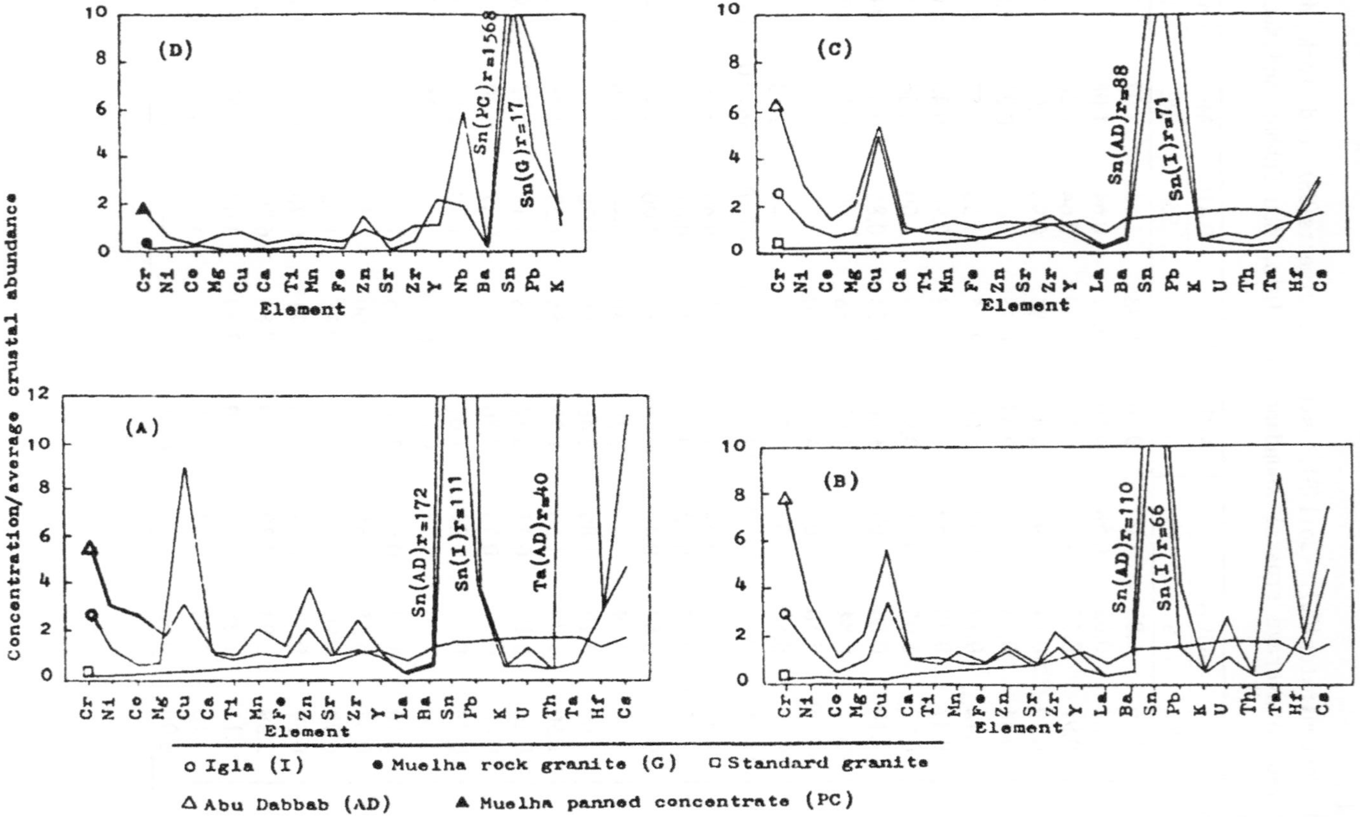


Fig.2.Element plots for A-panned concentrates, B-soil, C-stream sediment samples of the Igla and Abu Dabbab and D- panned concentrates and rock granite of Muelha based on normalized data to average crustal abundance.

TABLE 2
Mean element ratios for mineralized samples of Igla, Abu Dabbab and Muelha, compared with Egyptian (Abu Rusheid), Arabian and Nigerian mineralized granites.

| Type | Number of Samples | K:Rb | Rb:Ba | Rb:Sr |
|---|---|---|---|---|
| Panned Concentrates | 12 | 97 | 1.2 | 1.4 |
| Soil | 18 | 173 | 0.4 | 0.57 |
| Stream Sediments | 3 | 302 | 0.12 | 0.18 |
| Muelha Granite | 10 | 74 | 84 | 66 |
| Abu Rusheid Granite * | 58 | 48 | 8.3 | 46 |
| Arabian Granite ** | -- | 97 | 4.9 | 10 |
| Nigerian Granite *** | 10 | 53 | 10 | 267 |

* Mohamed, (1989).
** Jackson and Ramsay, (1986).
*** Imeokparia, (1983).

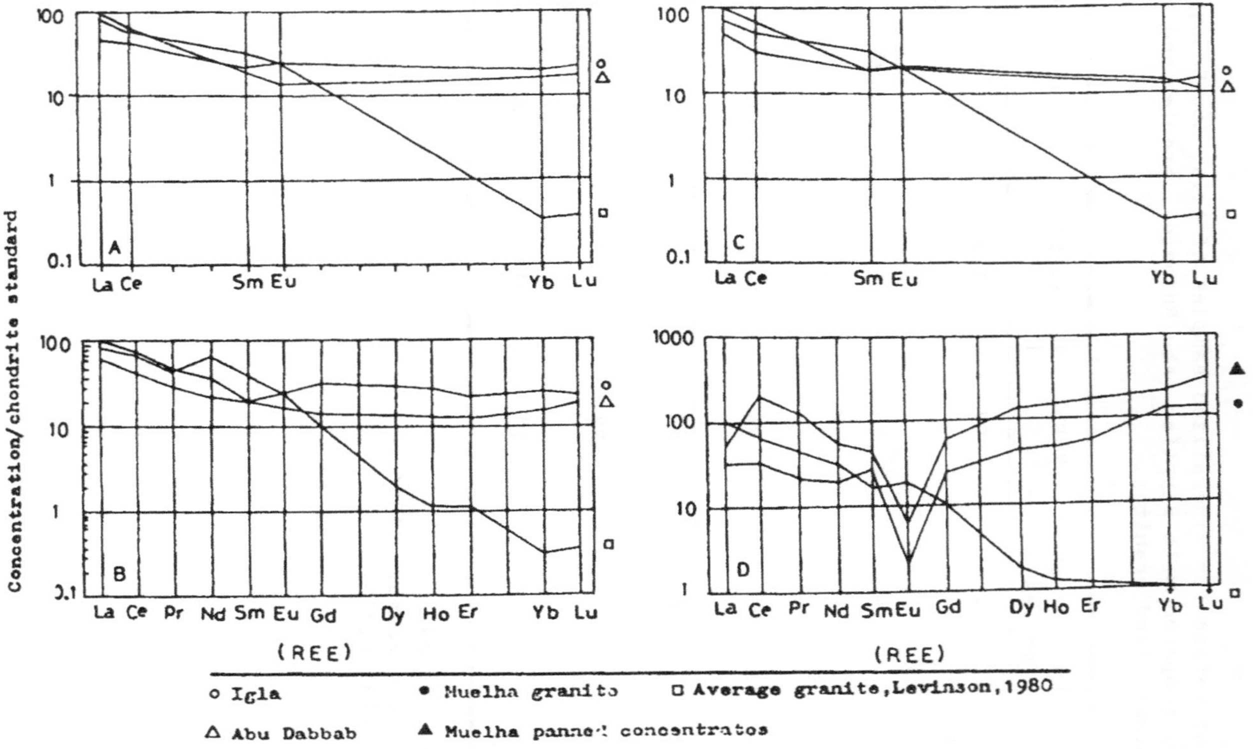

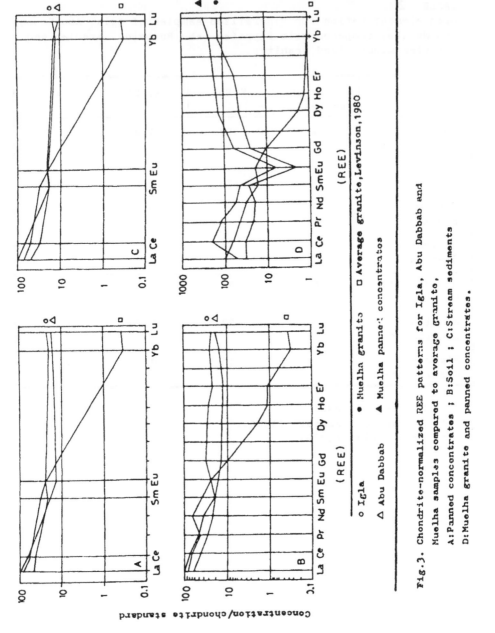

Fig.3. Chondrite-normalized REE patterns for Igla, Abu Dabbab and Muelha samples compared to average granite,
A: Panned concentrates ; B: Soil ; C: Stream sediments
D: Muelha granite and panned concentrates.

## FACTOR ANALYSIS

Factor analysis of the data sets for four groups was undertaken using the SPSS package [24], after transforming strongly skewed distributions (Sn, Rb, Zr, Ba, Sr, Zn, Co, Sc, V, Cr, Ni, Cu) to the corresponding $\log_{10}$ values. The first group includes the panned concentrates, the second group includes both the soil and stream sediments, the third group includes the rock samples of Muelha and the fourth group includes all the study sampling media. A 5-factor model was derived for the panned concentrates and a 4-factor model for the other groups. The factor loadings for these factor models are shown in Table 3. Each variable with a factor loading greater than 0.4 is shown in the left half of the respective box. Variables with factor loadings of less than -0.4 are shown in the right half of the boxes. The variable names are located on the base of the absolute value of their loadings on a scale from 0.4 (bottom of box) to 1.0 (top of box). In this way more important variables are shown near the bottom. At the base of each box, an interpretation and the percent variation explained are shown.

## PANNED CONCENTRATE DATA

The produced five factor model accounts for 91.7% of the data variability. Factor 1(Sc-Fe-V-Ti association) accounts for 53.9% of the variability of this model. There is also some contribution from Si. This metal association is known to occur together in mafic rocks, so grouping of these elements into factor 1 reflects the occurrence of the products of weathering of basic rock types in these panned concentrate samples. Factor 2 (Ca-Zr-Ba association) accounts for 14.7% of the data variability of this model. There is also some contribution from both Cu and Sr. This metal association is characteristic for granitic rocks and reflects their occurrences in the first type. Factor 3 (Sn,Rb-Na-K association) accounts for 11.1% of the data variability of this model. This factor model is considered as the Sn mineralization factor with the occurrence of pegmatitic association. Factors 4 and 5, which account for additional 7% and 5% respectively, are probably due to weathering effects in clay mineral adsorption and Mn oxide coprecipitation.

## SOIL AND STREAM SEDIMENT DATA

The 4 factor model produced accounts for 79.5% of the data variability. It gives similar pattern to that mentioned in the panned concentrates with the occurrence of mafic, granitic and pegmatitic associations with the presence of Sn mineralization in the third factor. The weathering effects are clear in both the third and fourth factors, which are responsible for the clay mineral and Mn oxide associations.

## ROCK GRANITE DATA

This factor model accounts for 71.6% of the data variability. Factor 1 (Zr-Ti-Ca-Sr-K association) and factor 2 (Si-Zn-Fe, Cu-Na-Sc association) account for 32.4% and 17.7% respectively of the variability of this model. The metal associations in both factors point towards granitophile elements and resemble the second factor in the panned concentrates

TABLE 3

Rotated factor loading for Sn-mineralized types in Igla, Abu Dabbab and Muelha samples

| Type | 1 | 2 | 3 | 4 | 5 |
|---|---|---|---|---|---|
| Panned concentrates n=12 | Sc Fe V Ti Si | Ca Zr Ba Cu Sr | Sn Rb Na K | Cr Ni Mg | Zn Mn Co |
| | mafic 53.9 | granitic 14.7 | minerliz + pegmatitic 11.1 | weathering -1- 7.0 | weathering -2- 5.0 |
| Soil and stream sediment n=21 | Sc Fe Ti V Sr | Si Ba Al Zn Ca | Co Ni Cr Cu Sn K | Rb Zr Mn Mg Na | |
| | mafic 32.6 | granitic 23.0 | weathering + mineraliz 14.6 | pegmatitic +weathering -2- 8.5 | |
| Rock granite muelha n=10 | Zr Ti Ca Sr K | Si Cu Zn Na Sc | Mg Co Ba | Sn Rb | |
| | granitic -1- 32.4 | granitic -2- 17.7 | mafic 15.9 | minerliz + Pegmatitic 5.6 | |
| All study samples n=43 | Zr Ti Rb Ba Sr K Ca V Cu Sc Ni Cr Fe | Mn Co Zn Mg | Al Si Na | Sn | |
| | granitic + mafic 60.2 | weathering 10.3 | albitization 6.9 | minerlization 5.8 | |

Key:

| Positive loading | Negative loading |
|---|---|
| 1.0 | −1.0 |
| 0.9 | −0.9 |
| 0.8 | −0.8 |
| 0.7 | −0.7 |
| 0.6 | −0.6 |
| 0.5 | −0.5 |
| 0.4 | −0.4 |

Variable names positioned in the boxes according to their loading values.

and soil and stream sediment data. Factor 3 (Mg-Co-Ba) accounts for 15.9% of the data variability and reflect mafic association. Factor 4 (Sn-Rb) accounts for 5.6% and shows the Sn mineralization with some pegmatitic association.

## TOTAL STUDY DATA

The 4 factor model produced explains the bedrock composition (mainly granitic and basic rocks) in the first factor, the weathering effects which include Mn oxide and clay mineral adsorption in the second factor, the metasomatic alterations in the third factor and finally the mineralization of tin is clear in the fourth factor.

From the above observations, it can be stated that the factors identified in the data from the individual four groups are very consistent and are restricted to four basic influences on the composition of the study samples: 1. Sn-mineralization, 2. bedrock, 3. metasomatic, and 4. weathering. The strength and rank of these factors vary from type to type depending on variation in bedrock composition, metasomatic and weathering effects, besides the presence of tin mineralization, which varies from type to type also, depending on the percentage of samples affected by mineralization and the strength of the expression of mineralization in the samples.

## DISCRIMINANT FUNCTION ANALYSIS

It is important to test the significance of the difference between the four sampling media by the use of discriminant function analysis using SPSS package [24]. This computes discriminant functions, incorporating the variables (analyzed elements) that most effectively distinguish the pre-defined types. Actually, in mineral exploration, discriminant analysis is primarily used to distinguish between anomalous and background situations. Provided that two or more population groups can be defined a priori, discriminant analysis is a useful tool to classify individuals into population groups on the basis of a number of measured variables [3, 9, 19, 23, 29]. Generally, the method used is a stepwise discriminant analysis [24]. The stepwise selection criterion is the overall multivariate F ratio for the test of difference among group centroids. The variable which maximizes the F-ratio also minimizes the Wilk's Lambda, a measure of group distinction.

Discriminant analysis was utilized for the classification of Arabian granitoids [21], which shows that the discriminating power of trace elements is less than that of the major oxide. In the present investigation, discriminant analysis was firstly carried out using the major elements (Si, K, Na, Mn, Fe, Ti, Ca and Mg) of the four sampling media (panned concentrates, soil, stream sediments and rock samples), in which 81.4% correctly classified samples were obtained. Then, the discriminant analysis was applied on the trace elements (Rb, Sn, Zr, Ba, Sr, Co, V, Cr, Ni and Cu), which resulted in 95.35% correctly classified samples. The use of both major and trace elements have succeeded in discriminating 100% of the classified samples. Consequently, discriminant analysis (DA) was carried out by the use of both the major and trace elements and two types of DA were used : direct and stepwise methods. The results of the direct method with the pooled within-groups correlations between discriminating variables and canonical discriminant functions show that the most characteristic variables in distinguishing between the four study types in the first function include, with the order of importance, (Rb, Ti, K) and for the second function, (Zn,Sn) and for the third function, (Na, Al). As the results obtained by this method, show only the direct effect of all the variables, it is tested to use the stepwise method, which show the individual effect of each variable and its importance in the classification. It is worthwhile noting that almost 67% of the study data were correctly classsified using only Rb, while another 4 elements (Sn, K, Zn and Na) raised the overall correct classification to 88%. Hence, the relative importance of the different elements on the classification of the total study samples is clearly seen. The variation of discriminant scores for the first function in the four types are shown in figure 4. It is clear that all the rock samples of Muelha are characterised by negative higher values, while the other types (panned concentrates, soil and stream sediments) show the positive values. This illustrates the possibility to use these discriminant scores in distingishing the study samples into two groups : primary (rock type) and secondary (weathering).

Another investigation was attemped to distinguish between tin mineralized and non-mineralized types. Twelve data analysis of non-mineralized grey granites from the Eastern Desert were selected from the work of El-Gaby [4] and compared with the above mentioned four mineralization types. The results of stepwise discriminant analysis to the five types show that 98.18% of the samples were correctly classified. Rubidium, Ba, Sn, K and Na are important discriminanting elements, while Co, Al, Mn and Sc are of little importance. The variation of discriminant scores for the first function in the five types are shown in figure 5. By using a critical value $R_1$ =7.0, the mineralized groups can be distinguished from the non-mineralized group. Also, by the use of the value $R_0$ =1.0, the secondary products types are separated from primary type.

## SUMMARY

The geochemical study of the tin mineralized localities in Igla, Abu Dabbab and Muelha can be summarized as follows:

1- There is a marked depletion of La, Th, Ba, P and K and enrichment of Cu, Cr, Zr, Zn, Pb, Hf and HREE in the Igla and Abu Dabbab data. 2- Tin is characteristically enriched in the three localities, Ta in Abu Dabbab, while Pb, Nb and Rb show high values in Muelha. 3- The panned concentrates, soil and stream sediments show enrichment in Fe, Ca, Mg, Sn, Zn, Co, Sc, Cr, Ni and Cu and a deficiency in Si, Ba and Sr. Rubidium is enriched in the panned concentrates. 4- The granitic rocks of Muelha are characterized by high values of Si, Na, Rb, Sn, Zn, Rb:Sr and low values of Mn, Fe, Ti, Ca, Mg, Zr, Ba, Sr, V, Cr, Cu and K:Rb. 5- Factor analysis was helpful in revealing the tin mineralization in the study samples with influences of metasomatic and weathering effects on the parent rocks. 6- Stepwise discriminant analysis on both major and trace elements was used in the correct classification of mineralized types and between primary and secondary types.

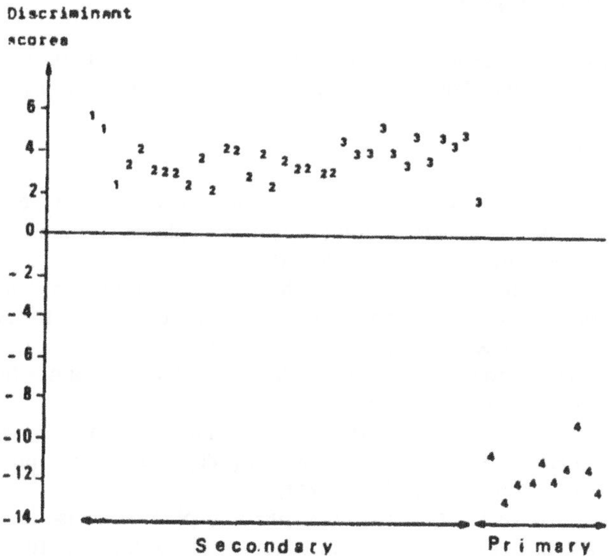


Fig. 4 Variation of discriminant scores for primary and secondary mineralized types 1-stream sediments, 2=soil, 3=panned concentrates, 4=rock granites

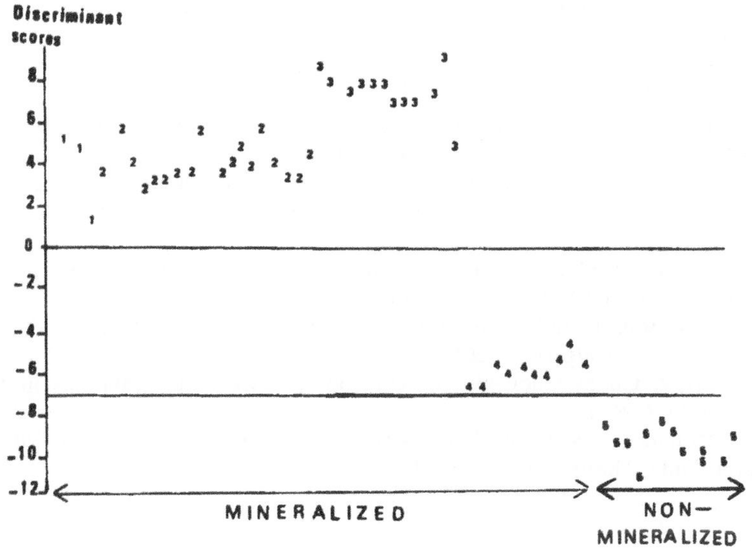


Fig. 5 Variation of discriminant scores for mineralized and non-mineralized types. 1=stream sediments, 2=soil, 3=panned concentrates, 4=Mueha granite, 5=grey granite.

ACKNOWLEDGEMENTS

The author wishes to express his appreciation to the staff of both the Technical University of Berlin and the Delft Nuclear Institute for their assistance during sample analyses.

REFERENCES

1. M.S. Amin. A tin-tungsten deposit in Egypt. *Econ. Geol.* 42. 637-671 (1947).
2. A.A. Beus and S.R. Grigorian. Geochemical exploration methods for mineral deposits. *Applied Publishing Co., Wilmette, Ill.* (1977).
3. F.L. Clausen and O. Harpoth. On the use of discriminant analysis techniques for classifying chemical data from panned heavy-mineral concentrates, central East Greenland. *J. Geochem.Explor.* 18. 1-24 (1983).
4. S. El Gaby, Petrochemistry and geochemistry of some granites from Egypt. *Neues Jb. Miner. Abh.*, 2, 147-189 (1975).
5. M.F. El Ramly, S.S. Ivanov and G.G. Kochin. Tin-tungsten mineralization in the Eastern Desert of Egypt. In: O. Moharram *et. al.*, (eds), studies on some mineral deposits of Egypt. *Geol. Surv. Egypt.* 43-52 (1970).
6. G.J.S. Govett. Rock geochemistry in mineral exploration. Elsevier. Amsterdam. (1983).
7. G.J.S. Govett, Bedrock geochemistry in mineral exploration. In: G. Garland (ed.), *Int. Conf. on Geophys. Geochem. explor. for minerals and groundwater, OGS spec.* volume 3. (1988).
8. G.J.S. Govett and P.R. Atherden. Application of rock geochemistry to productive plutons and volcanic sequences. *J. Geochem. Explor.* 30. 223-242 (1988).
9. R.J. Howarth. Empirical discriminant classification of regional stream-sediment in Devon and East Cornwall. *Inst. Min. Metall., Trans., Sect. B.* 8, 142-149 (1971).
10. R.J. Howarth. G.S. Koch. J.A. Plant and R.K. Lowary. Identification of uraniferous granitoids in the USA using stream sediment geochemical data. *Mineral. Mag.* 44, 455-470 (1981).
11. A.A. Hussein. Mineral deposits. In: R. Said (ed.). The Geology of Egypt. Balkama. Rotterdam 511-566 (1990).
12. E.G. Imeokparia. Lithogeochemical dispersion associated with Ririwai zinc-tin lode. Northern Nigerian. *J. Geochem. Explor.* 19. 643-661 (1983).
13. N.J. Jackson and C.R. Ramsay. Post-orogenic felsic plutonism. mineralization and chemical specialization in the Arabian shield. *Inst. Min. Metall., Trans. Sect. B* 95. 83-93 (1986).
14. G.G. Kochin and F.A. Bassiuni. The mineral resources of the U.A.R.. Part I: Metallic minerals. Internal Report. *Geol. Surv. Egypt* (1968).
15. A.A. Levinson. Introduction to exploration geochemistry. second edition. Applied Publishing Co., Calgary. (1980).
16. S.F. Lugov. On metallogenic specialization of Mesozoic granitoids in the Chukchi Peninsula. (1964).
17. F.H. Mohmed. Geochemical prospecting in Nugrus area. southern Eastern Desert. Egypt. Ph.D. Thesis. Alex. Univ. (1989).

18. M.A. Morsy and F.H. Mohamed. Geochemical characterization and petrogenetic aspects of Muelha tin specialized granite, Eastern Desert of Egypt. *Bull. fac. Sci., Alex. Univ.* 32 (A), 502-515 (1992).
19. Th. M. Pantzis and G.J.S. Govett, Interpretation of detailed rock geochemical survey around Mathiati, Cyprus. *J. Geochem. Explor.* 2, 25-36 (1973).
20. J.A. Plant, N. Breward, P.R. Simpson and D. Slater, Regional geochemistry and identification of metallogenic provinces: examples from lead-barium, tin-uranium and gold deposits. In: A.G. Darnley and R.G. Garrett (eds.). *J. Geolchem. Explor.* 39, 195-224 (1990).
21. C.R. Ramsay, J. Odell and A.R. Drysdall, Felsic plutonic rocks of the Midyan region, Kingdom of Saudi Arabia-II, Pilot study in chemical classification of Arabian granitoids. *J. Afri Earth Sci.* 4, 79-85 (1986).
22. A.A. Sabet, V. Tsogoev, L.B. Sarin, S.A. Azazi, M.A. El Bedewi and G.A. Ghobrial, Tin-tantalum deposit of Abu Dabbab. *Annals Geol. Surb. Egypt*, VI. 93-117 (1976).
23. O. Selinus, Factor and discriminant analysis to lithogeochemical prospecting in an area of central Sweden. In: G.R. Parslow (ed.). Geochemical Exploration 1982. *J. Geochem. Explor.* 19, 619-642 (1983).
24. SPSS. SPSS-x users guide, second edition, Mc Grow-Hill (1986).
25. S.R. Taylor, The application of trace element data to problems in petrology. *Phys. Chem. Earth* 6, 135- 213 (1966).
26. G. Tischendorf, Geochemical and petrographic characteristics of silicic magmatic rocks associated with rare element mineralization in metallization associated with acid magmatism, *Geol. Surv. Prague*, 2 (1974).
27. A.P. Vinogradov, Laws governing the distribution of chemical elements in the earth's crust. *Geochemistry*, 1. (1956).
28. A.P. Vinogradov, Average content of chemical elements in the major types of igneous rocks of the earth's crust. *Geochemistry*, 7. (1962).
29. R.E.S. Whitehead and G.J.S. Govett, Exploration rock geochemistry detection of trace element halos at Heath steel Mines, (N.B., Canada) by discriminant analysis. *J. Geochem. Explor.*, 3, 371-386 (1974).

*Proc. 30th Int'l. Geol. Congr.*, Vol. 19, pp. 241-256
Xie Xuejin (Ed)


# GOLD AND ASSOCIATED ELEMENTS IN THE LATERITIC REGOLITH AT JIM'S FIND SOUTH, TANAMI DESERT, NORTHERN TERRITORY, AUSTRALIA

C.L. Stott[1,4], L. Xu[1], C.R.M. Butt[2], G.M. Bailey[3] and J. C. van Moort[1]

1.Geology Department, University of Tasmania, GPO Box 252-79, Hobart, Tasmania 7001, Australia
2.CRC LEME, CSRIO Exploration and Mining, Private Bag, PO Wembley, WA 6014, Australia
3.Physics Division, ANSTO, Lucas Hights, Private Bag 1, NSW 2234, Australia
4.present address: WMC, PO Box 22, Leinster WA 6437, Australia

## ABSTRACT

The acid insoluble residue of rock pulps from deeply weathered terrain around the Jim's Find South gold prospect consists largely of quartz, with some minor sericite and feldspar. Original wall rock alteration patterns around the mineralisation can be observed and is characterised by increased Rb/Al and ratios. Arsenic values flare up from the mineralisation towards the surface. Ti/Zr ratios of the residue is of use to delineate original basalt flows between sandstones.

The composition of the raw rock pulps indicate strong leaching of Na, Ca and Sr, the retention of Si, Al, Fe, Mn and Co, and the near surface concentration of Zr, Hf and Ti.

## INTRODUCTION

The Granites-Tanami gold province is a relatively new gold producing area approximately 650 km NW of Alice Springs in the Northern Territory. Outcrop is sparse and the regional geology is largely interpreted from geophysical data. Detailed geological relations are only known at mine sites.

Prolonged chemical weathering spanning the Cretaceous to mid-Miocene produced a 90 m deep lateritic blanket, modified by subsequent drier climates. Although lithogeochemical differentiation of the weathered bedrock does not provide an alternative to lithology, bulk rock geochemistry appears to pick up alteration patterns related to mineralisation.

In order to find the most efficient way to define gold occurrences and wall rock alteration patterns in the regolith the results of an existing multi-element NAA-XRF study of *raw rock* pulps is compared with proton probe (PIXE/PIGME) analyses of the *acid insoluble residue* (silicate fraction) of the rock pulps.

## GEOLOGICAL INFORMATION

Regional Geology

The oldest exposed rocks are Palaeoproterozoic sediments and volcanics, which have been isoclinally folded, and cleaved and subjected to greenschist facies metamorphism at about 1960 Ma. Marked unconformities separate the Tanami complex basement from onlapping dominantly sedimentary packages, dated at about 1800 Ma. Widespread granite intrusion occurred at 1780-1710 Ma. The complex of these combined rocks forms an inlier between the Neoproterozoic and Phanerozoic rocks of the Canning Basin and the northern part of the Northern Territory, and the Mesoproterozoic shallow water sediments of the Birundu Group (Fig. 1).

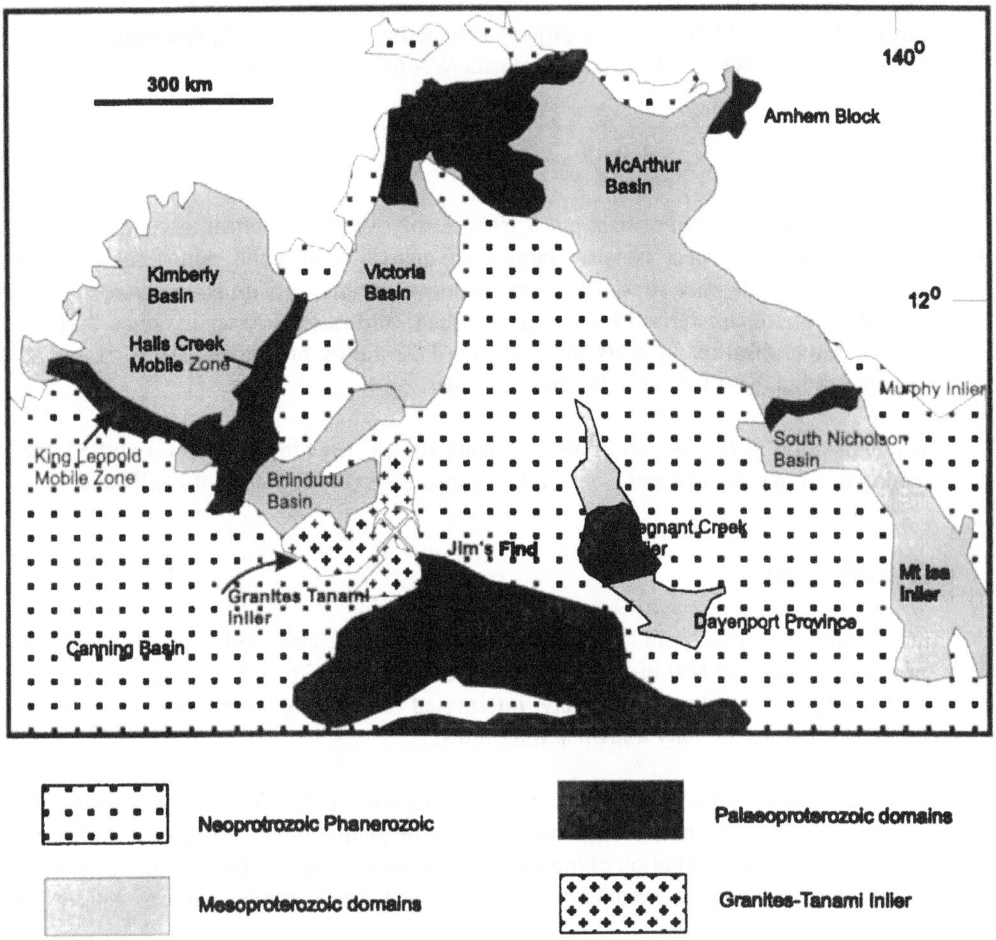


**Figure 1** The major Proterozoic inliers of Northern Australia (after Plumb 1990 and Tunks 1996)

Local Geology

Jim's Find is located on a NW/SE anticlinal structure at the southern end of the corridor between the Coomarie and Frankenia granite domes (Fig. 2). Of the five informal sedimentary units of the Tanami complex only the Charles Beds have been recognised in

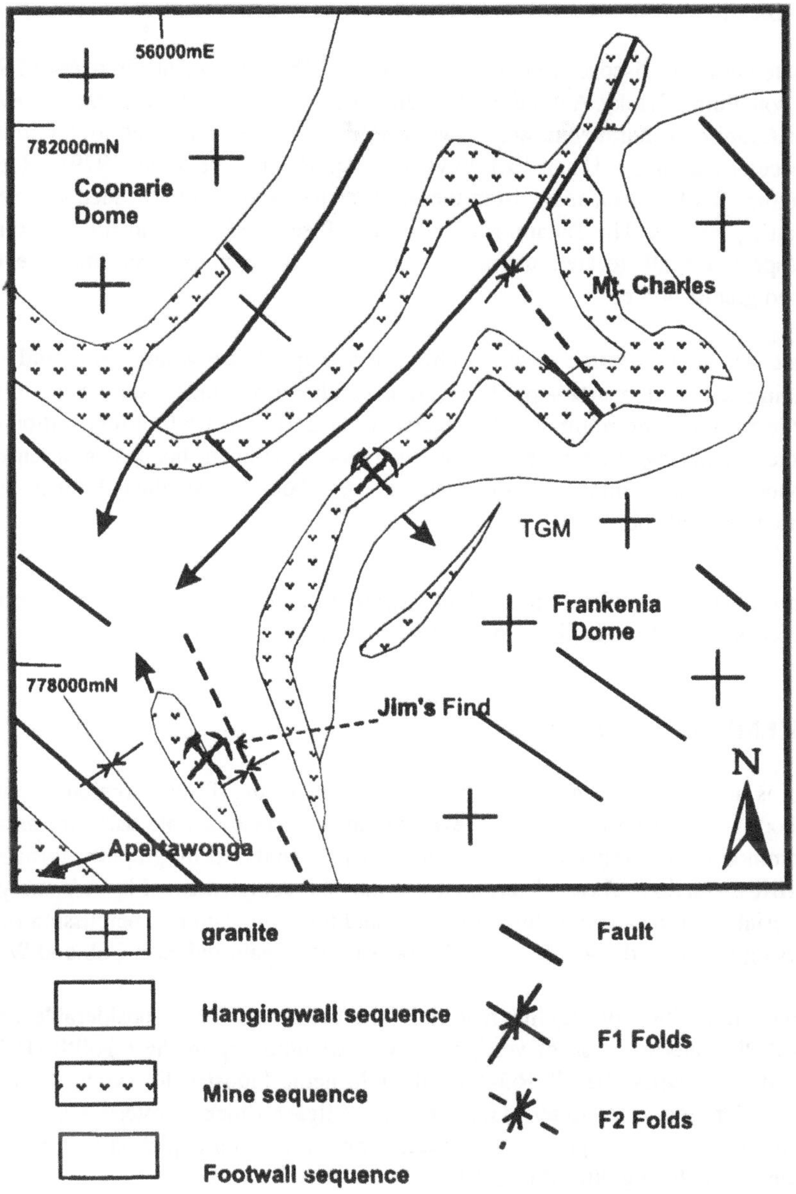


**Figure 2** Geological interpretation of the detailed study area, based on mapping and aeromagnetic data (after Tunks 1996)

the Jim's Find area. These comprise a sequence of fine to medium grained greywacke, phyllite, banded iron formation, chert, basalt and minor felsic volcanic rock. Exposure is confined to isolated low, rounded hills, with less than 10% outcrop.

Mineralisation

Particulate gold was discovered at Tanami in 1909. The subsequent discovery of 9.6 kg of rock containing 5.1 kg Au caused a small rush. Economic mineralisation recorded from the Granites-Tanami Mining Region have, to date, been restricted to Au, although minor occurrences of Cu, U and REE have been reported (Blake, *et al.* 1979). Recovery of the commodity has been solely from the Mt. Charles Beds of interbedded sedimentary and basaltic packages. The Tanami Mine, Redback Rise, Dogbolter and Jim's Find mines and prospect are all located along the structural corridor between the previously mentioned granite domes.

Economic Au mineralisation occurs within sheeted quartz-carbonate veins and in the surrounding sericite + quartz + pyrite ± carbonate alteration halos about steeply dipping and brittle fractures and minimum displacement faults (Tunks 1996). Mobilisation of Au within the weathering environment forms supergene enrichment blankets and lenses. A non-economic type of mineralisation occurs as stratabound disseminated Au in banded iron formations and cherts.

Ore mined at present grades 3 ppm. Plant operation in the mines is based on a carbon-in leach process. Jim's Find is likely to be opened as a mine in 1997.

## GEOCHEMICAL METHODS USED

Panning as an exploration technique did not prosper in the Tanami region because of water shortage. Conventional modern exploration of surface sample uses fire assay, or more commonly aqua regia soluble Au subsequently analysed by graphite furnace AA after MIBK extraction. These determinations may be complemented by AA analysis of As. Material from surface trenching, diamond and the more common percussion drilling is increasingly assessed by NAA and ICP for Au and its pathfinders As, Sb and W.

Multi-element analysis of regolith material has been used for a considerable time to assist with the interpretation of vestiges of original lithology in the regolith (Hallberg 1984). More recently the CSIRO-Australian Mineral Industry Research Association Project "Exploration for concealed Gold deposits, Yilgarn Block, Western Australia" has been using whole rock samples to characterise enrichment and depletion pattern of over forty elements in the regolith (Butt 1991).

Because of the often dispersed nature of gold occuring within the regolith, several special projects were set up in order to obtain better understanding of the nature and genesis of the lateritic regolith, of the supergene Au deposits within it and to provide data applicable to exploration for Au in and beneath the regolith. The most recent of thes studies is on Jim's Find South, a prospect immediately south of Jim's Find (Stott 1994). This study was followed by a PIXE/PIGME study (Stott et al. 1996)

## COLLECTION OF SAMPLES

The objective of the sampling program at the Jim's South anomaly was to obtain information on element distributions throughout the lateritic regolith. This information was collected in order to illustrate both the geochemical evolution of the regolith profile from saprock to lateritic horizons, and to determine the behaviour of the mineralisation system, including that of Au, its associated pathfinder elements and associated wallrock alteration assemblages. Samples were taken from drill holes along three section lines across the anomaly: 10150N, 10250N, 10400N (Fig. 3). Line 10150N was chosen to allow investigation of near surface depletion of gold and the presence of any immobile pathfinder elements. Lines 10250N and 10400N were chosen as they offered well constrained mineralised zones.

A total of one hundred and seventy four samples were collected for analysis. Collar locations of the drill holes from which the samples were taken are given in Fig. 3.

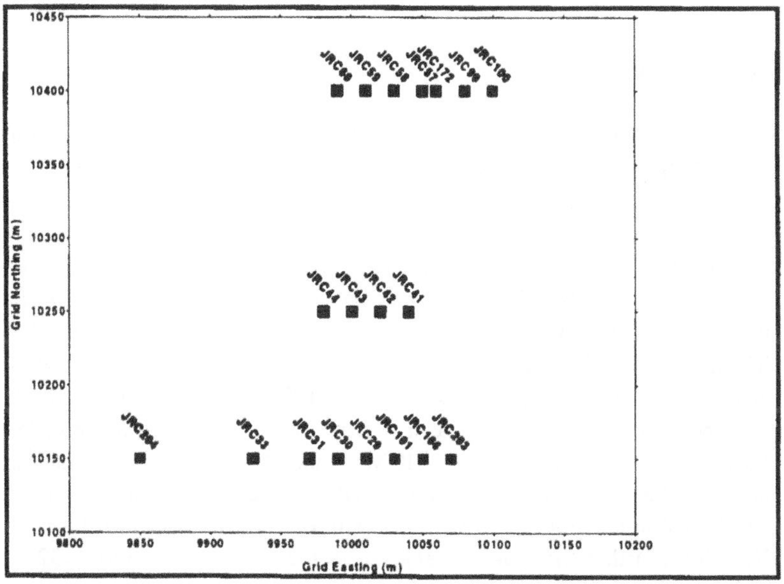


**Figure 3** The locations of drill holes used for Jim's south anomaly sampling program

## SAMPLE PREPARATION.

Reverse circulation and diamond core samples were jaw crushed to a maximum size of 4 mm. Crushed samples were then riffle split down to 1 kg, the excess being discarded. A

100-150 g fraction was subsequently split off the 1 kg sample for analysis, the excess being retained as a reference. Samples to be used for analysis were pulverised in a manganese steel mill to less than 75μm. This material was used for the NAA and XRF analysis of the initial study.

The PIXE/PIGME analyses were carried out on the same samples, but after they have been treated with hot aqua regia to destroy carbonates, oxides and sulphide minerals and subsequently with hot sulphuric acid to remove kaolinite. This acid treatment is stronger than any possible weathering process and removes all weathering products. The acid insoluble residue was subsequently cleaned with a hand held rare earth magnet. The acid insoluble residue of the Tanami samples consists of quartz, minor sericite and traces of feldspar, kaolinite and rutile (Fig. 4).

The acid insoluble residue of quartz veins, rocks and soils has been used as an exploration medium by the Geology Department for more than fifteen years, against the conventional wisdom of analysing the leach and throwing the leachate away. There is little difference between the chemical composition of the acid insoluble residue of weathered and unweathered rocks, although the residue of the weathered rocks is somewhat higher in rutile and zircon. ( van Moort et al. 1995). If this acid insoluble residue, which represents essentially silicates, is rich in eg. zinc or gold, mineralisation is nearby.

## INSTRUMENTAL ANALYSES

The neutron activation and XRF instrumentation has been widely used in the mineral exploration industry. Bulk rock analyses by proton probe, however, are rather rare, and consequently some details given below.

Bulk chemical analysis of most of the samples was carried out by simultaneous PIXE and PIGME spectroscopy using the ANSTO 3 MV Van de Graaff accelerator. For these analyses 400 mg powder samples were made into pressed pills with spectroscopically pure graphite. These were bombarded with a defocussed 2.5 MeV proton beam with typical beam currents of 200 nA; for each sample a dose of 100 μC was accumulated. A composite mylar-perspex pinhole filter was placed in front of the PIXE detector in order to reduce excessive X-rays of Si. The PIGME analysis provided elemental concentrations for Li, F, Na and Al relative to the USGS standard SGR 1 for Li and F and NBS standard 278 (obsidian) for Na and Al, all with a minimum detection limit of several ppm. PIXE allows further simultaneous analysis of the elements $Z=13$ to $Z=92$.

Simultaneous PIXE/PIGME analysis was chosen because of its speed and the wide range of elements which can be determined simultaneously, including Li, F and Na, its flexibility, precision and detection limit. Similar results may also be obtained by XRF spectrometry, which method however cannot provide information on the concentration of the elements Li and F.

## DETECTION LIMITS

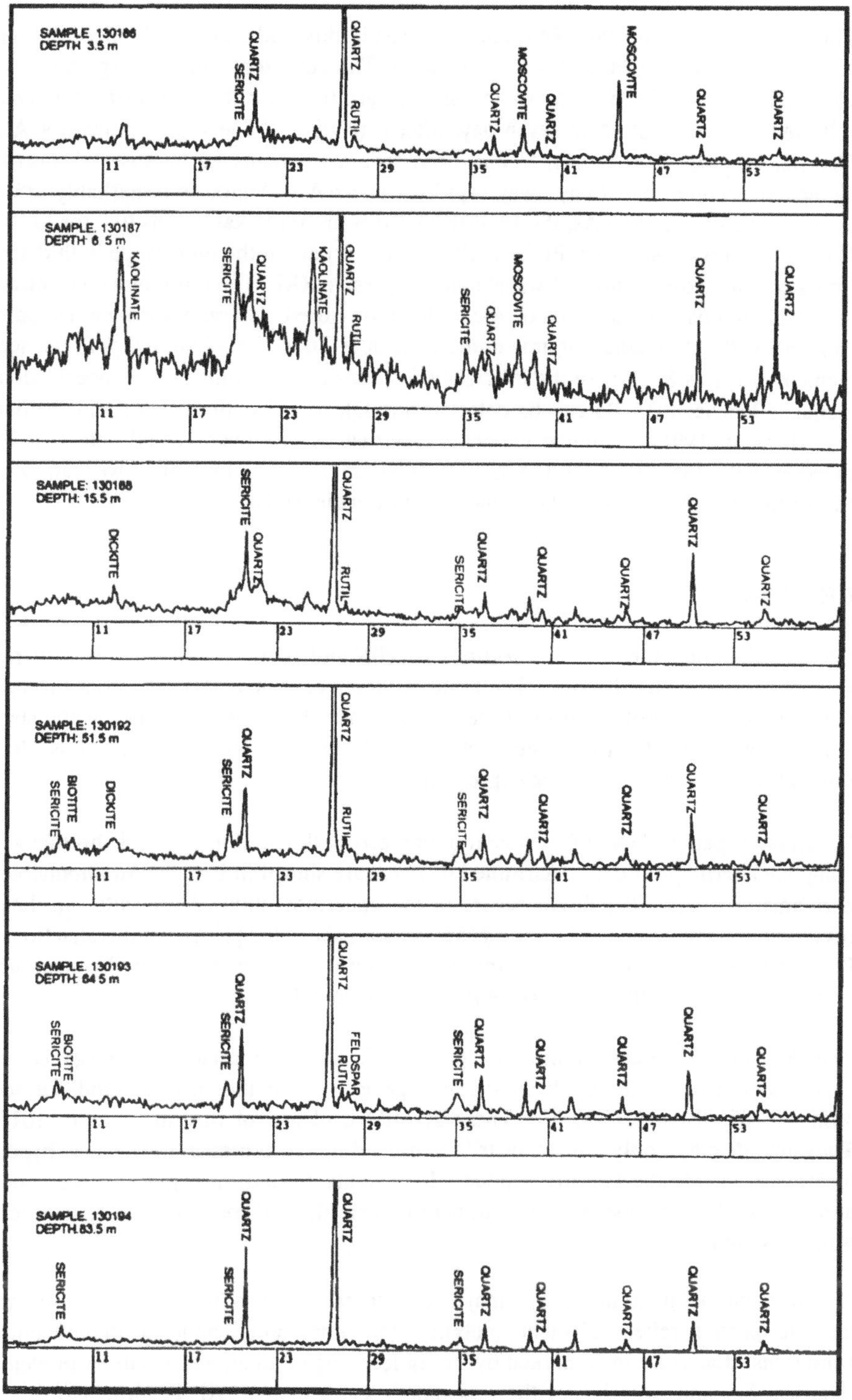


Figure. 4. X-raydiffraction data of drill core JRC 030 from section 10150N

The detection limits for the INAA analyses used in this study are mainly those of thermal NAA ie using thermal neutrons for activation. The detection limits for Ag, Au, Ba, Co, Cr, Fe, K, La, Na, Sb, Sc, Th, W and also Zn are those for thermal neutron activation, followed by a count at about seven days after activation. The values shown for Al, Cl, Cu, Ga, Ge, Mg, Mn, Sr, Ti, V and need separatye counts eg because of short-lived elemental isotopes that decay away rapidly. The NAA method is remarkably sensitive for elements like Au and As, the PIGME analyses are remarkably sensitive for Li. F and Na. The (energy dispersive) PIXE analyses provide data in the ppm range with detection limits coming close to those of wavelength dispersive XRF. The minimum concentrations detectable in this work depends on the pinhole filter used, and ranges from a few ppm for elements with intermediate atomic numbers such as Zn to a few hundred ppm for heavier elements such as Sn, where the efficiency for Kα-rays is diminished. Concentrations of elements heavier than Ba can be detected through their Lα lines (Bird and Williams , Russell et al. 1996). Common elements analysed by this method and detection limits together with the INAA and XRF are show in table 1, in which we can see the PIXE/PIGME is very good at both major and trace elements..

## RESULTS

Gold distribution patterns across sections 10250N and 10400N define highly constrained subvertical mineralised zones. The trend of these zones is visibly persistent throughout the entire regolith profile with well defined Au, As, Sb and W anomalies at the surface. On the other hand the large section 10150N exhibits a broader less defined mineralisation with little surface expression.

Distribution patterns for gold concentrations across the section 10150N throughout the complete laterite profile, as determined by NAA are shown in Fig. 5. An anomalous Au enrichment of only 70 ppb Au occurs in the uppermost mottled saprolite, the lateritic residuum and the overlying lateritic generally soil, above the primary mineralisation. The PIXE Au distribution is similar to the NAA pattern , but does not show the near surface enrichment, because the near surface Au is acid insoluble.

The As and Sb values in the *raw* pulps indicate the formation of large cones of enrichment which flare up from the hypogene zone to the modern land surface in sections 10250N and 10400N. There is little evidence of this in section 10150N. Tungsten displays only minor distribution tending to preserve its former hypogene location even into the lateritic residue. K contents around the hypogene ore zone are anomalously high compared with unmineralised basalts and correspond to sericite related mineralisation

The distribution patterns of the major and trace elements throughout the complete lateritic profile reflect element mobility and immobility under different climatic conditions. Laterisation is marked by strong leaching of alkali and alkali earth elements (Na, Ca, Sr), the retention of the less mobile elements (Si, Al, Fe, Mn, Co) and the residual near surface concentration of the immobile elements (Zr, Hf, Ti). Modification of the profile under arid conditions is marked by the accumulation of alkali and alkaline earth elements in the upper parts of the profile (Na, Sr) and by the redistribution of the semi-mobile elements (Si, Al and REE).

**Table 1.** Typical detection limits for INAA, PIXE/PIGME and XRF for geological materials (ppm)

| Element | INAA* | PIXE/PIGME** | XRF*** |
|---|---|---|---|
| Ag | 5 | 80 | 2 |
| Au | 5ppb | 7 | 3 |
| Al | | 400 | 100 |
| As | 1 | 3 | 23 |
| Ba | 100 | 50 | 30 |
| Ca | 10,000 | 45 | 100 |
| Cl | | 75 | 20 |
| Co | 1 | 3 | 20 |
| Cr | 5 | 13 | 10 |
| Cu | | 2 | 4 |
| Fe | 200 | 5 | 40 |
| Ga | | 2 | 3 |
| Ge | | 2 | 3 |
| K | 2000 | 35 | 10 |
| La | 0.5 | 600 | 2 |
| Mg | | 1500 | 100 |
| Mn | | 7 | 50 |
| Na | 100 | 40 | 250 |
| Nb | | 5 | 4 |
| Ni | 50 | 4 | 25 |
| P | | 40 | 2 |
| Pb | | 4 | 5 |
| Rb | 20 | 3 | 5 |
| S | | 140 | 50 |
| Sb | 0.2 | 25 | 2 |
| Sc | 0.1 | 40 | 2 |
| Si | | 1100 | 20 |
| Sn | 100 | 100 | 1 |
| Sr | 500 | 3 | 5 |
| Th | 0.5 | 10 | 2 |
| Ti | | 17 | 60 |
| V | 2 | 27 | 5 |
| Y | 1 | 6 | 5 |
| W | 2 | 25 | 2 |
| Zn | 100 | 2 | 5 |
| Zr | 500 | 5 | 5 |

* Baquerel Laboratories, Lucas Heights, NSW, Australia
** Physics Division, Lucas Heights, NSW, Australia
*** CSIRO, Wembly, W.A., Australia

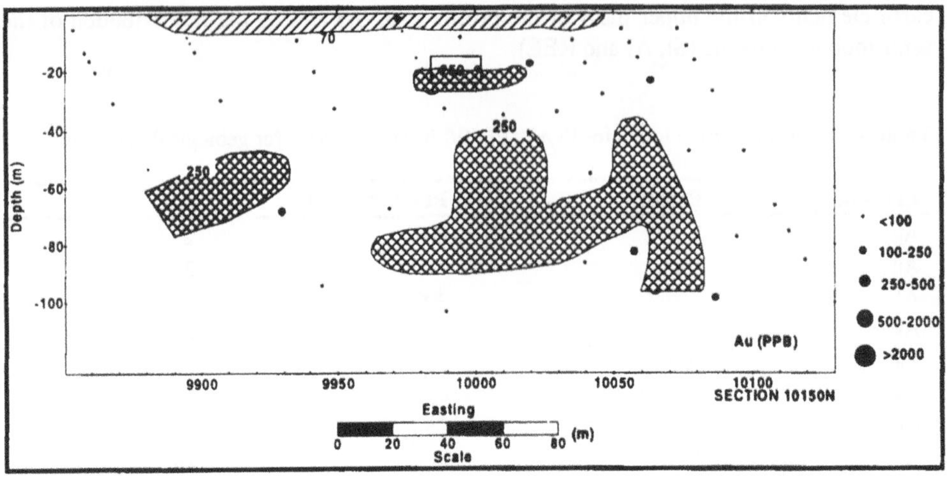


**Figure 5** The NAA distribution map of Au along section 10150N. The surface expression is weak at only 70 ppb.

The *raw* pulp data do not permit lithogeochemical differentiation of the weathered bedrock using standard immobile element ratios, including ratios of Ti to Zr. Bedrock geology remains most accurately defined by diamond drill core control.

The chemical data of the *acid insoluble* residue give three factor analysis clusters. The first cluster comprises the elements F, Al, K, V, Mn, Fe, Ga, Rb, W and Au and is essentially related to sericite wallrock alteration. The second cluster consists of Ti, As, Y, Zr and Nb and is largely related to resistate minerals. The third cluster consists of Na, Ca and Sr and is interpreted to comprise elements in weatherable minerals such as feldspar and thus represents weathering intensity. It appears also to represent wallrock alteration.

The results do not confirm the earlier conclusion that the Au is remobilised into lenses in the deeper part regolith. The Au trapped in the silicates of the acid insoluble residue displays the same distribution pattern as the Au in the rock prior to treatment. In retrospect we feel that the Au distribution in the deeper part of the regolith reflects the original distribution pattern.

Arsenic values in the acid treated material flare upward to the surface from the underlying areas of high Au values. See Fig. 6.

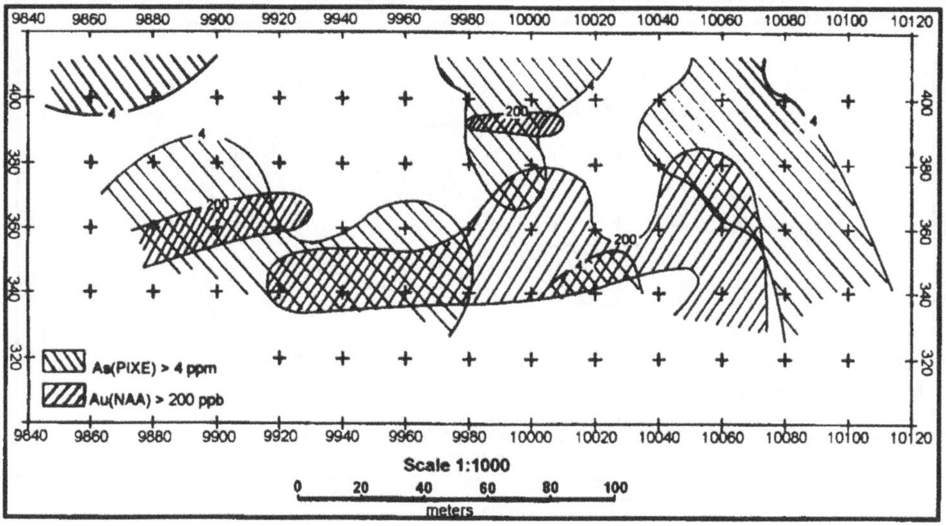


**Figure 6** The distribution of As in the acid insoluble residue along 10150N

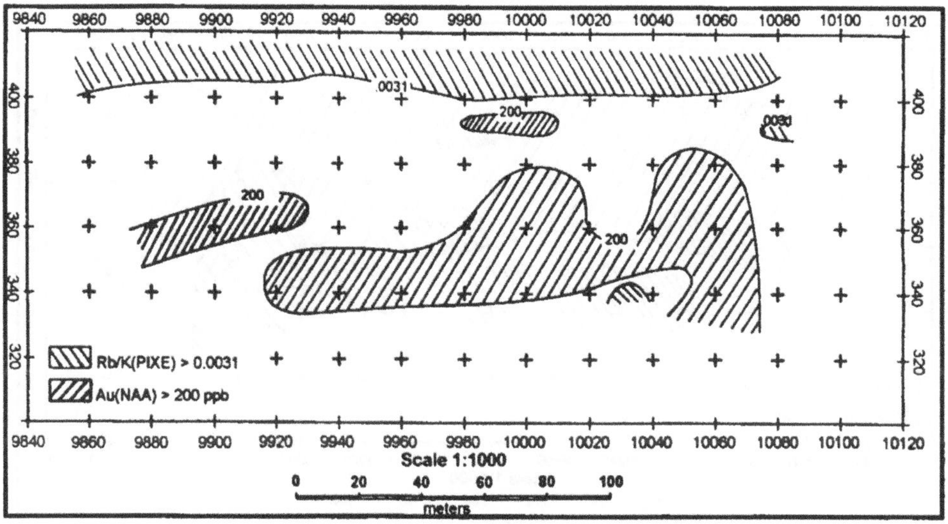


**Figure 7** The distribution of Rb/K in the acid insoluble residue along section 10150N

The Rb/K ratio is high at the upper level of the regolith, over the Au occurrences (see Fig. 7). There is in the acid insoluble residue an indication that Ni, Cu, Zn, As and Pb are enriched in the upper level of the regolith, and more precisely above the Au occurrences. This trend is very much enhanced by use of the multiplicative indices of As×Cu×Ni and Cu×Pb×Zn, as shown in the Figs. 8 and Fig.9.

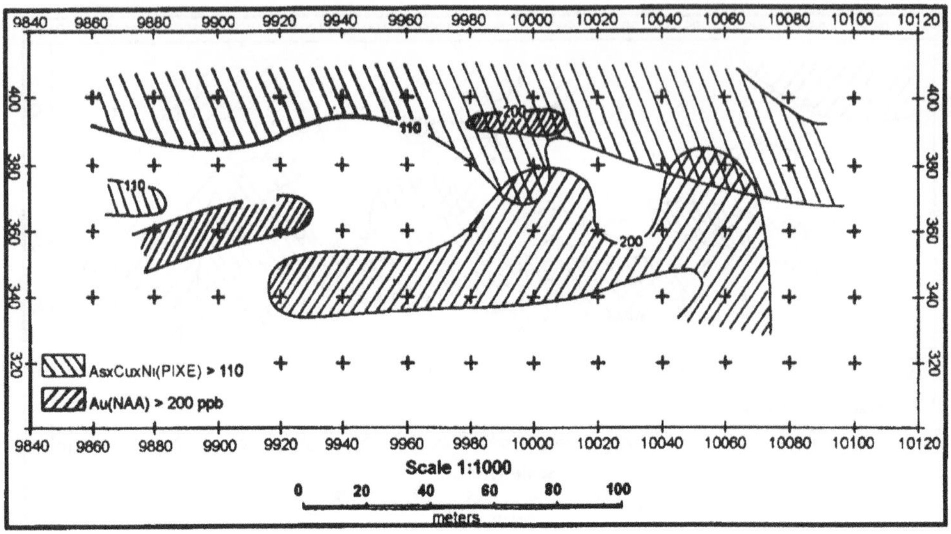


**Figure 8** The distribution of As×Cu×Ni in the acid insoluble residue along the section 10150N

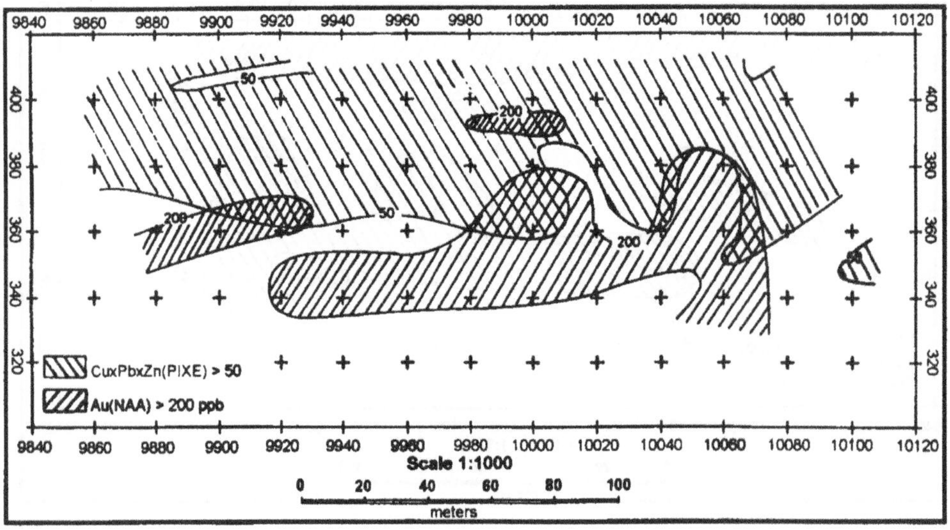


**Figure 9** The distribution of Cu×Pb×Zn in the acid insoluble residue along the section 10150N

The areas of high Au values are surrounded by a halo of high K and Rb values. Of the various possible multiplicative and divisional ratios the Rb/Al (Fig. 10) ratio appears to reflect best the presumed areas of primary Au mineralisation.

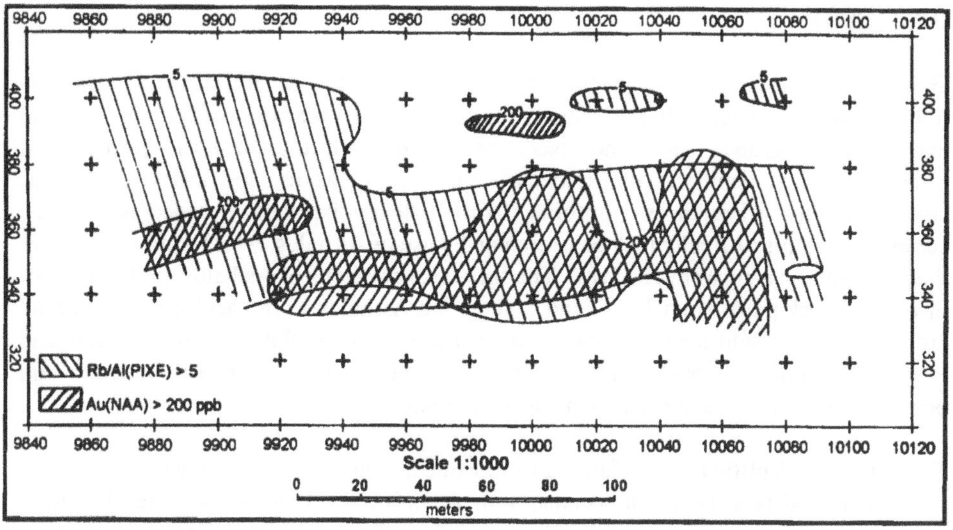


**Figure 10** The distribution of Rb/Al in the acid insoluble residue along the section 10150N

Na, Ca and Sr are depleted near the surface and also around the areas of high Au.

The Ti/Zr (Fig. 11) ratios of the acid insoluble residue of the regolith allows distinction of the original basalt flows in the sandstones. This trend cannot be discerned in the untreated samples.The average values of 10,000 ppm Ti and 100 ppm Zr appear to indicate Fe-tholeites (Hall and Plast 1992).

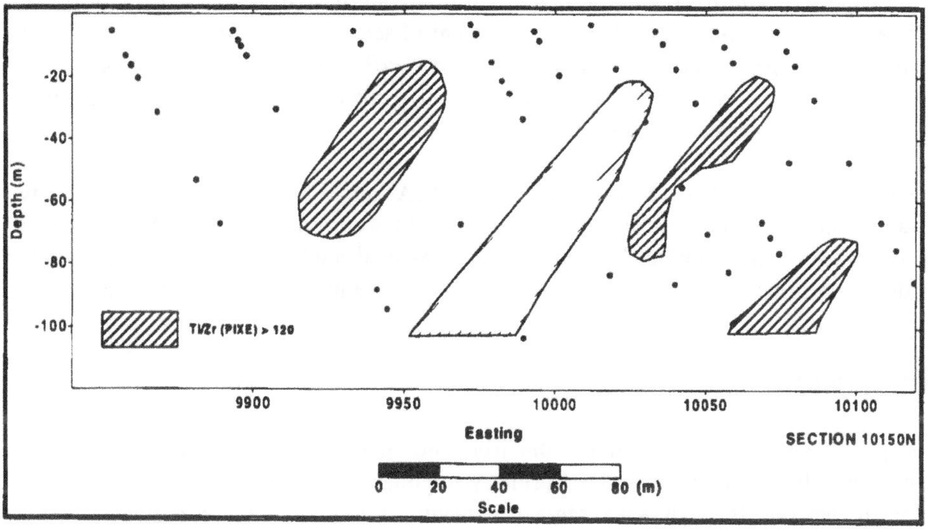


**Figure 11** The distribution of Ti/Zr in the acid insoluble residue along section 10150N

## DISCUSSION

Gold enrichment in the upper levels of the regolith, as occurring at Tanami, is common (e.g. Porto and Hale, 1995). Such enrichments are easily located in regional exploration programs. It is however usually difficult to locate the primary Au mineralisation at depth, as the secondary enrichment has in most cases no obvious continuation toward depth (as in profile 10150N), being displaced or eroded away. To a certain extent this difficulty may be overcome by use of As halos, which tend to be larger than the Au occurrences themselves. The surface expressions of As halos tend to be poor in the raw rock pulps from Jim's Find South. The As halos in the acid treated material are better defined and break through to the surface. The absolute As values in the acid insoluble residue are smaller than the raw material as both secondary arsenates and primary minerals like arsenopyrite have been removed.

Tunks (1996) identified in the Tanami Gold Mine a total of five vein stages. The fourth stage veins and breccias are associated with Au mineralisation and sericite alteration. Consequently recognition of wall rock alteration patterns in the Tanami region is important. The bleached sericite alteration halos that surround auriferous quartz-carbonate veins, are typical of many shear zone hosted Au deposits in sub-greenschist facies host-rocks (Mikucki and Groves 1990), as well as Carlin style Au mineralisation (van Moort et al. 1995). This expresses itself in increased K content around the orebody, better observed in the PIXE and XRF data than in the NAA data (as neutron activation is a poor method to analyse K).

The Rb/K ratio of the sericite at Jim's Find South is increased, as is consistent with the trend at Carlin (van Moort and Xu 1995), porphyry copper deposits in Chile (Armbrust et al. 1977) and Sn-W deposits and greisen (Govett 1983), and possibly many other types of ore deposits. The Rb/K ratio is highest in the regolith above the ore body, near the surface. This expresses probably an enrichment of sericite in the near surface regolith relative to K-feldspar. The use of As×Cu×Ni and Cu×Pb×Zn as surface indicators of the gold mineralisation is not well understood, as all secondary minerals containing these elements appear to have been washed out by the acid.

Another expression of the sericitisation is the K/Al and Rb/Al ratios. The ore body shows distinct halos of these ratios around the gold occurrences. The K/Al and Rb/Al are of no use in near surface location because residual kaolinite in the acid insoluble residue causes increased aluminium values and consequently decreased Rb/Al ratios.

## CONCLUSION

It is possible to remove effectively the iron and manganese oxides and hydroxides, carbonate, sulphates and remnants of primary sulphides from regolith material (and rock pulps in general) through aqua regia treatment. Kaolinite may be removed by a subsequent sulphuric acid treatment. The acid insoluble residue consists mainly of quartz with minor sericite and some heavy minerals like rutile and zircon.

The chemical composition of the residue can de used to describe original wall rock alteration patterns around the Au mineralisation through a set of parameters: K and Rb enrichment, increased Rb/K and Rb/Al ratios and a general decrease in Na, Ca and Sr depletion through feldspar destruction. Arsenic in the acid insoluble residue appears to be of specific use to define surface expression of hidden Au mineralisation. Ti/Zr ratios can be used to characterise original rock composition.

**Acknowledgements**

The PIXE/PIGME analysis of this study were financed by the Australian Institute of Nuclear Science and Engineering (AINSE) Grant No 96/076.

**References**

Armbrust, G.A., Oyarzun, J., and Aria, J., 1977, Rubidium as a guide to ore in Chilean porphyry copper deposits: *Economic Geology*, v. 3, p. 347-357.

Bird, J.R., and Williams, J.S., 1989, *Ion Beams for Materials Analysis*: Sydney, Academic Press, 719 p.

Blake, D.H., Hodgson, I.M., and Muhling, P.C., 1979, Geology of the Granites-Tanami region, Northern Territory and Western Australia, Bureau of Mineral Resources Geology and Geophysics, p. 197.

Butt, C.R.M., 1991, Dispersion of gold and associated elements in the lateritic regolith mystery zone, Mt. Percy, Kalgoorlie, Western Australia, Wembley, WA, Division of Exploration Geoscience, CSIRO.

Govett, G.J.S., 1983, *Rock geochemistry in mineral exploration*: Handbook of exploration geochemistry: Amsterdam, Elsevier.

Hall, G.E.M., and Plant, J.A., 1992, Application of geochemical discrimination diagrams for thetectonic interpretation of igneous rocks hosting gold mineralisation in the Canadian Shield: *Chemical Geology*, v. 95, p. 157-165.

Hallberg, J.A., 1984, A geochemical aid to rock type identification in deeply weathered terrain: *Journal of Geochemical Exploration*, v.20, p. 1-8.

Mikucki, E.J., and Groves, D.I.G., 1990, Mineralogical constraints, *in* Ho, S. E., Groves, D. I., and Bennett, J. M.,( eds.), *Gold deposits of the Archaean Yilgarn Block: Western Australia: Natue, Genesis and Exploration Guides*, Perth, Geology Department and University extension, The University of Western Australia, p. 212-220.

Porto, C.G., 1995, Gold redistribution in the stone line lateritic profile of the Posse Deposit, Central Brazil: Economic Geology, v. 90, p.308-321

Russell, D.W., Bailey, G.M., van Moort, J.C., and Cohen, D.D., 1996, Chemical composition and paramagnetism of vein quartz from the Tasmania gold mine, Beaconsfield, Northern Tasmania: *Nucl. Instr. and Meth. in Phys. Res.*, p. 598-600.

Stott, C., 1994, *A regolith study of the Jim's South gold anomaly, Tanami Desert, Northern Territory*: Unpub. Honours thesis, University of Tasmania 137 p.

Stott, C., Xu, L., Butt, C.R.M., Bailey, G.M., and van Moort, J.C., 1996, Gold and associated elements in the lateritic regolith at Jim's Find Tanami Desert, Northern Territory, Australia: *30th International Geological Congress*, p. 64.

Tunks, A.J., 1996, *Geology of the Tanami Gold Mine, Northern Territory*: Unpub. Ph.D. thesis, University of Tasmania 214 p.

van Moort, J.C., Hotchkis, M.A.C., and Pwa, A., 1995, EPR spectra and lithogeochemistry of jasperoids at Carlin, Nevada: distinction between auriferous and barren rocks: *Journal of geochemical Exploration*, v. 55, p. 283-299.

van Moort, J.C., and Xu, L., 1995, Preliminary results of the research on the development of paramagnetic and geochemical exploration technique for hidden gold deposits in highly weathered terrain, Hobart, Geology Department, University of Tasmania.

*Proc. 30th Int'l. Geol. Congr.*, Vol. 19, pp. 257-273
Xie Xuejin (Ed)


# The Study of Surfacial Geochemical Dispersion Around a Polymetallic Mineralization and Geochemical Exploration in Permafrost and Forest-swamp Terrain, Da-Hinggan Mts., Northeastern China

WANG MINGQI[a], LIU YINHAN[a], REN TIANXIANG[a] and LIU SHUXING[b]
[a]*Institute of Geophysical and Geochemical Exploration, Langfang, Hebei, 065000, P.R. China*
[b]*Geophysical Exploration Team, Shuangcheng, Heilongjiang 150105, P.R. China*

Abstract

Forest-swamp terrain covers over 400,000 $km^2$ of northeast China. Soil is poorly developed in the terrain. In contrast with A horizon soil , most elements including Cu, Pb, Zn, Ag are strongly enriched in B+C horizon. To compared with bedrock, however, these elements are enriched in both soil horizons. In the soils above the mineralization, Cu and Zn are mostly in the sulphides and crystalline iron oxyhydroxides, Pb is mainly in the sulphides and Ag is mainly in the amorphous Fe oxyhydroxides.

The drainage is not developed in the upper reaches of the valleys. The clastic stream sediments can be collected only in the low reaches of the valleys in the bottom of or sometimes near-waterline bank of streams. In the -5 mm fraction clastic stream sediments, coarse grains (+0.2mm) represent a large portion of clastic stream sediments (over 90%). Metal elements tend to increase in the fine fractions. The sequential extraction results show that the partitioning of Cu, Pb, Zn, Ag in clastic sediments corresponds closely to that of soils. Organic stream sediment (peat) can be collected at either side of streams, in which the concentration of organic carbonate is between 3~16%. In peat, both the total of many elements and the organic bound partition of Cu, Pb, Zn, Ag are markedly increased in contrast with clastic stream sediments. The study of partitioning of metals in stream sediments indicates that clastic sediment is a good medium for pinpointing the mineralization in forest swamp area..

Water plays an important role in surfacial geochemical process in the terrain and pH of water and organic matter of soil are the key control factors in migrating elements. To consider the dispersion pattern of elements in the landscape, we suggest that,clastic stream sediment (-0.9mm) at the density of one sample/4$km^2$ is the best alternative approach for the Regional Geochemistry National Reconnaissance (RGNR) project in the forest-swamp terrain.

Keywords: geochemical dispersion, geochemical exploration, forest-swamp terrain, permafrost

## INTRODUCTION

Forest and swamp terrains are developed mostly in high latitude areas like some part of Northern Europe, Northern America and Northeastern Asia. Organic stream sediments (peat) were highly recommended as the media for the regional geochemical surveys in the area in North Europe and former Soviet Union in the past [1, 2, 3]. In China, Forest-swamp terrain covers over 400,000 $km^2$ of northeast China with northeastern Inner Mongolia and northern Heilongjiang representing very typical examples of this landscape (Fig. 1) . Due to the lack of detailed geochemical studies of the secondary environment and methodologies for geochemical exploration, Regional Geochemistry National Reconnaissance project (RGNR) in the past was carried out in the areas by exclusively applying the approaches in China proper terrains (sample density: one sample/km$^2$. 39 elements were analysed on composite samples which is the combination four adjacent samples resulting in a density of 1 composite sample/4 km$^2$). After more than a decade implementation, the project was close to be completed. However, there had not been any significant successes in the application of geochemistry to mineral exploration in this terrain and some perplexing phenomena, for example, some strong regional anomalies disappear after follow-up and the concentrations of metal elements decrease significantly from stream sediments to soil to bedrock, were observed during the follow-up stage. In order to establish a applicable approach for regional geochemical exploration in the forest-swamp terrain, an orientation study was carried out in July of 1991 to investigate secondary chemical dispersion in that region. This study involved collection of clastic stream sediments, peat, soil, rock, heavy minerals and stream water.

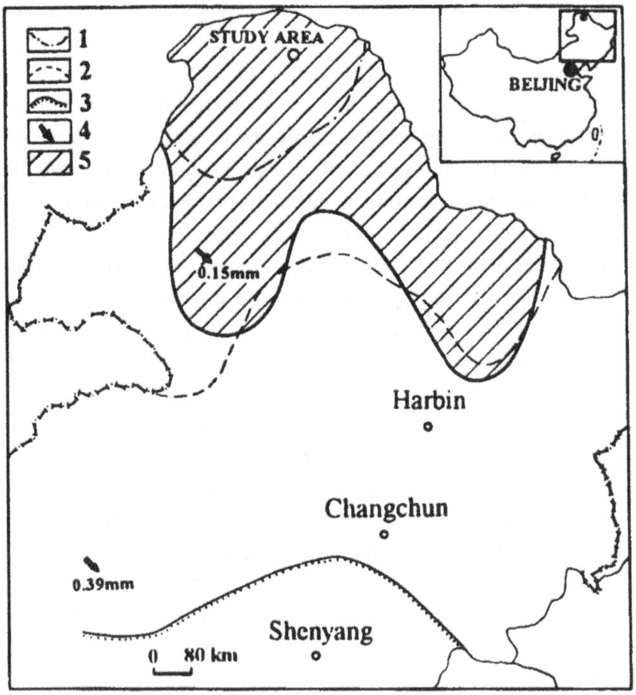


Fig. 1. Distribution of forest-swamp terrain and periglacial geomorphology in Northeastern China (the data of periglacial geomorphology from Wang [4] and Xie [5]. (1) Permafrost region. (2) Permafrost island. (3) Boundary of permafrost in Pleistocene epoch. (4) Wind direction and the median of loess grain size in Pleistocene epoch. (5) Forest swamp area.

**PREVIOUS WORK**

Low density of regional geochemical surveys covering whole Helongjiang Province were conducted at a density of one sample per 55 $km^2$ prior to the RGNR project [6]. Marked multielement anomalies including Cu and Zn were delineated in the area (Fig. 2) and Bailushan polymetallic mineralizations were soon discovered by following up the geochemical anomalies.

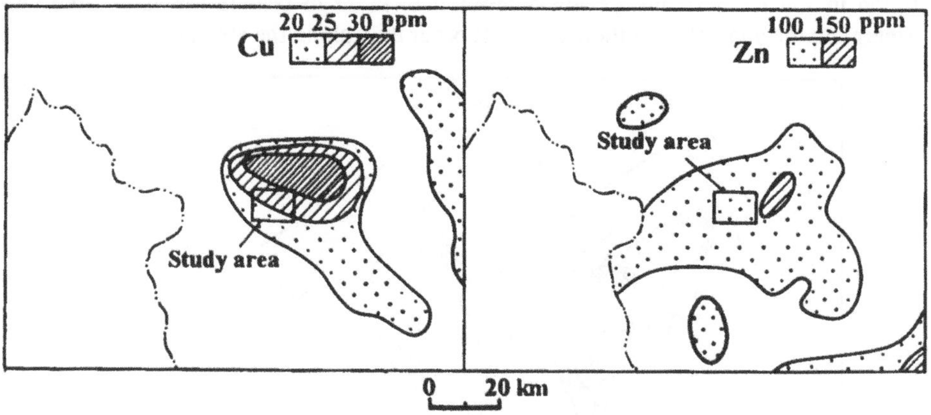


Fig.2. Anomalies of Cu, Zn from stream sediment survey at a density of one sample /55$km^2$. Simplified from the unpublished maps by Heilongjiang Geophysical Exploration Team [6].

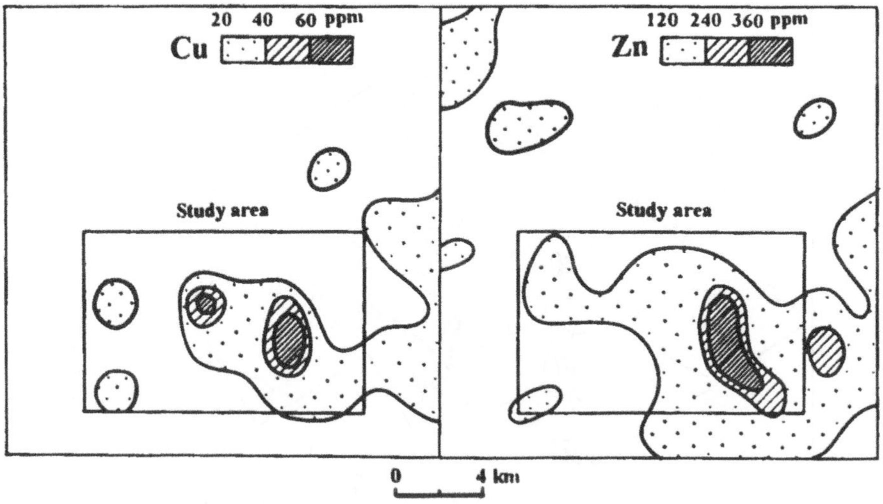


Fig. 3. Regional organic stream sediment anomalies around study area from RGNR project. Simplified from maps by Heilongjiang Geophysical Exploration Team [7].

A regional organic stream sediment including valley peat survey was conducted at the density of 1 sample/km$^2$ in the study area in 1988 as part of the RGNR project [7]. Anomalous concentrations of Cu, Zn and other elements covering a surface area of more than 200 km$^2$ were outlined (Fig. 3). To follow-up the anomalies, a detailed soil (mostly A horizon soil or peat in seepage zones and swampy valleys) survey on a 50×200 m grid [8]. Difficulties were encountered not only with sample collection owing to poorly developed soil and hard access, but also in interpreting the results. The concentrations of Cu, Zn in soil (A horizon) are much lower than those in peat in the same area and furthermore,the anomalies of these elements occur only along the stream

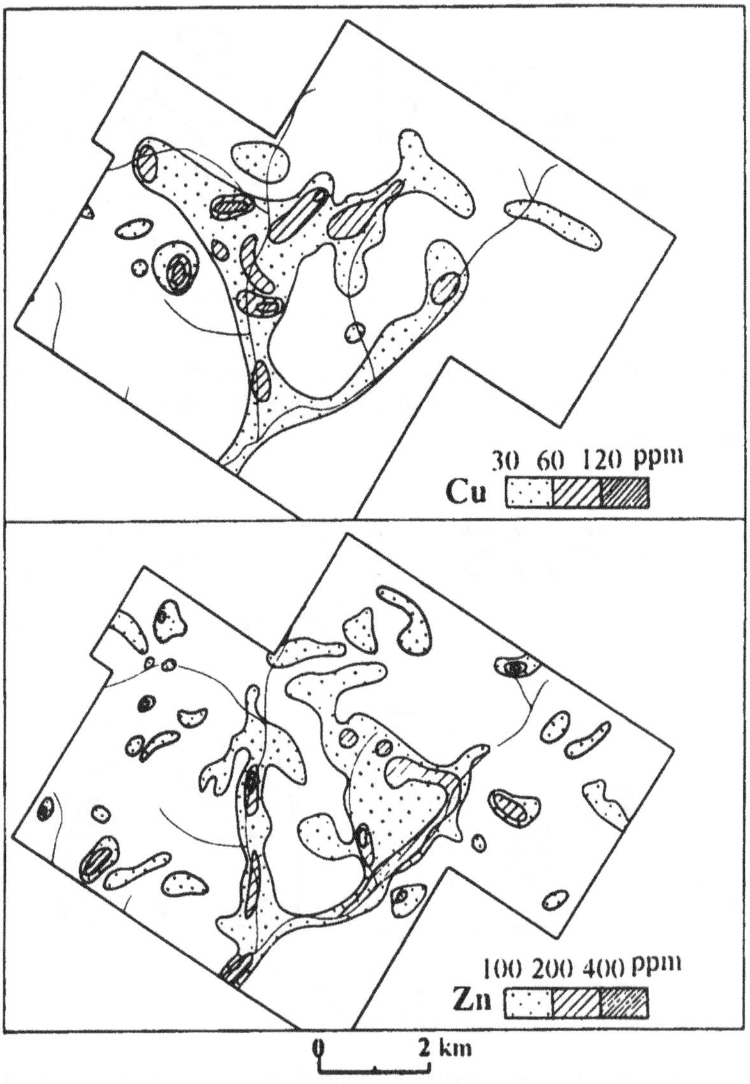


Fig. 4. Cu and Zn anomalies by detailed top soil survey on a 50 × 200 grid. Simplified from the maps by Heilongjiang Geophysical Exploration Team [8].

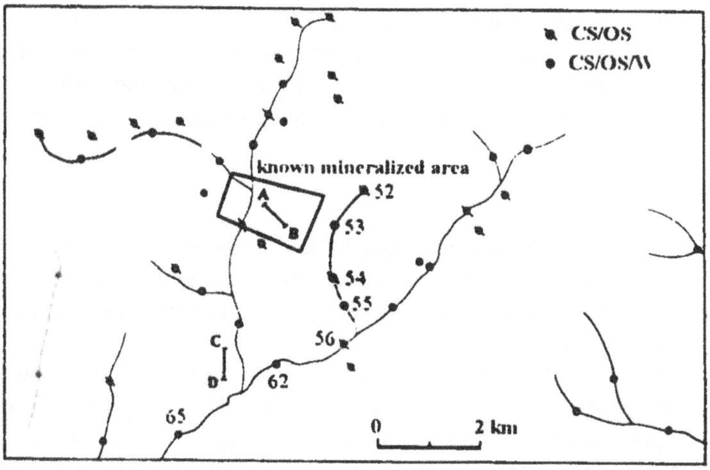


Fig. 5. Location of sample sites in study area. A-B, C-D = location of soil profiles. CS = clastic stream sediment. OS = organic stream sediment. W = water.

in 0.25 M HCl); (5) Crystalline Fe oxide (1.0 M $NH_2OH \cdot HCl$ in 25% $CH_3COOH$); (6) sulphides ($KClO_3/HCl/HNO_3$); (7) residual ($HF/H_2O_2/HNO_3/HCl$). Cu, Pb, Zn, and Ag were determined in each extract by AAS.

## GEOLOGY

The study area is situated in the northern part of the Da Hinggan uplifted region. Da Hinggan mountains rose continuously and has been subject to erosion since late Cretaceous period. Basaltic volcanic eruptions took place during late Tertiary period or early Pleistocene epoch and the basalt covered a lot of area [4]. Forest-swamp landscape has been developed for over 10 000 years (Table 1).

TABLE 1
Evolution of vegetation and climate in Da Hinggan mountains in Holocene Epoch*

| Time | Vegetation | Climate |
|---|---|---|
| post-glacial Period (about 12000) | mostly birch, small amount of conifer | turned warm and humid |
| 8000-6000 | broadleaf trees | the warmest climate period |
| late Holocene to date | coniferous trees increased, broadleaf trees decreased | turn cold and did not change much till present |

* Simplified from Guo [12].

Bailushan mineralized area is underlain largely by Cambrian sedimentary and volcanic rocks, consisting mostly of sandstone, schists, rhyolite and tuff, and by intrusive rocks including granite and diorite [13]. Other volcanic and subvolcanic rocks, including valley (Fig. 4). This suggests that those elements be leached out of the soil profile and precipitate in the organic material. A new project was funded by bureau of Geological survey, MGMR to continue our study on geochemical dispersion and establish an effective approach for follow-up of regional geochemical anomalies in the terrain.

## LANDSCAPE

Forest-swamp areas in China were mostly located in the Periglacial perimorphology regions with high latitude, high organic matter content of water, cold climate, low pH in media and heavy vegetation cover [4], and permafrost and permafrost island are well developed in the terrain (Fig. 1).

The soil profile is very thin, and blockmeer and stone slope were well developed on hillsides and hilltops in the area. The terrestrial moss blanked covers almost all the surface. In the upper reaches of the valleys, the drainage is not developed and it is impossible to collect transported clastic sediments. Swamp, covered with scrubs and carex, forms zones tens to hundreds of meters wide; In the low reaches of the valleys, the drainage is better developed and clastic stream sediments can be collected only in the bottom or sometimes near-waterline bank of streams. swamp, covered with carex and moss, forms zones over hundreds of meters.

## METHODS

Soils and bedrock from two traverse lines over mineralization and background areas were collected through trenching at 50 m sample interval. Clastic stream sediments were sampled using long handle spades and 5 mm sieve at the stream bed at a density of one sample/3-4 $km^2$. Stream peat (organic stream sediment) sampling was undertaken at either side of stream waterline at about one sample/2-3 $km^2$ density. Sampling locations are shown in Fig. 5. Water sampling, preprocessing, analysing methods and pH measurements were described in great detail by Wang [4].

All solid samples were air-dried. Soil and clastic stream sediment were sieved (40 mesh) and ground into -200 mesh powder. In addition, 6 soil samples (B+C horizon) weighting among 2 and 3 kg were separated into four fractions from -0.076 mm to 5 mm and 10 clastic sediment samples (4-5 kg) were sieved into eight fractions from -0.076 mm to 5 mm to study the distribution of grain size weight and elements in soil and clastic stream sediment. 18 elements including Cu, Pb, Zn, Ag were determined by AAS. Sieved peat samples (60 mesh) were separated into two equal parts to understand the impacts of organic matter on analytical results: One was directly ground and the other was ignited at 450℃ for over 5 hours and then ground . The peat samples were analysed by XRF for 12 elements including Cu, Pb, Zn, Fe, Mn.

Typical samples of soil, rock and stream sediments were subjected to a 7-step sequential extraction scheme based on procedures developed by Filipek and Chao [9, 10, 11]. The scheme was designed to separate individual hydromorphic and biogenic fractions from resistant crystalline minerals in the following order: (1) absorbed, exchangeable and carbonate (HOAc);(2) Mn oxides (0.1 M $NH_2OH \cdot HCl$); (3) organic compounds (0.1 M $Na_4P_2O_7$); (4) amorphous Fe oxyhydroxide (0.25 M $NH_2OH \cdot HCl$

basalt, andesite, granite-porphyry, and dacite are also present (Fig. 6). The Cu-Pb-Zn-Ag-Mo mineralization, hosted in rhyolite and tuff, is located at the center of the study area. Due to dense vegetation cover and difficult access condition, detailed geological mapping is very difficult to undertake, no detailed geological data is available at present.

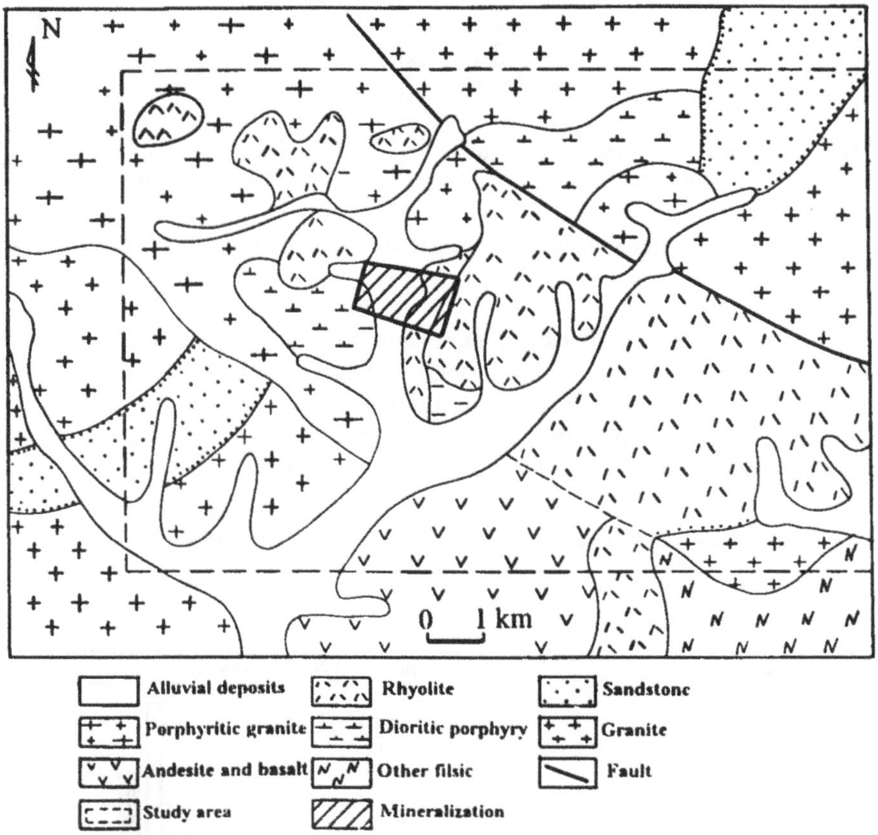


Fig. 6. Geologic map of study area. Simplified from Second Geological Survey in Heilongjiang Province [13].

## GEOCHEMICAL DISPERSION OF METALS IN THE AREA

1 Distribution of elements in soil and bedrock

Soil profile is poorly developed and the weathering layer comprising mainly rock chips (>3cm) is only 20 -50 cm on the hillsides and ridges in the area. A horizon soil is composed mostly of low maturity humus matter and small amount of leached clay. The residual layer is located under freeze-thaw debris and it is impossible to draw a line between B and C horizon. In the -5 mm size fraction, only less than 20% was in the fractions of -0.18 mm. The Pb concentrations in different size fractions are very close and Zn, Cu, Ag tend to be enriched in fine fractions, especially in the -0.9+0.18 mm fraction (Fig. 7).

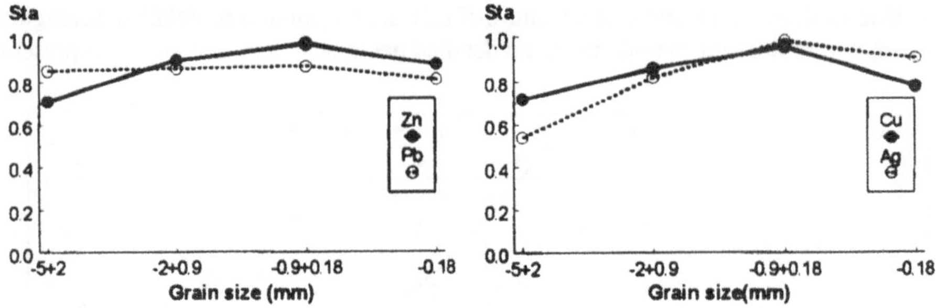


Fig. 7. Distribution of elements in different size fractions in residual soil in study area (n=6).

In contrast with weathered bedrock, most elements including Cu, Pb, Zn, Ag both in mineralized and background area are strongly enriched in soils, particularly in residual layer (Fig. 8). At the same time, the concentrations of the elements except Zn in B+C horizon are much higher than in A horizon in mineralized area, whereas the elements except Pb in background area remain almost the same in both layers .

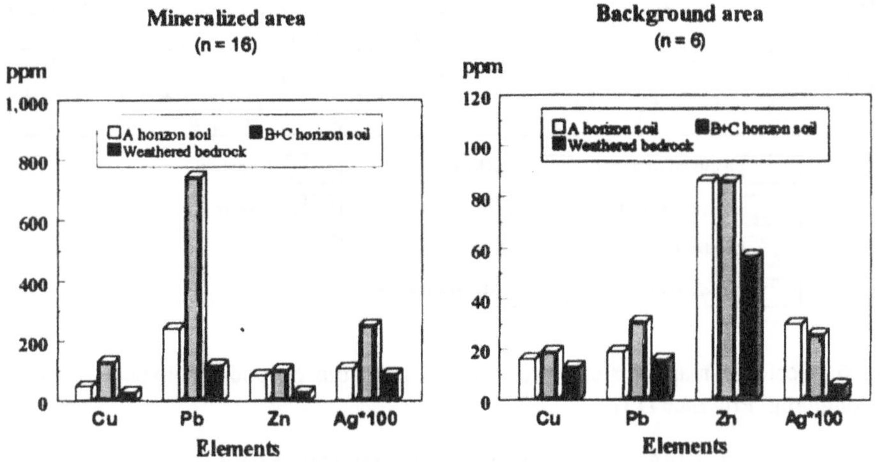


Fig. 8. Concentration of trace elements (geometric mean) in soils and rock in Bailushan area.

The results of sequential extraction show that in the soils above the mineralized zone, Cu are mostly in the forms of crystalline Fe oxides, sulphides; Zn in the crystalline Fe oxides, sulphides, and residues; Pb in the forms of sulphides, and Ag in the forms of amorphous Fe oxides (am Fe ox) and organic bound (Table 2); Although high organic matter is highly enriched in A horizon soil (organic carbonate 2-8%), lower amounts of Cu, Pb, Zn except Ag than expected appear to be complexed by the organic matter because of low maturity of organic material and strong acid condition . An association exists between the percentage of metals extracted in each step in soils and that in the weathered bedrock.

Table 2
Percentages of Cu, Zn, Ag, Pb extracted sequentially in the seven steps in soils and weathered bedrock in mineralized zones.

| Medium | Element | AEC | Mn ox | organic | am Fe | cry Fe | sulphides | residues |
|---|---|---|---|---|---|---|---|---|
| soil A | Cu | 1-2 | 0-1 | 4-9 | 2-8 | 33-39 | 39-50 | 4-5 |
| soil B+C | Cu | 1-2 | 0-1 | 5-8 | 3-5 | 31-44 | 44-48 | 2-6 |
| w-bedrock | Cu | 2-3 | 0-4 | 3-7 | 10-13 | 27-42 | 35-42 | 4-5 |
| soil A | Zn | 2-8 | 1-2 | 4-8 | 3-6 | 27-45 | 22-27 | 15-24 |
| soil B+C | Zn | 0-1 | <1 | 4-7 | 3-5 | 34-56 | 20-33 | 14-22 |
| w-bedrock | Zn | 1-5 | 0-2 | 0-18 | 3-14 | 3-32 | 18-42 | 14-42 |
| soil A | Pb | 2-7 | 5-9 | 6-11 | 8-12 | 7-10 | 42-65 | 2-14 |
| soil B+C | Pb | 2-4 | 2-4 | 5-9 | 8-11 | 6-12 | 50-65 | 4-11 |
| w-bedrock | Pb | 1-4 | 3-5 | 4-9 | 6-9 | 8-15 | 49-71 | 6-7 |
| soil A | Ag | 0 | 2-5 | 8-30 | 40-60 | 13-20 | 9-13 | 1-14 |
| soil B+C | Ag | 0 | 2-4 | 18-37 | 40-61 | 15-22 | 10-15 | 4-6 |
| w-bedrock | Ag | 0 | 0-3 | 4-10 | 28-40 | 19-26 | 36-46 | 2-3 |

AEC = absorbed/exchangeable/carbonate; Mn ox = Mn oxides; am Fe = amorphous Fe oxyhydroxides; cry Fe = crystalline Fe oxides.
soil A = A horizon soil; soil B+C = B+C horizon soil; w-bedrock = weathered bedrock.

2 Distribution of elements in stream sediments

*Concentrations*

As result of heavy vegetation cover, grain substance in high position is hardly to be moved away to the streams unless the geological bodies is being eroded by the streams. The lack of enough supply of grain matter leads to scarcity of fine grain in the streams and in the size fraction of -5 mm clastic stream sediments, over 90% were in the fractions of >0.18mm coarse grains (Fig. 9).

Zn, Cu, Ag tend to be enriched in the fine fractions (Fig. 9), perhaps due to the absorption effect. Because of lower chemical mobility of Pb in hypergenic the zone, smallest changes are observed among Pb concentrations in different size fraction and highest amounts of Pb occur in -9+0.45 and -0.45+0.18.

The elements in different size fractions along a stream are similar to each other in the anomaly persistency and extension although metal elements with Zn, Cu and Ag in particular are highly enriched in the fine fractions (Fig. 10). This indicates that most of

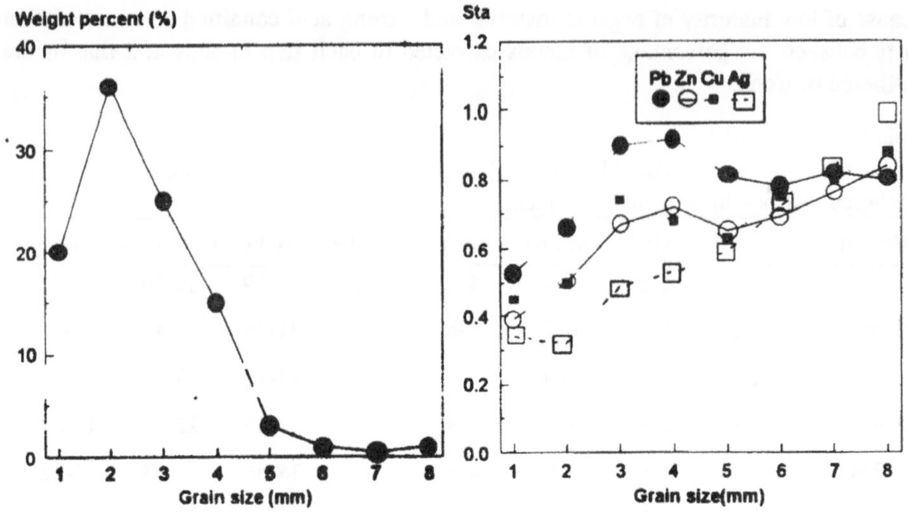


Fig. 9. Distribution of metal elements in different size fractions in clastic stream sediments. Grain size (mm): (1) -5+2. (2) -2+0.9. (3) -0.9+0.45. (4) -0.45 +0.18. (5) -0.18 +0.125. (6) -0.125+0.098. (7) -0.098+0.076. (8) -0.076.

the ore-forming heavy minerals have not yet been separated from the rock fragments and there is a stronger absorption effect in the clastic sediment, especially in fine fractions. Considering the distribution of grain size frequency, element concentrations in different fractions and anomaly persistence, -0.9 mm is a proper sample size for the RGNR project in the terrain.

Organic stream sediment consist of dead organic debris in various stages of humification and different amount of inorganic material. Metal elements are highly enriched in the organic stream sediment in contrast with the clastic sediment (Table 3).

Table 3
Concentrations of Cu, Pb, Zn and Ag in stream sediments in Bailushan area.

| | Cu | | Pb | | Zn | | Ag | |
|---|---|---|---|---|---|---|---|---|
| | Mean | S.D. | Mean | S.D. | Mean | S.D. | Mean | S.D. |
| Clastic stream sediment (-5+0.9mm) | 39.44 | 54.44 | 74.2 | 59.6 | 127.3 | 127.3 | 0.24 | 0.17 |
| Clastic stream sediment (-0.9mm) | 56.11 | 52.71 | 84.2 | 48.9 | 251.9 | 160.2 | 0.41 | 0.27 |
| Stream or valley peat (-0.18mm) | 79.60 | 46.10 | 119.5 | 44.4 | 407.2 | 357.3 | 2.97 | 1.51 |

N=42; Mean = geometric mean; S.D. = standard deviation.

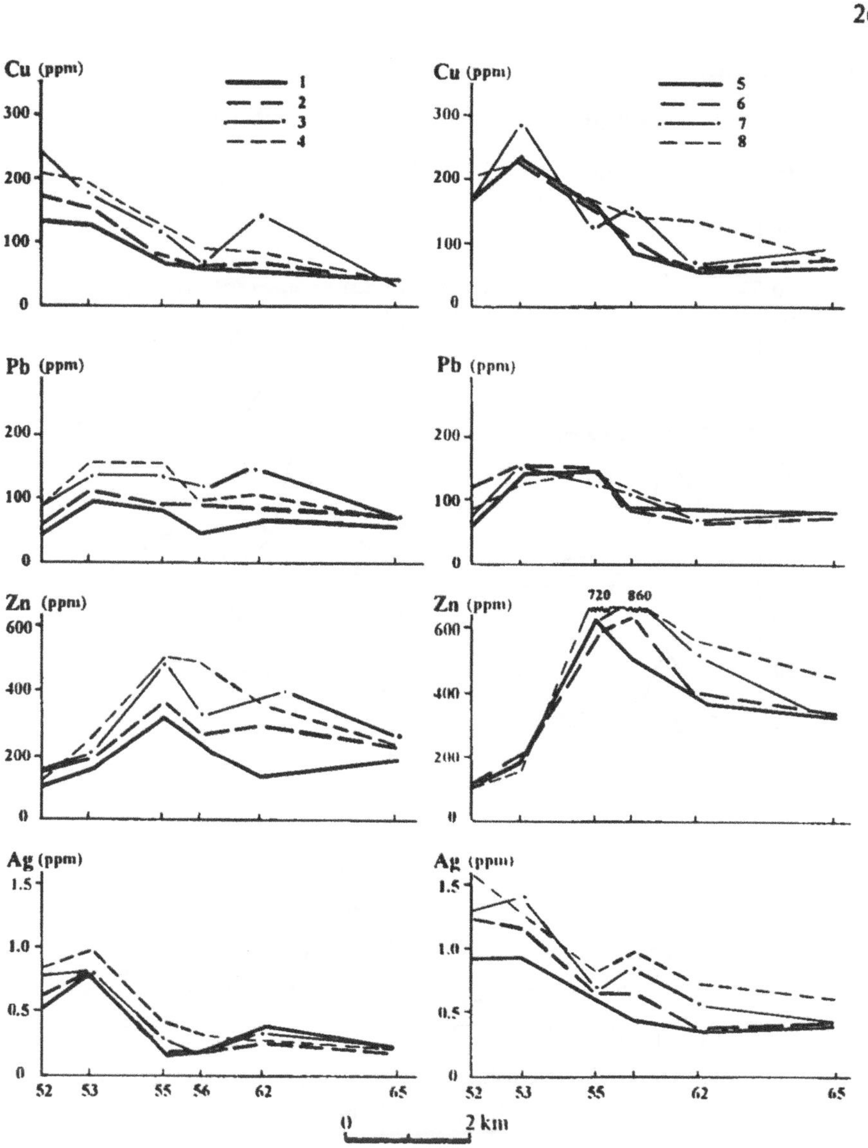


Fig. 10. Persistance of metal elements in different size fractions of clastic stream sediments along a stream drainage in study area. Grain size (mm): (1) -5+2. (2) -2+0.9. (3) -0.9+0.45. (4) -0.45 +0.18. (5) -0.18 +0.125. (6) -0.125+0.098. (7) -0.098+0.076. (8) -0.076. Location of numbered sample sites shown on Fig. 5.

*Partitioning*

The partitioning of Cu, Zn, Pb, Ag of clastic sediments corresponds to that of soils (Table 2 and 4). In comparison with clastic sediment, significant difference lies in which both the total and organic bound fraction of metal elements are markedly increased in peat, which shows that considerable amount of the elements is complexed by organic material.

environment changes.

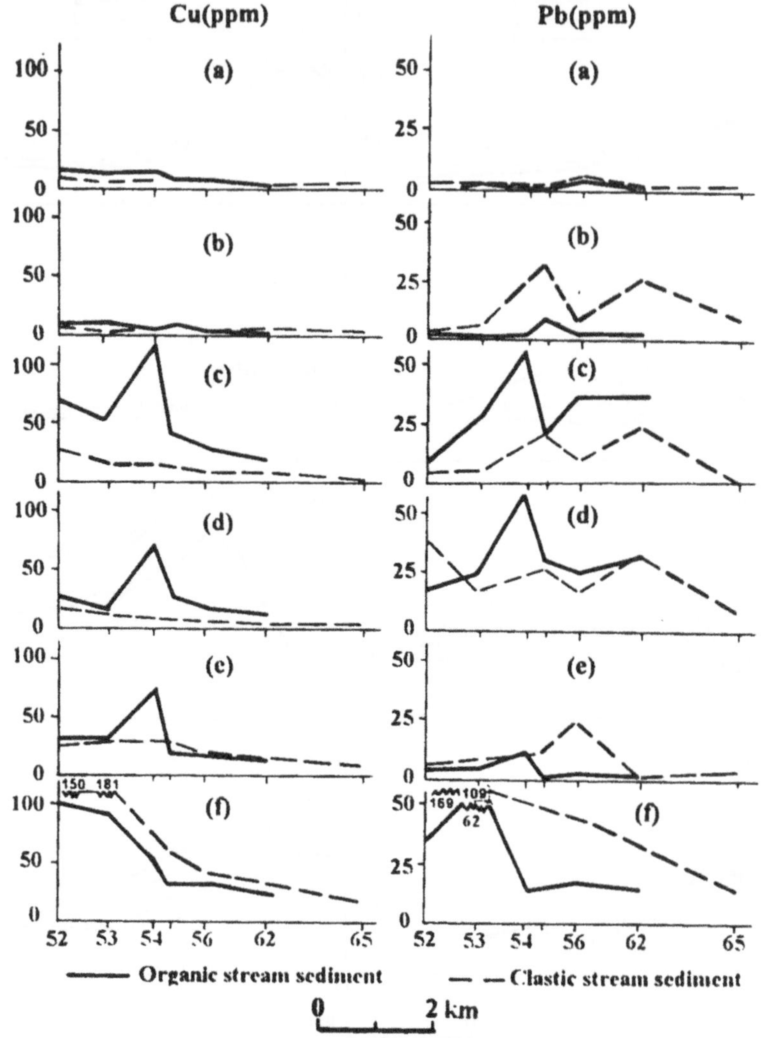


Fig. 11. Distribution of Cu, Pb in the various phases of stream sediments along a stream drainage in study area. (a) Adsorbed/exchangeable/carbonate (b) Mn oxides (c) Organic (d) Amorphous Fe oxides (e) Crystalline Fe oxides (f) Sulphides. Location of numbered sample sites shown on Fig. 5.

4 Dispersion pattern

The above results show that water plays an important role in surfacial geochemical process in the terrain. When rain and melton ice water enters the top soil (A horizon), it can leach out most of elements because of high organic matter and low pH values (Table 4) in the soil. For the elements leached from A horizon soil, some of them precipitate in the B + C horizon as a result of pH and temperate increases and organic carbon decreases in the horizon and some of them enter surface run-off with water,

The distribution of ore-forming elements varies greatly in the various phases of stream sediments along a stream drainage in study area (Fig. 11). For Cu and Pb in clastic sediment, only the sulphides fraction forms markedly anomalous trains along streams; Cu, Pb in sulphides in organic stream sediments also forms anomalies in the drainage, but the anomalous values are much lower than that in clastic sediment and other forms of Cu fluctuate greatly with hydromorphic forms including organic bound and amorphous Fe oxyhydroxides in particular.

Table 4
Percentages of Cu, Zn, Pb, Ag extracted sequentially in the seven steps in stream sediments in mineralized zones

| medium | Element | AEC | Mn ox | organic | am Fe | cry Fe | sulphides | residues | T(ppm) |
|---|---|---|---|---|---|---|---|---|---|
| CSS | Cu | 2.5-10 | 2-6 | 6-15 | 5-11 | 10-24 | 30-71 | 2-5 | 45-252 |
| OSS | Cu | 2-7 | <1-5 | 22-34 | 8.1-24 | 11-22 | 15-42 | 2-7 | 79-346 |
| CSS | Zn | 1-17 | <1-16 | 3-9 | 6-17 | 14-43 | 20-39 | 11-29 | 170-526 |
| OSS | Zn | 2-19 | 0.5-16 | 8-15 | 5-10 | 10-36 | 16-43 | 13-26 | 88-1324 |
| CSS | Pb | 2-7 | 3-23 | <1-17 | 11-35 | 1-20 | 24-67 | 3-20 | 51-163 |
| OSS | Pb | <1-3 | 1-4 | 12-47 | 23-41 | <1-8 | 9-47 | 6-8 | 79-171 |
| CSS | Ag | <1 | 1-21 | 4-10 | 15-46 | 10-16 | 28-50 | 4-13 | 0.2-1.0 |
| OSS | Ag | <1-2 | <1-3 | 8-23 | 50-70 | 5-15 | 6-13 | 4-8 | 0.9-3.0 |

AEC = absorbed/exchangeable/carbonate; Mn ox = Mn oxides; am Fe= amorphous Fe oxyhydroxides; cry Fe = crystalline Fe oxides; T = total
CSS = -20# clastic stream sediment (n=6); OSS = organic stream sediment (n=8).

3 Distribution of elements in surface water

Surface water is widely spread in the area during May-August. Cu, Zn, Fe, Mn in stream water increase markedly and pH values diminish significantly around the mineralization [4]. Fig. 12 shows that dissolved organic carbon in water and pH exhibit an inverse correlation and control the mobility of the elements in water. Copper, Zn and Mn anomalies occur mainly in water at pH <6 and DOC >40; their contents in water drop sharply at pH values 6. These elements are significantly enriched in peat primarily in the hydromorphic forms including AEC, organic complexes, Mn and Fe oxides. Fe is transported over shorter distances than the other elements because it begins to precipitate at about pH =4 and is almost completely deposited at pH=6.

We can see the distribution patterns of elements in the media even clearly in Fig.13. The pH values in peat decrease greatly around the mineralization, and Cu and Fe anomalies in peat correspond to the Cu and Fe anomalies in water. However, geochemical lows in peat for Mn and Zn were outlined within areas of water anomalies for the same metals and correspond closely to pH lows. This also indicates that metals in the mineralizations can be leached out chemically in the upper reaches of streams in the acid environment and precipitated in lower reaches of streams and seepage zones due to transport a certain distance in the upper reaches of streams, precipitate in the seepage

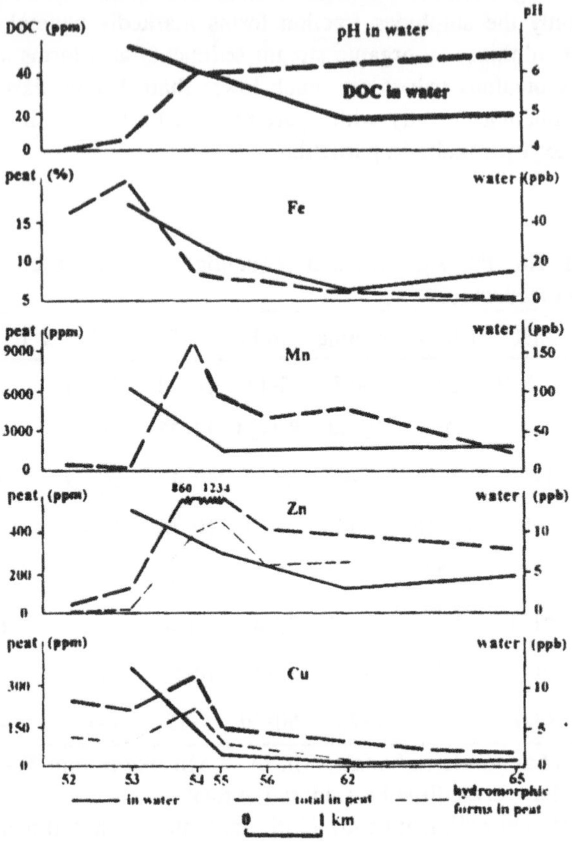


Fig. 12. Variation of water pH, dissolved organic carbon in water (DOC) and various elements in water and organic stream sediment (peat). Location of numbered sample sites shown on Fig. 5.

Table 4
**pH values in media in study area**

| Medium | Mineralized area | Background area |
|---|---|---|
| A horizon soil[a] | 2.2-3.1 | 3.2-4.2 |
| B+C horizon soil[a] | 3.1-4.0 | 4.0-5.1 |
| Stream peat[a] | 2.5-4.6 | 4.4-5.8 |
| Surface water | 4.0-6.0 | 5.8-6.8 |

[a] in 0.1 N KCl.

zone or the low reaches of streams also as a result of variation of the environment and form displaced anomalies of the elements in the peat (Fig. 14).

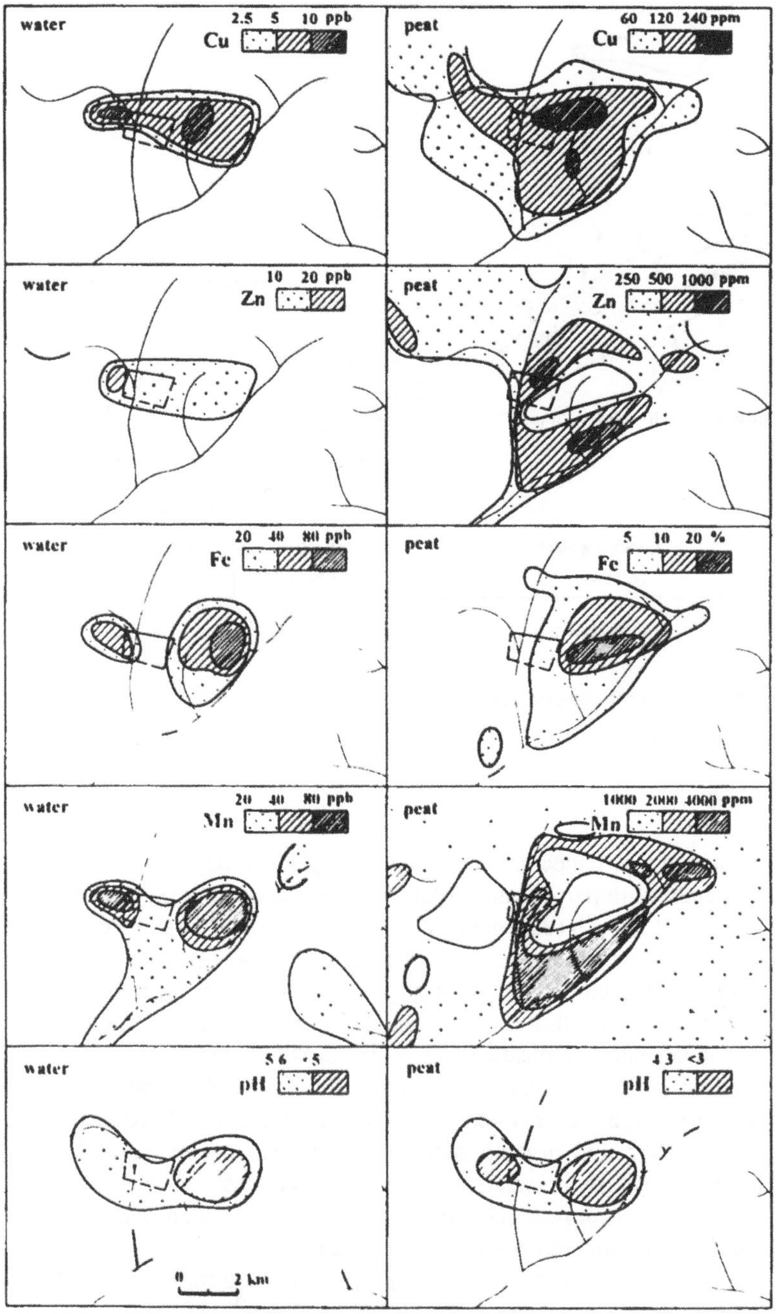


Fig. 13. Distribution of matal elements and pH in peat and water around the mineralization in Bailushan area.

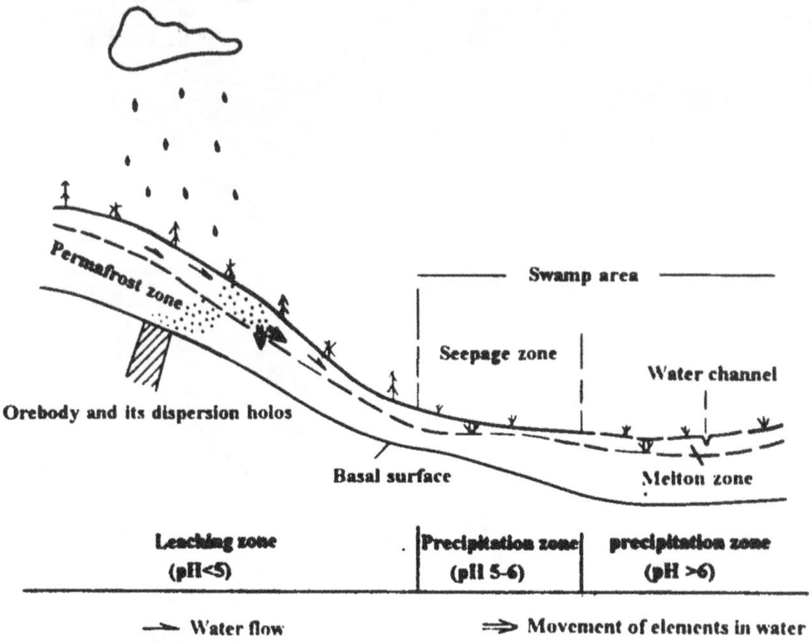


Fig. 14. The chemical dispersion pattern of elements in forest-swamp terrain, DaHingan Mountains.

## CONCLUSION

Surface water in this temperate-cold study area plays a major role in the secondary dispersion of elements, in part because mechanical migration is greatly hindered by thick vegetation. DOC, pH and Eh (data not available) of water determine the concentration at which ions can remain in solution; changes in any of these parameters is sufficient to cause precipitation. Vegetation can be not only a mechanical barrier to halt migration of coarse grain materiel downhill but also a chemical activator (DOC) to intensify transportation of elements chemically when it dies and decomposes.

Clastic stream sediment is an ideal medium for the RGNR project in the forest-swamp terrain. Although the marked anomalies of most mineralized elements can be formed both in clastic and organic sediments, the elements in sulphides in clastic sediment decrease regularly downstream from the mineralization and have a higher contrast than that in stream peat. This indicates that clastic stream sediments are more effective in pinpointing mineralizations in the terrain.

The alternative approaches for the RGNR project in the terrain are to collect clastic stream sediment (-20#) at a density of one sample/4 $km^2$. After two years test, the approaches are suitable for the RGNR project and may play an important role in mineral exploration in the terrain.

Organic stream sediment and valley peat can be used as a sampling medium for regional and detailed geochemical surveys, however, some problems were come across in sample processing and analyzing, and particularly because hydromorphic anomalies (displaced anomalies) are often found in the seepage zone and stream swamp, great difficulty exits in the data interpretation and following-up of anomalies.

Considering poorly developed soil and heavy vegetation, stream water is a good medium at the follow-up stage for quickly locating base and precious metal mineralization.

Acknowledgments

We are deeply appreciative for the help of Zhang Wenhua, Sen Xiaoling, Lu Yinxou and the Central Lab of IGGE for conducting the analyses. We are grateful to Mrs. Song Wei for preparing the figures. Thanks are also directed to Gao Hui and Mi Shouchao for cooperation during field work. Finally, the authors would like to express their gratitude to Professor Xie Xuejing for his constructive review of the draft manuscript.

**REFERENCES**

1. Brundin, N. H., 1972. Alternative sample types in regional prospecting. J. Geochem. Explor., 1:7-46.
2. Borovitskii, V.P., 1976. The application of bog-sampling in prospecting for ore deposits in perennial frost regions. J. Geochem. Explor., 5:67-70.
3. Larsson, J.O., 1976. Organic stream sediments in regional geochemical prospecting, Precambrian Pajala district, Sweden. J. Geochem. Explor., 6: 233-249.
4. Wang, M., Ren, T., Liu, Y. and Liu X., 1995. Stream water, an ideal medium for rapidly locating polymetallic mineralization in forest-swamp terrain: a case study in Da Hinggan Mts., northern Heilongjiang, China. J. Geochem. Explor., 55: 257-264.
5. Xie, Y., 1981. Periglacial landforms and regionalization in Northeast China. J. Glaciology and cryopedology, 3: 17-23.
6. Li, X., 1988. 1980. Unpubl. geochemical maps. Of low density of stream sediment survey in northern Da Xinggan Mts. (in Chinese).
7. Hao, J., 1988. Unpubl. geochemical maps of the 1/200,000 Donfanghong quadrangles (in Chinese).
8. Geophysical Exploration Team of Heilongjiang Province, 1984. Unpubl. geochemical maps. of detailed geophysical and geochemical surveys in Huzhong region, Da Hinggan Mts., Heilongjiang (in Chinese).
9. Filipek, L. H. and Theobald, P.K., 1981. Sequential extraction techniques applied to a porphyry copper deposit in the basin and range province. J. Geochem. Explor,. 14:155-174.
10. Chao, T.T. and Zhou L.,1983. Extraction techniques for selective dissolution of amorphous iron oxides from soil and sediments. J. Soil Sci. Soc. Am. Proc., 47:225-232.
11. Chao, T.T.,1984. Use of partial dissolution techniques in geochemical exploration. J. Geochem. Explor., 20: 101-135.
12. Guo, D., 1981. Preliminary approach to the history and age of permafrost in northeast China. J. of Glaciology and Cryopedology, 3: 2-16.
13. Second Geological Survey of Heilongjiang, 1986. Unpubl. Rep. of 1/50,000 geological survey in Xiji-neshan area (in Chinese).

# ISOTOPE GEOCHEMISTRY

*Proc. 30th Int'l. Geol. Congr.*, Vol. 19, pp. 277-287
Xie Xuejin (Ed)


# Os, Sr and Nd Isotope Research on the Shizhuyuan Polymetallic Tungsten Deposit, South Hunan, China

LI HONGYAN, MAO JINGWEN
*(Institute of Mineral Deposits, Chinese Academy of Geological Sciences, Beijing 100037, China)*
DU ANDAO, ZOU XIAOQIU and SUN YALI
*(Institute of Rock and Mineral Analysis, Chinese Academy of Geological Sciences, Beijing 100037, China)*

**Abstract**

The Shizhuyuan polymetallic tungsten deposit was formed in the contact zone between the Qianlishan granite stock and Devonian limestone. The Qianlishan granite stock is composed of pseudoporphyritic biotite granite, equigranular biotite granite and granite porphyry, from early to late. The Rb-Sr whole rock isochron ages of these three phases of granitic rocks are 152Ma, 137-136Ma and 131Ma, respectively. There are close spatial-temporal relationships between the multiple phases of granitic rocks and multiple periods of mineralization in the deposit. Massive skarn ores of the first period are associated with the pseudoporphyritic biotite granite. Using the Re-Os isochron method on molybdenite, we directly measured the ore-forming age of skarn W-Sn-Mo ore related to the first-phase granite, and obtained a Re-Os age of 151.0±3.5Ma. This proves that polymetallic tungsten mineralization can occur in the early phase of granite series. Isotopic evidence also indicates the granites and the ore have characteristics of both continental crust and mantle. The granitic rocks are enriched in ore-forming elements and, therefore, are considered to have been the main source of the ore-forming materials.

*Keyword: Re-Os, Rb-Sr, Sm-Nd, polymetallic tungsten deposit, granite*

## INTRODUCTION

The Shizhuyuan polymetallic tungsten deposit, which is situated in the south Hunan depressive region of the South China Folded Belt and located 13 km southeast of Chengchou City, Hunan Province, is a famous supergiant deposit of nonferrous metals in the world. It has attracted a lot of researchers' attention since it was discovered in 1970's for it's feature of large reserves, multi-types of mineralization as well as complex material compositions. Wang Changlie et al.[1,2], Yang Chaoqun[3], Wang Shufeng et al.[4] have described its geological setting and metallogenesis. Zhang Ligang[5] studied the genetic relationship between the Qianlishan granitic stock and the Shizhuyuan polymetallic tungsten deposit with oxygen and sulfur isotopes methods. However, the data of rock-forming and ore-forming ages are still ambiguous.

Previously, the mineralizing age of a deposit was usually deduced by the age of the igneous rocks or altered wall rocks near the deposits, which can only give an approximate age, and also probably results in error conclusions to those deposits with a large time gap between the ore-forming and rock-forming events. In this study, based on the new isotopic data obtained for both the Qianlishan granite and the Shizhuyuan polymetallic tungsten deposit, we focus on the study of the relationship between the rock-forming and ore-forming ages, as well as their material sources.

## Geological Setting

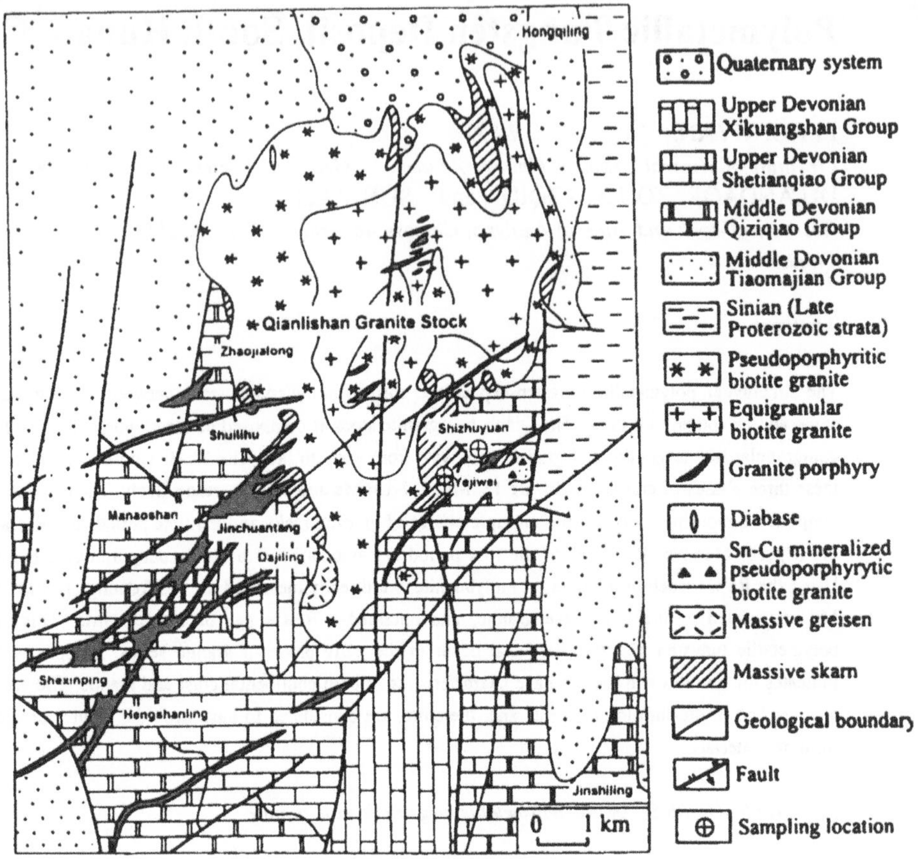


Fig. 1 Schematic map of the Qianlishan granite stock
(modified after Wang Changlie et al., 1987)

The Shizhuyuan supergiant polymetallic tungsten deposit was formed in the contact zone between the Qianlishan granite and limestone of Devonian. The country rocks of the Shizhuyuan deposit include Precambrian metagraywacke, Middle Devonian sandstone of the Tiaomajian Group, dolomitic limestone of the Qiziqiao Group and Upper Devonian limestone and marbles of the Shetianqiao Group, and dolomitic limestone of Xikuangshan Group. The Devonian formation, especially the Shetianqiao Group, is the main wall rock for the mineralizations.

The Qianlishan granite stock, whose outcrop being about 10 $km^2$, is a composite intrusion. From early to late, it consists of pseudoporphyritic biotite granite (phase I), equigranular biotite granite (phase II) and granite porphyry (phase III). Pseudoporphyritic biotite granite is the result of the first and the largest emplacement event in the Shizhuyuan district, the pseudoporphyritic minerals are mainly perthite and quartz, and plagioclase is the most common matrix mineral. The equigranular biotite granite usually shows a medium-coarse grained structure, except locally a fine-grained structure, especially in the margin of the granite body. The granite porphyry occurs as dike swarms, along NE-striking faults and cutting through the earlier granites.

All the three phases of granitic rocks are enriched in $SiO_2$, $Al_2O_3$. Petrochemical data show that they belong to peraluminous granite series. Their differentiation indices are all higher than 92, and the first two phases have sea-gull shaped chrondrite-normalized REE pattern, the last one belong to the right-dipped type. Accessory minerals of phase I and phase II granites are different from those of phase III, with zircon, monazite, xenotime and thorite in the former and allanite, apatite, zircon, rutile and monazite in the latter. There are no obvious evolutionary relationships in petrochemistry between the first two phases and the last one[6]. These granites are specially enriched in F, Li, Rb, Be, Ga and radioactive elements such as uranium and thorium, show the characteristics of high heat production granites.

A close spatial-temporal relationship exists between the Qianlishan granite and the Shizhuyuan polymetallic tungsten deposit. W-Sn-Mo-Bi-Be mineralization is related to the first two phases of granite, while the granite porphyry is frequently associated with lead, zinc and silver mineralizations.

Massive calcic skarn develops in the contact zone between the pseudoporphyritic biotite granite (phase I) and Devonian limestone. The massive skarn mainly consists of garnet, pyroxene, vesuvianite and wollastonite. During the retrograde alternation of the massive skarn, garnet and pyroxene were replaced by salite, fluorite and magnetite. The massive skarn is enriched in scheelite, wolframite, cassiterite, bismuthinite, molybdenite and fluorite. The emplacement of the equigranular biotite granite (phase II) caused the development of the big stockwork greisen type and fine stockwork greisen type polymetallic tungsten mineralizations, with the former is superimposed on the massive skarn and the latter on marble and pseudoporphyritic biotite granite. Granite porphyry dikes (phase III) are accompanized by manganese skarn, which is related to Pb-Zn-Ag mineralization.

## The Rb-Sr isotopic ages of the Qianlishan granites

The ages of the Qianlishan granite stock have been studied with K-Ar, Rb-Sr isotopic methods since the 1960's. Owing to the difference of analysis precision, these results show strong divergence, ranging from 63Ma to 172Ma[2 4]. Yet up to now, no Rb-Sr or K-Ar isochron has been reported. Wang Changlie et al.[1 2], Zhao Yongxin[7] considered the Qianlishan granite stock as products of multiple-step emplacements of a mono-magmatic activity. Shen Weizhou et al.[8], Zhang Ligang[5] proposed that the primitive magma of the Qianlishan granite were formed from remelting of crust materials. Wang Shufeng et al.[4] collected some Rb-Sr isotope data of the Qianlishan granite stock, and deemed that the strontium isotope compositions of the Qianlishan granite stock have a large variation, half of the initial $^{87}Sr/^{86}Sr$ rations are 0.71205~0.73282, half are 0.70322~0.70477.

We collected 21 granitic rock specimen from both ground outcrops and undergraund mining tunnels for Rb-Sr isotopic analysis. Sm-Nd isotopic compositions are also analyzed to research their materials sources for these samples. They were done in Institute of Geology, Chinese Academy of Geological Sciences, and in Institute of Geology, Chinese Academy of Sciences. The results are presented in the Table 1. The method for analysis of Rb-Sr has been described by Zhao Zijie et al.[9], and method for Sm-Nd described by Huang Xuan[10].

Rb-Sr isochron age of pseudoporphyritic biotite granite (phase I), is 152±9Ma with correlative

Table 1 Rb-Sr and Sm-Nd isotopic data of granitoids in the Qianlishan granite stock

| No. | Sample | Rb (ppm) | Sr (ppm) | $^{87}Rb/^{86}Sr$ | $^{87}Sr/^{86}Sr$ | $\varepsilon_{Sr}$ | Sm (ppm) | Nd (ppm) | $^{147}Sm/^{144}Nd$ | $^{143}Nd/^{144}Nd$ | $\varepsilon_{Nd}$ |
|---|---|---|---|---|---|---|---|---|---|---|---|
| Q-3 | | 723.93 | 53.71 | 39.3157 | 0.79231 | 43.1 | 10.51 | 44.12 | 0.1440 | 0.512258 | -6.39 |
| Q-4 | Pseudopor- | 577.76 | 71.29 | 23.5670 | 0.76060 | 76.0 | 11.79 | 49.25 | 0.1448 | 0.512228 | -6.99 |
| Q-5 | phyritic | 597.26 | 59.11 | 29.4114 | 0.77079 | 41.4 | 7.34 | 28.42 | 0.1561 | 0.512222 | -7.33 |
| Q-6 | biotite | 633 27 | 43.54 | 42.4523 | 0.80021 | 59.0 | 8.34 | 32.01 | 0.1576 | 0.512241 | -6.99 |
| 490-46 | granite | 745.32 | 57.59 | 39.7505 | 0.79348 | 107.7 | 8.07 | 30.60 | 0.1595 | 0.512212 | -7.59 |
| 490-18 | | 945.53 | 22.40 | 125.2856 | 0.97225 | 339.9 | 30.19 | 90.02 | 0.2566 | 0.512312 | -7.41 |
| 490-19 | | 916.90 | 9.70 | 289.5376 | 1.30381 | 506.2 | 11.56 | 33.70 | 0.2074 | 0.512322 | -6.35 |
| 490-20 | | 664.76 | 15.57 | 126.5452 | 0.95812 | 104.5 | 15.81 | 36.85 | 0.2595 | 0.512313 | -7.44 |
| 490-21 | | 918.30 | 8.87 | 317.6460 | 1.32806 | 73.4 | 16.49 | 35.52 | 0.2808 | 0.512341 | -7.27 |
| 490-22 | Equigra- | 941.78 | 12.58 | 225.8068 | 1.14096 | -43.91 | 6.36 | 59.47 | 0.0647 | 0.512288 | -4.25 |
| 490-28 | nular | 946.20 | 18.67 | 151.1954 | 1.02954 | 314.8 | 15.49 | 38.53 | 0.2430 | 0.512302 | -7.34 |
| 490-29 | granite | 837.70 | 16 50 | 151.4687 | 1.02996 | 313.0 | 15.50 | 37.39 | 0.2506 | 0.512312 | -7.28 |
| XN-1 | | 1061.79 | 18.15 | 174.6716 | 1.03590 | 86.5 | 24.22 | 52.49 | 0.2791 | 0.512344 | -7.16 |
| XN-4 | | 748.20 | 45.11 | 48.5073 | 0.81890 | 295.3 | 7.46 | 24.82 | 0.1817 | 0.512033 | -11.55 |
| 490-31 | | 921.90 | 17.45 | 156.5534 | 1.00892 | -130.4 | 13.44 | 35.22 | 0.2308 | 0.512304 | -7.10 |
| 490-32 | | 740.16 | 55.26 | 39.0755 | 0.79465 | 170.3 | 3.83 | 13.33 | 0.1739 | 0.512232 | -7.53 |
| 490-5 | | 436.27 | 60.84 | 20.8363 | 0.75558 | 176.6 | 9 64 | 54.92 | 0.1061 | 0.512224 | -6.56 |
| 490-6 | Porphyry | 400.42 | 100.65 | 11.5431 | 0.73828 | 176.6 | 9 52 | 51.02 | 0.1128 | 0.512301 | -5.17 |
| 490-7 | granite | 383.57 | 80.73 | 13.7928 | 0.74252 | 177.3 | 9.55 | 50.32 | 0.1147 | 0.512125 | -8.64 |
| 490-8 | | 458.78 | 75.24 | 17.7108 | 0.74994 | 179.1 | 8.38 | 44.65 | 0.1135 | 0.512124 | -8.64 |

coefficient r=0.9934. The ages of the medium grained equigranular biotite granite and the fine grained equigranular biotite granite (both belong to phase II) are 137±7Ma and 136±6Ma, respectively. The granite porphyry (phase III) has an age of 131±1Ma.

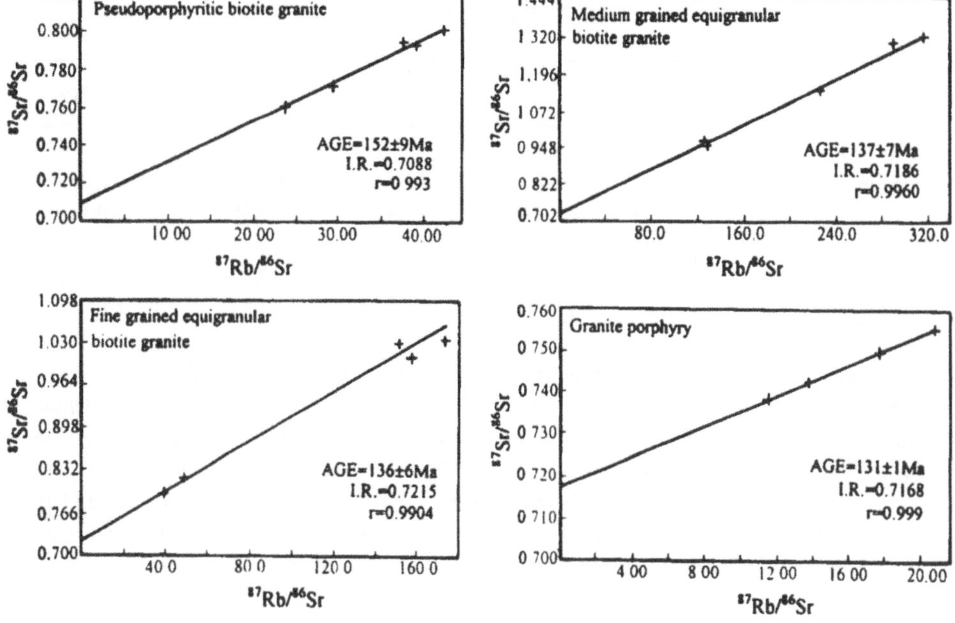


Fig.2 Rb-Sr isochrons of the granitoids in the Qianlishan granite stock

## Ore-forming age of the Shizhuyuan polymetallic tungsten deposit

There is a close spatial-temporal relationship between the Qianlishan granite stock and the Shizhuyuan polymetallic tungsten deposit in the field. Previously, the age of mineralization were only supposed to be equal to or later than that of the granitic rocks near the deposit. So, there is no accurate age data formerly. However, the development of Re-Os isotope test technique leads us to obtain accurate ore-forming age data directly.

Three massive skarn ore samples, with relation to the pseudoporphyritic biotite granite, were collected at 560 tunnel and Taipingli producing area (Fig. 1). Sample TP-16 and TP-17, the massive vesuvianite-garnet skarn ore, contain a great amount of molybdenite, bismuthinite and flourite. Sample 560-9, the massive pyroxene-garnet skarn ore, is enriched in scheelite, wolframite, molybdenite and bismuthinite. We separate six molybdenite samples for Re-Os isotope analysis from the three ore samples. The grain of molybdenite and bismuthinite is very small and shows poicilitic texture, so it is difficult to separate them completely. These molybdenite have fine lepidoblastic texture and belong to $2H_1$ type. Their average chemical composition is 58.32% molybdenum, 0.14% tungsten and 39.32% sulfur. Their molecular formula are $(Mo_{0.999}W_{0.001})_{0.99}S_2$.

The molybdenite samples were ground to finer than 100 mesh and measured by isotope dilution-ICP/MS in the Re-Os Isotopic laboratory of Institute of Rock and Mineral Analysis, Chinese

Academy of Geological Sciences. The analysis methods and procedures were described by He Hongliao et al.[11]. The blanks of Re and $^{187}Os$ are 0.07ng and 0.01ng respectively. The ages were determined with the precision of about 3%.

Table 2 Results of Re-Os analysis

| Sample | Re ($\times 10^{-6}$) | Os ($\times 10^{-9}$) | $^{187}Os$ ($\times 10^{-9}$) | $^{186}Os$ ($\times 10^{-9}$) | $^{187}Re/^{186}Os$ | $^{187}Os/^{186}Os$ |
|---|---|---|---|---|---|---|
| TP-16(1) | 1.22 | 3.93 | 2.00 | 0.0310 | 24670(1420) | 64.6(3.9) |
| TP-16(2) | 1.34 | 13.35 | 2.49 | 0.1740 | 4811(456) | 14.3(0.9) |
| TP-17(1) | 1.22 | 8.58 | 2.09 | 0.1040 | 7325(363) | 20.0(1.1) |
| TP-17(2) | 1.32 | 18.72 | 2.33 | 0.2630 | 3138(340) | 8.9(1.1) |
| 560-9(1) | 1.04 | 6.19 | 1.84 | 0.0698 | 9327(784) | 26.4(2.4) |
| 560-9(2) | 1.16 | 164.60 | 7.30 | 2.5300 | 287(16) | 2.9(0.2) |

The results of Re-Os analysis are presented in Table 2. The rhenium content of the six molybdenite samples range from $1.04\times 10^{-6}\sim 1.34\times 10^{-6}$, which are similar to the $2H_1$ type molybdenites in typical skarn or greisen deposit in the world[12]. Except for 560-9(2), the $^{187}Os$ contents of these samples are low, range from $1.84\times 10^{-9}$ to $2.94\times 10^{-9}$, and show the regularity of positive correlation between rhenium and $^{187}Os$ values. Sample 560-9(2) has higher osmium value of $164.6\times 10^{-9}$, and its $^{187}Os$ value is evidently higher than that of the other samples. This indicate that sample 560-9(2) has normal osmium. This might be caused by bismuthinite which is not separated from molybdenite.

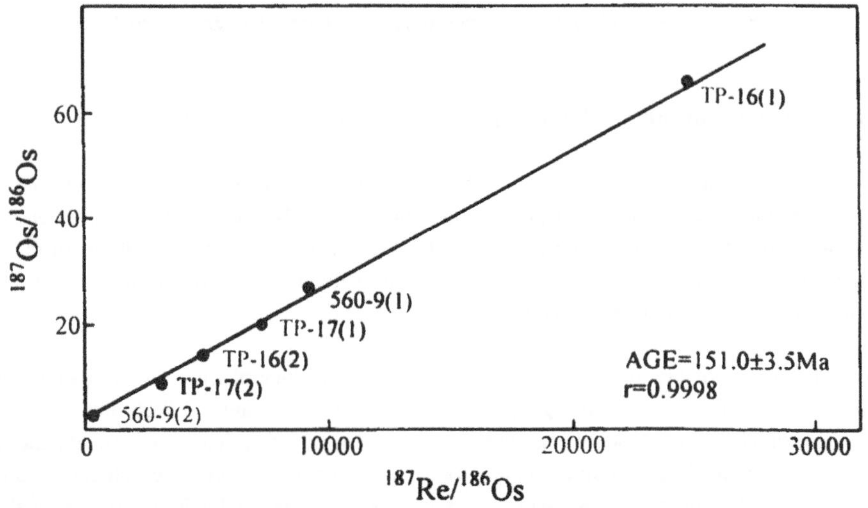


Fig.3 Re-Os isochron of molybdenites in Shizhuyuan polymetallic tungsten deposit

On the Re-Os isochron of molybdenites (Fig. 3), six samples are plotted onto a line with a correlative coefficient r=0.9998. The ore-forming age is 151.0±3.5 Ma. The polymetallic tungsten deposits extensively develop in South China, and have a close genesis relationship with the Yanshanian granite. In general, the granitiods associated with tungsten and tin deposits frequently emplaced as granitic rock series. The W-Sn mineralization were usually considered to be related to the latest phase of granitic series[13,14,15]. Using the Re-Os isochron method of molybdenite, we directly measured the metallogenic age of massive skarn W-Sn-Mo ore, and got a result of 151.0±3.5Ma. This result corresponds to the age of the pseudoporphyritic biotite granite (phase I), and is quite different from the later phases (137-131Ma), hence proves that the first stage of mineralization occurred shortly after the intrusion of the pseudoporphyritic biotite granite. This result is also in good agreement to geological evidences in the field. Therefore, this deposit should be an example which shows polymetallic tungsten mineralization occur in the early period of granite series.

## Discussion

### *Petrogenesis of the Qianlishan granites*

The initial $^{87}Sr/^{86}Sr$ ratios of the three phases granitic rocks, from early to late, are 0.7088±0.0045, 0.7187±0.0127 and 0.7215±0.0073, and 0.7168±0.0002. These data are relatively lower than those of typical S-type granites of South China, such as Darongshan granite (0.7249) in Guangxi Province, Lincang granite (0.7240) in Yunnan Province and Shanjiang granite (0.7208) in Jiangxi province[16].

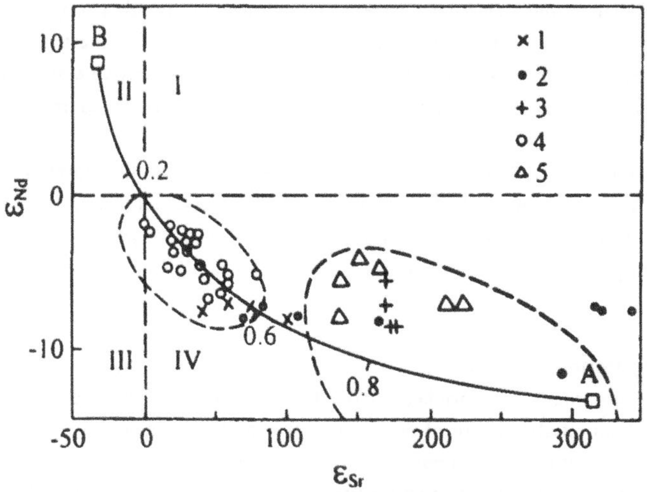


Fig. 4 The $\varepsilon_{Sr}$-$\varepsilon_{Nd}$ diagram of Yanshanian granitoid in South China (according to Liu Changsi et al.[17])

1 pseudoporphyritic biotite granite; 2 equigranular biotite granite; 3 granite porphyry; 4 syntectonic series granites in South China; 5 continental crust anatexis granites in South China; A the endmember of the upper crust of South China; B the endmember of the depleted mantle

On the $\varepsilon_{Sr}$-$\varepsilon_{Nd}$ diagram of granitoids of Yanshanian in South China (Fig. 4), except two phase II samples, the samples of the Qianlishan granite stock are all located in the district of $\varepsilon_{Nd} < 0$, $\varepsilon_{Sr} > 0$, and nearly plotted onto a level line. This indicates a great quantity of crustal materials took part in the genesis process of these granitic rocks[18,19]. The phase I granite, plotted in the left side of quadrant IV, have 50~65% upper crust materials and 35~50% depleted mantle materials. They partly plot into the syntectonic series granites in South China. Liu Changsi et al.[17] calculated the upper crust materials (UC) and depleted mantle materials (DM) compositions of the granites in South China, and got the conclusion that the granites containing > 67% UC belong to crust-source type, the higher is the UC componet, the shallower is the level of the source; granites with 50~

67% UC were generated from deep-source crustal materials, lower crustal mafic granulite and some mantle materials. Granite porphyry (phase III), plotted in the right side of quadrant IV, is located in the area of continental crust anatexis granite and has > 70% UC. The spots of the phase II granite have large divergence. Two spots were plotted in the area of $\varepsilon_{Nd} < 0$, $\varepsilon_{Sr} < 0$. Carter et al.[18] reported that Lewisian granulite samples usually have characteristics of $\varepsilon_{Nd} < 0$ and $\varepsilon_{Sr} < 0$. Some samples of the phase II granite have similar $\varepsilon_{Nd}$, $\varepsilon_{Sr}$ value to the phase I granite, some are close to the granite porphyry, and some show evidently high $\varepsilon_{Sr}$ values. This implies that granitic magma of the second phase of the Qianlishan granite stock has a same source of the phase I granite, and was seriously contaminated by upper crust materials in uplift process.

From Table 1, we can see that the Sm/Nd ratios of pseudoporphyritic biotite granite and granite porphyry are of 0.24~0.26 and 0.18~0.19. The crustal rocks have 0.16~0.24 Sm/Nd ratio, which also indicate that rock-forming materials of pseudoporphyritic biotite granite include mantle materials. The equigranular biotite granite has great changes of Sm/Nd ratios (0.11~0.46) and Rb/Sr ratios (13.4~103.5), which suggest that this phase of granite formed in a relative open system and might have experienced strong differentiation and contamination.

According to petrochemical data, accessory mineral assemblage and Nd-Sr isotope characteristics of the three phases of the Qianlishan granitic rocks, we can conclude that the pseudoporphyritic biotite granite and equigranular biotite granite are products of two successive emplacements from the same magma chamber. Granite porphyry seems to have a different magma sources.

The Nd model ages of the pseudoporphyritic biotite granite and granite porphyry, calculated according to single-stage evolutional model, are 1949~2632 Ma with an average of 2307 Ma and 1284~1578 Ma with an average of 1434 Ma, respectively. Hence the phase I granite might be formed from remelting of a Early Proterozoic source, while phase III from the Middle Proterozoic source.

*The source of ore-forming materials*

The Qianlishan granitic rocks are enriched in ore-forming elements[4,6]. The concentration of tungsten, tin, bismuth and molybdenum of the Qianlishan granitic rocks are not only much higher than the clarke values for the acidic igneous rock, but also higher than the average values of the granites of South China and the Yianshanian granites. Except for tungsten, the Qianlishan granitic rocks have higher tin, bismuth and molybdenum contents than the Devonian carbonate rocks. Zhang Ligang[5] studied the H, O isotope characteristics of the Qianlishan granitic rocks and related orebodies, and concluded that the first stage ore-forming fluid came from the reequilibrium magmatic water of the first phase granite. So it can be deduced that the Qianlishan granites should be the most important source of the ore-forming materials.

Re-Os isochron method can be useful not only in the studies of the mineralization age but also in genesis of hydrothermal deposits. Although molybdenites should contain almost pure radiogenic osmium[20], because of impurity such as bismuthinite or scheelite in molybdenite samples, we obtain a initial $^{187}Os/^{186}Os$ ratio of 2.17±0.08 for the molybdenite samples, which we consider to represent the characteristic of the ore-forming materials. Allegre and Luck[21] put forward initial osmium isotopic $^{187}Os/^{186}Os$ ratio evolution diagram for the continental crust in the case of various models (Fig. 5). In a model where the continental crust is being continuously derived from the mantle, the initial $^{187}Os/^{186}Os$ ratio in granitoids will have values close to the mantle evolution curve which is around 1. On the other hand, if the continental crust is a continuously reworking primitive crust differentiated at 3.8 b. y., the initial $^{187}Os/^{186}Os$ ratio will have much higher values, up to 26 for young granitoids. The data of 2.17±0.08 is plotted above the mantle evolution curve, and is lower than the initial $^{187}Os/^{186}Os$ ratio of the old recycled crust. The average Nd model age of the pseudoporphyritc biotite granite are 2307 Ma. So we can deduce that the ore-forming substances are from a mixture of 2307 Ma crust materials, which producted by old continental crust together with mantle material, and new mantle material. The initial $^{87}Sr/^{86}Sr$ ratio of the phase I granite is 0.7088±0.0045, which shows a conformable feature.

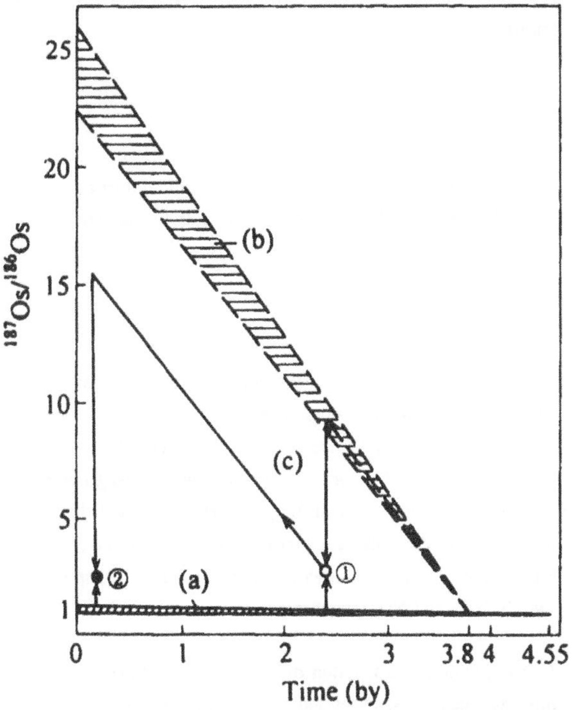


Fig. 5 The initial osmium isotopic $^{187}Os/^{186}Os$ ratio evolution diagram for the continental crust in the case of various models

a. continental crust continuously derived from mantle material; b continental crust continuously recycled from a primitive crust differentiated at 3.8 b y.; c mixing model in (1)formation of a granite from a mixture of old continental crust and new mantle material; in (2)formation of a granite from a mixture of new crust formed in (1) with new mantle material

As described above, the Qianlishan granite stock is a composite intrusion of Yanshanian. The first two phases of granitic rocks, especially the pseudoporphyritic biotite granite (phase I), was formed in deep-seated region. The magma, enriched in ore-forming elements, volatile components and radioactive elements, intruded into limestone of Devonian, and induced the W-Sn-Mo-Bi-Be mineralization with skarnization and greisenization.

## Conclusion

(1) The Qianlishan granite stock is a composite intrusion which consists of pseudoporphyritc biotite granite, equigranular biotite granite and granite porphyry, from early to late. The Rb-Sr

whole rock isochron ages of the three phases granitic rocks are 152Ma, 137-136Ma and 131Ma, respectively.

(2) With Re-Os isochron method for molybdenites, a 151Ma metallogenic age of massive skarn W-Sn-Mo ore has been obtained. This result corresponds to the age of the pseudoporphyritic biotite granite (phase I), hence polymetallic tungsten mineralization can occur in the early period of granite series.

(3) The initial $^{87}Sr/^{86}Sr$ ratios and initial $^{187}Os/^{186}Os$ ratio show that the material sources of the ore and the phase I granite have characteristics of both crust and mantle. The ore-forming substances are mainly from the granite.

## Acknowledgments

This research is supported by China Association of Geological Foundation. The Authors would thank Wang Changlie and Xiu Youzhi for their help during the field work.

## REFERENCES

1. Wang Changlie, Xu Youzhi, Xie Ciguo and Xu Wenguang. The geological characteristics of Shizhuyuan W-Sn-Mo-Bi deposit. In Symposium on Tungsten Geology, Jiangxi, China. Hepworth, J V and Yu Hongzhang (Ed.), Beijing: Geological Publishing House, 413-426 (1982)
2 Wang Changlie, Luo Shihui, Xu Youzhi, Sun Yihong, Xie Ciguo, Zhang Zhongming and Xu Wenguang The geology of the tungsten polymetallic deposit. Beijing: Geological Publishing House, 29-48 (1987).
3 Yang Chaoqun. Mineralization of the composite greisen-stockwork-skarn type W-(scheelite and wolframite) Bi-Mo deposit of Shizhuyuan, Dongpo, southern Hunan, China. In Symposium on Tungsten Geology, Jiangxi, China. Hepworth, J. V and Yu Hongzhang (Ed.), Beijing: Geological Publishing House, 503-520 (1982).
4 Wang Shufeng and Zhang Qiling. Introduction of geology of the Shizhuyuan W-Sn-Mo-Bi ore deposit. Beijing: Beijing Publishing House of Sciences and Technology, 30-45 (1988).
5 Zhang Ligang Stable isotope geochemistry of the Qianlishan granites and tungsten-polymetallic deposit in Dongpo area, Hunan. Journal of Guilin College of Geology 9, 259-267 (1989)
6. Mao Jingwen and Li Hongyan Evolution of the Qianlishan granite stock and its relation to the Shizhuyuan polymetallic tungsten deposit. International Geology Review 37, 63-80 (1995).
7. Zhao Yongxin. Ore-forming mechanism of Qianlishan granite stock discussed from the relationship of Shizhuyuan W-polymetallic deposit with the stock Earth Science Journal of China University of Geosciences. 13 (2), 155-162 (1988).
8 Shen Weizhou. Stable isotope geology. Beijing: Atomic Energy Press. 304-336 (1987).
9. Zhao Zijie and Yuan Zhongxin. Geology of granitoids in Nanling region and their petrogenesis and mineralization. Beijing Geological Publishing House. 226-248 (1989)
10. Huang Xuan. The study on the granitoids in Shanxi Province by Nd-Sr isotopes. Advances in Geosciences, 2, 212-219. Institute of Geology, Acadenia Sinica. China Ocean Press (1992).
11. He Hongliao, Du Andao, Zou Xiaoqiu, Sun Yali and Yin Ningwen. A study on rhenium-osmium isotope systematics by using Inductively Coupled Plasma Mass Spectrometry and its application to molybdenite dating. Rock and Mineral Analysis. 12 (3), 161-165 (1993).
12. Han Yinwen. Molybdenite polytypes and the mechanism of polytype formation. Earth Science—Journal of China University of Geosciences. 13 (4), 385-394 (1988).
13. Groves, D. I. and Mccarthys, S. Fractional crystallization and the origin of tin deposits in granitoids.

Mineral Deposit. **13**, 11-26 (1978).

14 Mao Jingwen. The igneous rock series and tin polymetallic minerogenetic series in the Tengchong area, Yunnan. Acta Geological Sinica **63** (2), 175-187 (1989).

15 Lehmann, B Metallogeny of tin Springer-Verlag, 211 (1990).

16 Shen Weizhou, Wang Dezi and Liu Changsi. Geochemical and isotopic geological study of the tin-bearing porphyry in Yin Yan, Guangdong Province Journal of Nanjing University (Earth Sciences). 6 (2), 159-165 (1994)

17 Liu Changsi, Zhu Jinchu, Shen Weizhou and Xu Shijing. Classification and source materials of continental crust trasformation series granitoids in South China. Acta Geological Sinica. 64 (1), 43-52 (1990).

18. Carter, S. R., Evensen, N. M., Hamilton, P. J. and O'Nions, R. K. Neodymium and strontium isotope evidence of crustal contamination of continental volcanics. Sciences. **202**, 743-747 (1978).

19. Allegre, C. J. and Othman, D. B. Nd-Sr isotopic relationship in granitoid rocks and continental crust development: A chemical approach to orogenesis. Nature. **286**, 335-342 (1980).

20. Luck, J. M. and Allegre, C. J The study of molybdenites through the $^{187}$Re-$^{187}$Os chronameter. Earth Planet. Sci Lett. **61**, 291-296 (1982).

21 Allegre, C. J. and Luck, J M. Osmium isotopes as petrogenetic and geological tracers. Earth Planet. Sci. Lett. **48**, 148-154 (1980)

Printed and bound by CPI Group (UK) Ltd, Croydon, CR0 4YY
23/10/2024
01778248-0007